TAKE-OFF

Technical English
for Engineering

Teacher's Book

Fiona McGarry
and Nicholas Regan

Published by

Garnet Publishing Ltd.
8 Southern Court
South Street
Reading RG1 4QS, UK

First published 2008.
Reprinted 2013.

ISBN: 978 1 85964 975 6

British Library Cataloguing-in-Publication Data
A catalogue record for this book is available from the British Library.

Production

Programme management:	Doug Mackie, Synergy Total Business Solutions Ltd
Project manager:	Richard Peacock
Project consultants:	Fiona McGarry, Rod Webb
Editorial team:	Carol Rueckert, Emily Clarke, Matthew George
Design:	Henry Design Associates, Robert Jones, Christin Helen Auth, Bob House
Illustration:	Doug Nash
Typesetting:	Nick Asher
Photography:	Clipart.com; Corbis; Getty Images; Stockbyte; www.boeingimages.com.
Audio production:	Motivation Sound Studios, Silver Street Studios

Garnet Publishing wishes to thank Chris Murray and the staff of Saudi Development and Training (SDT) for their assistance in the development of this project.

Every effort has been made to trace the copyright holders and we apologize in advance for any unintentional omissions. We will be happy to insert the appropriate acknowledgements in any subsequent editions.

Printed and bound

in Lebanon by International Press: interpress@int-press.com

Contents

Book map

Unit	Topics	Skills	Language
1 **Design and innovation**	• Properties of materials • Design rationale • Aircraft specifications	**Listening and reading:** • Develop sub-skills: skimming and scanning • Find key information from different sources **Speaking and writing:** • Descriptions and comparisons • Note completion	• Numbers and measurement • Language for description, e.g., material properties • Question forms • Reason and purpose • Tenses: past, present and future
2 **Manufacturing techniques**	• Functions of hand and machine tools • Manufacturing processes	**Listening and reading:** • Identify key information • Transfer information, e.g., from text to table or diagram **Speaking and writing:** • Explanation of components and processes • Contrast and comparison	• Definitions and descriptions • Verbs for manufacturing operations • Imperatives for instructions • Prepositions of movement • Word-building • Parts of speech
3 **Frameworks**	• Aircraft structure • Principles of flight • Assembly line jobs and processes	**Listening and reading:** • Practise extracting key information • Recognise 'clues' such as signposts in a text **Speaking and writing:** • Restate and summarise information • Practise note-taking and using notes for a talk	• Movement and manoeuvres 1 • Forces and stresses • Direction and location • Passive voice for processes • Speculation and prediction • Word combinations • Vowel sounds/sentence stress
4 **Control systems**	• Hydraulics • Hydraulic applications • Braking and landing systems • Aircraft control surfaces	**Listening and reading:** • Intensive comprehension: follow descriptions of procedures and systems **Speaking and writing:** • Describe, evaluate and compare systems • Practise reducing and summarising information	• Hydraulics, control surfaces and linkages • Movement and manoeuvres 2 • Adverbs of manner, degree and frequency • Compound nouns • Word stress and vowel strong/weak forms
5 **Engine and fuel systems**	• Engine parts and how engines work • Engine types and specifications • Engine overhaul • Fuels and fuel systems	**Listening and reading:** • Familiarisation with different types of text, e.g., data sheets and tables • Focus on text organisation **Speaking and writing:** • Discussion and speculation using notes, tables and diagrams • Practise note-taking using reduced forms and abbreviations	• Engine parts and operations • Verbs for engine problems and repair • Fuel and air movements • Mathematical concepts, symbols and abbreviations • Passive with *by* + agent • Reduced passive forms in notes
6 **Review unit**			

Unit	Topics	Skills	Language
7 **Safety and emergency**	• Risks and hazards • Emergency procedures and equipment	**Listening and reading:** • Familiarisation with longer and multiple texts • Work out meaning from context **Speaking and writing:** • Complete tables and reports with notes • Practise giving safety information and explanations	• Chemicals and elements • Safety warnings, equipment and systems • Nouns and adjectives for damage and dangers • Imperatives and modals of obligation • Language of purpose • Syllables and stress
8 **Air and gas**	• Pneumatics • Heating and cooling systems	**Listening and reading:** • Familiarisation with different text types, e.g., instruction manuals and advertisements • Interpret graphs and diagrams **Speaking and writing:** • Discuss and interpret diagrams and schematics • Form and table completion	• Measurements/calculations • Language for changes of physical state, e.g., *condense* • Technical verb/noun collocations • Compounds and complex noun phrases • Defining relative clauses • Discourse markers • Rhythm and stress
9 **Electrical systems**	• Electrical maintenance and repair • Electric motors and batteries • Diagrammatic representation	**Listening and reading:** • Interpret/follow explanations of charts and formulae • Recognition of words and sounds in a flow of connected speech **Speaking and writing:** • Explain and discuss electrical problems and procedures • Form-filling and detailed note-taking	• Electrical parts and systems • Electrical abbreviations and symbols • Multi-word verbs • Comparison and recommendation • Past and past perfect tenses • Consonants and clusters
10 **Communication**	• Avionics • Cockpit instruments and navigation systems • Technical drawing and PCBs	**Listening and reading:** • Practise skimming and scanning more complex texts • Follow instructions and detailed information **Speaking and writing:** • Develop editing skills	• Technical verb/noun collocations • Language for electrical faults and problems • Linking words • Fact, tendency and possibility: zero and 1st conditional sentences • Modal verbs for speculation
11 **Maintenance**	• Forms and certification • Maintenance procedures	**Listening and reading:** • Interpret and understand various forms • Ask questions and give explanations about maintenance systems **Speaking and writing:** • Discuss possible maintenance scenarios/problems • Practise writing clear instructions	• Language for position, assembly and disassembly • Language for mechanical damage • Word-building • Hypothetical situations: 2nd and 3rd conditional sentences • Phrases for explanations and definitions
12 **Review unit**			

Introduction

Who is *Take-Off* for?

Take-Off is aimed at people who need to study technical English. It was written primarily for engineering students whose English is intermediate level (European Framework B1) or above, but it is suitable for pre- and in-service technicians in all MRO areas (NVQ Level 2 and above). It covers general engineering topics, but has an aeronautics focus, so is also particularly suitable for anyone working in the aeronautics industry from co-MRO operatives to supervisors, management and pilots.

The approach

Take-Off assumes that students will have a basic grounding in English grammar and some knowledge of technical terms for their subject, but that they need to improve their listening, speaking, reading and writing skills. Consequently, the book takes a *skills-building approach*, but will also help students develop their range of vocabulary and grasp of language structures relevant to their daily lives. It aims to develop students' 'survival' English so that they can pursue technical courses in further or higher education institutions, or start working as a technician or engineer in an English-speaking environment.

Take-Off also assumes that English teachers may have little or no technical background or specialist knowledge of the aeronautics industry. Therefore, authentic technical texts are adapted and exploited so that they are accessible, and tasks are designed to focus on language development, rather than test technical knowledge.

Features of the course

- Language boxes: provide short, visual explanations of grammatical rules and syntactic patterns. Students are encouraged to 'notice' the language patterns that are used in the texts, so that new language is always presented in context.
- Skills boxes: focus awareness on the useful sub-skills and strategies that will help learners in the real world. These include advice on different ways of reading, strategies for listening and advice on how to take notes and edit written work.
- Authentic texts: include instruction pages and diagrams from BA Systems manuals and certification and report forms. The use of this type of material aims to motivate students, prepare them for the work environment and promote credibility.
- Learner-centred tasks: encourage learners to draw on their own experience and contribute their own ideas in many activities. A lot of the tasks in *Take-Off* involve pair or group work, including discussions, working out problems, presenting and swapping information. This maximises practice opportunities and facilitates a positive group dynamic.
- Phonology work: addresses specific difficulties which research indicates are critical. This includes awareness – particularly for listening – of sentence stress, connected speech, vowel quality and length and consonant clusters.

- Word lists: include a list of key vocabulary for each lesson as well as an alphabetical list of all vocabulary at the back of the book. The lesson-by-lesson list can be exploited in several ways, for example, teachers may wish to present the words during the lesson or learners may check the words using technical dictionaries as preparation for the lesson.

- Glossary: a list of key technical terms is included at the back of the book. This is a useful resource which provides teachers and learners with simple explanations and definitions as well as clarifying common abbreviations and electrical symbols.

- The Workbook: provices additional reading texts for comprehension practice as well as writing tasks and extra vocabulary and grammar-focused exercises. It may be used in class to provide extra written practice and language consolidation. Alternatively, teachers may wish to concentrate on new vocabulary work and skills-building during class time, and set the Workbook activities for self-study.

- Independent learning: is encouraged through the use of tasks that require students to research topics in reference books and on the Internet. The Teacher's Book suggests weblinks for many of the lessons.

- The Teacher's Book: includes full step-by-step procedures for each lesson as well as a full answer key and model answers where students are asked to write their own texts. It also includes advice for teachers on how to clarify problem areas, suggestions for extra activities and links to useful websites.

- End-of-unit tests: Please contact your local Garnet Education representative for information about how to access these resources.

The syllabus

Vocabulary

Vocabulary is taught systematically in clear contexts with a focus on collocation and word-building as well as spelling and pronunciation. High-frequency vocabulary has been selected from technical word lists and semi-technical vocabulary is given particular prominence. Vocabulary tasks encourage students to group and categorise lexis according to topic and to focus on how vocabulary is used in a range of texts.

Vocabulary tasks include dictionary work, matching words with definitions, production of personalised sentences, and labelling diagrams and schematics. It is assumed that the students will have their own technical dictionaries and/or access to class sets of dictionaries.

Grammar

Take-Off focuses primarily on language that is used frequently in a technical environment. Therefore, it gives special emphasis to the following: imperatives, the passive voice (including reduced passive forms) and modal verbs.

Students are also encouraged to notice and learn 'chunks' of language and notional/functional exponents for key areas such as: talking about cause and effect, stating work completed, giving warnings, explanations and instructions, discussing how to deal with problems, calculations, measurements and numbers.

Listening

Instructors commonly report that spoken communication with maintenance crews is very difficult when English is their L2. This is often due to the instructors' lack of awareness of their own speech (possibly highly colloquial and/or delivered with a regional accent), and the operative's lack of listening skills and experience. Listening skills and strategies are clearly crucial to continued training throughout trainees' careers, and to the safe and correct maintenance of aircraft.

Three or four lessons in each unit are generally listening-focused. Recordings introduce learners to different types of English (including American and Australian varieties) and a range of regional accents.

Students are encouraged to develop their ability to listen for gist and to pick out specific information. *Take-Off* also helps students to develop microskills; these listening tasks help students to pick out keywords and signposts, recognise use of intonation, rhythm and pauses, and identify stressed words and syllables.

Speaking

Take-Off aims to develop students' confidence as well as their competence in speaking. Speaking activities are a central part of every lesson, and it is essential that students are encouraged to engage and participate in pair and group work. Tasks often require a degree of critical thinking by asking students to offer opinions, deductions and suppositions. This will be vital to their further studies and training.

Tasks include discussions, role plays, information-gap exercises, giving short talks involving explanations and instructions, and solving problems. These are all useful preparation for the engineering workplace.

Speaking exercises should be coupled with a focus on pronunciation: common L2 speaker problems such as word and sentence stress, vowel length and consonant clusters are given attention in more controlled speaking exercises.

Reading

Take-Off helps students to develop skills required to deal effectively with all the text types they will come across in their training and workplace. These skills include: dealing with numbers, extracting data from dense technical text, working with bullet-pointed and numbered manual-style instructions and reading and interpreting diagrams, flow charts, graphs and labelled drawings.

Writing

Writing tasks are devised to meet the needs of students in both training and workplace situations. Tasks are scaffolded so that weaker students can use the prompts and model texts to help them produce their own texts. Task-types include: note-taking, including abbreviations; writing numbers and bulleted notes; form-filling; transferring data from listening text to pictures, from diagram to notes and instructions, and from written text to speaking.

Objectives

To raise awareness of the need to listen in different ways and practise listening for gist and specific information.

To practise describing products and technologies.

Language

Language of description, e.g., *It's really* + adj. *It can* + verb, numbers and measurements.

Introduction

- Introduce the topic of innovation by eliciting tools and/or household appliances that students or their families use every day. Elicit which technologies they feel are most important/have changed people's lives the most.

Listening

A Match nouns to form compounds and label pictures

- Point out that compound nouns are often used for tools and appliances. Elicit a few compounds that students are familiar with, e.g., *washing machine, chain saw*.

- Set the task for individual work and pair checking. Encourage students to guess any combinations they are not familiar with rather than pre-teaching them. The emphasis should be on students working out the answers from what they already know, rather than focusing on getting the answers correct.

- Draw attention to the fact that students can check the meaning of many new terms connected with engineering and aeronautics (e.g., rotary engine) in the Glossary at the back of the book. Emphasise that students

will also need to look up new vocabulary in a good technical dictionary.

Answers

1 bagless – (vacuum) cleaner

2 MP3 – player

3 high-performance – engine*

4 air cushion – vehicle

5 flat screen – television

6 vertical take-off – jet

* Students may point out that this is a Wankel rotary engine.

B Listen for gist: number pictures according to the order they are discussed (CD1 Track 1)

- Set for individual work and pair-checking. Monitor, but do not spend too long feeding back unless there are specific problems.

Answers

1	5	4	6
2	4	5	1
3	2	6	3

C Listen for specific words in the conversations (CD1 Track 1)

- As students do the exercise, monitor to check that they can pick out most of the key words.

- Conduct feedback. Elicit words that were used in combination with the circled words, e.g., *massive memory, air base*. Point out that *motor* is used twice, i.e., both for the electric motor of the vacuum and the car engine (or the car itself).

Answers

1 picture (flat screen television)

2 boat, water (air cushion vehicle)

3 memory, music (MP3 player)

4 base (vertical take-off jet)

5 machine, motor (bagless vacuum cleaner)

6 motor, motorway (high-performance engine)

D 💿 Complete the table with numbers and quantities (CD 1 Track 1)

- This table completion exercise is good preparation for listening and note-taking in later units. It might be a good idea to see how much of the table students can fill in before they listen a third time.

- Encourage pair-checking before you begin whole-class feedback onto the board or OHP. Check that students know what the abbreviations *kph* (kilometres per hour) and *bhp* (brake horse power) mean and elicit other abbreviations that could be used: *k, m, gb* and *l*.

Answers

product	measure	unit	how much?
flat screen television	weight	kilos	18
air cushion vehicle (hovercraft)	height	metres	1
MP3 player	memory	gigabytes	60
vertical take-off jet	speed	kph	1,000
bagless cleaner	capacity	litre	2
high-performance engine	power	bhp	230

E Recall information about the products

- Elicit what students recall about the product. Prompt with questions if necessary, e.g., *The hovercraft is 1 metre above the water, but how fast is it? (120 kph) What can hold 2 litres? (the bin of the cleaner)*

- Encourage students to reflect on the different ways they listened in Exercises B, C and D. Then ask them to read the Skills Box and ask questions if necessary.

Speaking

A Describe products and technological advances

- The focus of the exercise is on fluency; nevertheless, draw attention to the useful language in the box and encourage students to use it to scaffold their discussion.

- Set the exercise as pairwork followed by whole-class feedback. Allow personalisation of the topic: students may be interested in discussing more modern technological developments such as nanotechnology, smart textiles, etc.

Answers

Answers depend on the students.

B Group discussion: new technologies in different areas

- Divide the class into groups of three or four. If necessary, start them off by eliciting or suggesting technologies yourself, e.g., use of alcohol, sugar cane or grasses as fuel; biometric identification systems such as iris recognition, etc.

- If students have problems coming up with ideas for this, you could choose one area to go through with the whole class as an example.

Answers

Answers depend on the students.

C Give a talk to persuade the group to spend money on a technological idea

- Check that students understand the context. Pre-teach the word *grant* and clarify/discuss how a company might choose how to spend a research and development grant.

- Check that each group member has chosen a different idea or invention. You may need to allow time for students to plan their talks before giving them to their group. Alternatively, you may prefer to set the preparation stage for homework and allow time in a subsequent lesson for the talks. If students need ideas for the topics they are going to discuss, you could direct them to the following websites:

  www.livescience.com
  www.marketlaunchers.com

- Monitor the groups as they do the exercise and feed back on the ideas that were chosen by each group.

- **Note:** Further vocabulary work: there is a list of key vocabulary for this lesson on Course Book page 266.

Answers

Answers depend on the students.

- Now have the students do the Workbook exercises for Unit 1, Lesson 1.

Workbook answers

Language: quantities

A Write quantities

weight	825 mg 19 kilos 999 g
height	950 mm 1,060 cm 0 mm 2 ft 3 ins 0.5 m
memory	6 gb 1 mb 550 kb 728,046 bytes
speed	95 fps 1,110 kph 220 bhp 75 mph 33.3 rpm 15 fps
capacity	250 cc 2-litre
power	410 kw 5.2 mw

B Check understanding of abbreviations

Answers depend on the students.

Language: word order

A Look at examples of production descriptions

The Wankel rotary engine <u>has</u> neither a crankshaft nor a flywheel. This means that it <u>is</u> lighter and <u>can</u> produce more power than a normal engine of the same size.

The ACV (hovercraft) <u>flies</u> on a cushion of air a short distance above the ground. This means that it <u>can</u> move easily on land or water.

The order is: subject - verb - (object).

B Rewrite sentences with the correct word order

1 My colleague came with me.
2 It's got a big motor.
3 It can travel at 1,000 km per hour.
4 It weighs 18 kilos so it's really easy to lift.
5 You can play music on it.

Product description

Objectives

To practise the reading sub-skill of prediction and focus on key (active) vocabulary in a reading text.

To raise awareness of how pronouns are used in a text and to practise pronoun referencing.

Language

Vocabulary connected with the production process; pronouns: *it*, *they*, *this*, and *these*.

Reading

A Discuss what students already know about the design process

- Use the visuals to clarify the different stages of the design process: *drawing*, *model* and *prototype*.

- Encourage students to record their ideas about the design process in any way that suits them, e.g., drawings, flow chart, list, etc.

- After they compare answers in pairs, elicit a few ideas, but do not confirm answers at this point.

Answers

No answers at this point.

B Read to find and note stages

- **Note:** It is not necessary to pre-teach the key words that appear in bold in the text unless you feel that students will really not be able to work them out from their context.

- Students read to find the stages of the design process in the text and compare them with their own ideas.

- **Note:** It would be an idea to set a time limit so that students skim the text and do not spend too long worrying about unknown words.

- Feed back, bringing out the fact that there may be variations on the suggested order, e.g., the design brief may precede the specifications.

Suggested answer

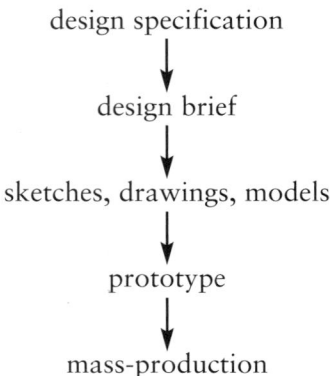

design specification

↓

design brief

↓

sketches, drawings, models

↓

prototype

↓

mass-production

C Compare lists with partners

- Have students check their answers with a partner before whole-class feedback.

D Match key words from the text with definitions

- Draw attention to the bracketed letters in front of the definitions. Elicit that (*n*) refers to *noun* and (*npl*) refers to *plural noun*.

- Set the exercise for individual work and pair-checking before whole-class feedback.

- Refer students to Language Box 1. Emphasise that, in the future, they should choose and record active vocabulary from texts independently. Point out that it is useful to choose words that form a set (all words are connected to the same topic) and that it is important to record the part of speech along with the word.

Answers

1	mass production	**6**	specification
2	model	**7**	regulations
3	conditions	**8**	drawing
4	sketches	**9**	component
5	prototype	**10**	materials

Language

A Identify pronouns in the text

- Set for individual work and brief whole-class feedback.

- Point out that learners frequently confuse *it* and *this*. You might want to elicit the difference between them before referring students to Language Box 2.

Answers

it: line 9, line 10, line 11, line 12, line 20

these: line 13, line 18

this: line 14

they: line 21

B Identify pronoun references

- Set for individual work. If students are unclear about the exercise, do an example with the whole class, e.g., write on the board: *The answers to questions like these (line 13)* and elicit what these refers to (*the questions asked in lines 7–12*).

Answers

1 it (line 10): the product

2 this (line 14): the design brief

3 these (line 18): the sketches, drawings and models

4 it (line 20): the prototype

5 they (line 21): the engineers

C Correct pronoun use in product descriptions

- Set for pairwork discussion and whole-class feedback. Emphasise that it is important for students to check their own work and correct errors such as these.

- Elicit the fact that in these sentences *this* refers to the information in the whole of the preceding sentence.

Answers

1 The Wankel rotary engine has neither a crankshaft nor a flywheel. <u>This</u> means that <u>it</u> is lighter and can produce more power than a normal engine of the same size.

2 The ACV (hovercraft) flies on a cushion of air a short distance above the ground. <u>This</u> means that <u>it</u> can move easily on land or water.

- Now have the students do the Workbook exercises for Unit 1, Lesson 2.

Workbook answers

Writing: descriptions

A Complete the paragraphs

Microwave ovens <u>use</u> high-frequency waves of heat to cook food. This means <u>they</u> can <u>cook</u> faster and <u>use</u> less power <u>than</u> ordinary cookers.

Digital cameras <u>don't</u> <u>use</u> photographic film. They <u>have</u> <u>a</u> memory chip instead. This <u>means</u> <u>they</u> <u>are</u> lighter and can store more photos <u>than</u> <u>ordinary</u> <u>cameras</u>.

B Write paragraphs about other products

Answers depend on the students.

Language: describing products

A Write key words from the text

1 component		**6** regulations	
2 drawing		**7** sketches	
3 specification		**8** models	
4 materials		**9** prototype	
5 conditions		**10** mass production	

B Complete the sentences with key words

1 mass production	**6** sketch
2 model	**7** Materials
3 conditions	**8** drawings
4 component	**9** prototype
5 regulations	**10** specification

Lesson 3   Materials and properties

Objectives

To review and extend language for describing appearance and properties of materials.

To read and summarise the contents of a text.

Language

Vocabulary connected with material properties, question forms.

Vocabulary and speaking

A Identify materials in the classroom

- Lead in by giving an example description of a familiar item for students to guess. Use some of the exponents in the Language Box.

- Set part *a* for group work or do it together as a class.

- Encourage students to look up the materials in a dictionary, and draw their attention to the phonemic transcriptions. Even if students are not familiar with phonemic script (specifically the International Phonetic Alphabet (IPA)), they should be able to guess the words in part 2. Explain how knowledge of IPA can help students find out how to pronounce new words, because phonemic transcriptions are used in learner dictionaries.

- **Note:** It is not necessary for students to study and memorise the whole alphabet, and they will not be tested on it. However, they should become familiar with phonemes as they encounter them during the course. You may want to point out that IPA refers to received or southern standard pronunciation and that there may be regional differences, e.g., with the length of vowel sound in words like *glass*.

Answers

Answers depend on the students.

B Match questions and answers

- Set for individual work and pair-checking. Point out that the pattern for question *2* can be used with other verbs, e.g., *Can it be used for repairing/moving/making/things?*

Answers

1	c	**4**	f
2	e	**5**	d
3	b	**6**	a

Object: door handle

C Discuss materials used in car and aircraft manufacture

- Set the task for pair or group work. Prompt students if necessary, but encourage them to paraphrase rather than pre-teaching the names of unknown materials.

- Conduct whole-class feedback, going over any new vocabulary.

Suggested answers

1 aluminium, steel, (reinforced) glass, synthetics, e.g., plastics, fibreglass

2 aluminium, steel, (reinforced) glass, synthetics, e.g., plastics, fibreglass

Reading

A Choose a sentence to summarise a text

- Set the task for individual work and pairwork discussion.

- Feed back on the task. You may also wish to discuss the materials that are mentioned in the text and whether they are the same ones students thought of in Speaking Exercise C.

- **Note:** It is an important skill for students to be able to extract the main ideas in a text as well as understand the text at sentence and word level.

Answer

1 is wrong because although modern jets are discussed in paragraph 2, it is not the topic of the whole text

2 is probably the best summary

3 is inaccurate because some traditional materials are still used, e.g., steel

B Match questions to material properties

- If students are already familiar with the terms for material properties, set the first part of the task for pairwork and monitor for problems. If students have problems

with the lexis, you may need to elicit or explain the meaning of the terms before they do the task; alternatively, students could look up the terms in a dictionary or in the Glossary.

Answers

1 electrical resistivity

2 brittleness

3 ductility

4 conductivity

- Elicit suggestions for additional questions for some of the other properties before asking students to work in pairs or groups to come up with their own ideas.

- Conduct whole-class feedback, eliciting ideas from different groups and suggesting questions where necessary.

- **Note:** Reformulate incorrect questions or encourage students to self-correct, but don't spend too long going over problems with syntax at this point. The focus should be more on checking comprehension than on testing grammatical knowledge.

Suggested answers

How much can its length increase? (coefficient of linear expansion)

Can the surface be damaged easily? (hardness)

Is it suitable for the pistons of a car engine? (toughness)

Can it be moulded into different shapes? (malleability)

Can it withstand wear and tear? (toughness)

Does it return to its original shape after stretching? (elasticity)

Does it turn liquid when it is heated? (melting point)

Does it float in water? (density)

How much heat is needed to raise its temperature by 1 degree Celsius? (specific heat capacity)

Can it withstand heavy loads? (strength)

- Now have the students do the Workbook exercises for Unit 1, Lesson 3.

Workbook answers

Language: properties of materials

A Match properties of materials to the definitions

hardness	the ability of a material to withstand abrasion or surface scratches
melting point	the temperature at which material changes from a solid to a liquid state
electrical resistivity	the degree to which a material opposes an electrical current running through it
elasticity	the ability of a material to return to its original length
malleability	the ability of a material to be moulded into other shapes by bending, hammering, rolling, pressing or pulling
ductility	the ability of a material to be stretched until it is long and thin like a wire or string
brittleness	the tendency of a non-elastic material to crack and break rather than bend
conductivity	the ability of a material to conduct heat or electricity
toughness	the ability of a material to withstand the natural elements

B Use prompts to make questions

1 Can it absorb heat?
2 How heavy is it?
3 At what temperature does it melt?
4 Does it corrode easily?
5 Can it be flattened?
6 Does it go back to its original shape?
7 How well does it conduct electricity?

C Decide which property is referred to

1 specific heat capacity
2 density
3 melting point
4 toughness
5 malleability
6 elasticity
7 conductivity

Language: word-building

A Complete the table

noun	adjective
heat	hot
conductivity	conductive
resistivity	resistive
density	dense
toughness	tough
elasticity	elastic
strength	strong
brittleness	brittle
malleability	malleable
ductility	ductile

B Find words that fit the stress patterns

Suggested answers

Oo	toughness
Ooo	brittleness
oOo	conductive
oOoo	ductility
ooOoo	conductivity
oooOoo	malleability

C Check your pronunciation

Answers depend on the students.

Lesson 4   An amazing material

Objectives

To practise listening and note-taking.

To give a short talk using notes.

Language

Vocabulary connected with material properties, question forms.

Listening and note-taking

A 🔊 **Listen for gist: identify the topics that are mentioned in the listening text (CD1 Track 2)**

- Use the visual to lead in to the topic. Elicit what students know about spider webs.

- Allow time for students to read through the list of topics and check that they understand key terms such as *tensile strength* and *synthetic*.

- Set the activity for individual work and pair-checking before conducting brief feedback.

Answers

2 the tensile strength of steel

4 properties of spider silk

5 elasticity

6 difficulties for commercial use

7 synthetic spider silk

8 how spiders use their silk

B 🔊 **Listen and take notes (CD1 Track 2)**

- Explain that students will listen again and take notes. Elicit any problems students have with note-taking before referring them to the strategies in the Skills Box. You might want to remind them of the abbreviations used for quantities in Lesson 1 and elicit examples of other useful abbreviations.

- Students should write the headings in the appropriate space with gaps between each heading for their notes.

- Play the recording without pausing while you monitor to check that students are writing notes. If students are having difficulties, you could play the recording a third time, pausing it after each topic to give time to write.

Answers

Answers depend on the students.

C **Compare and edit notes**

- Give students an opportunity to compare, discuss and add to their notes before feeding back. Go around and monitor as they do this.

- Feed back by building up a model set of notes on the board as you elicit ideas from the class. Alternatively, show the model notes below (also in the Workbook) on OHP after students have discussed their own ideas and elicit any suggestions for changes and additions.

Suggested answers

1 **How spiders use their silk**

- wrap small creatures in

- make shelters

- lifelines

2 **Properties of spider silk**

- adhesive: v. sticky

- strength: up to 5x stronger than steel

- tensile strength: t s of radial thread = > 1,000 Pa c/f mild steel = 400 Pa

- weight: v. light – 25% lighter than synth plastics

- keeps strength down to v. low temps (-40°)

3 Difficulties for commercial use

– almost imposs. to produce in useful quantities

– can't farm spiders!

4 Synthetic spider silk

– Canada: scientists trying to produce spider silk from goats' milk

Speaking

A Make notes about a material

- Students should prepare their notes individually, but can plan individual presentations or work in pairs. Encourage students to choose a wide range of materials. These could be traditional, e.g., types of metal, wood, etc., manufactured, e.g., stainless steel, nylon, or even patented brands, e.g., Gore-tex, Velcro. See the materials section of **www.wikipedia.org** for information and ideas.

- Monitor and help with ideas where necessary.

Answers

Answers depend on the students.

B Practise presenting information

- Emphasise the importance of practising the presentations, because it is not always easy to present information using notes effectively.

- If students lack confidence, you could scaffold the activity by suggesting how to expand their notes, e.g., *It is very* + adj. *It contains … It has a high degree of … In the past it was used for … It is now used for … It could be used for … in the future.*

- **Note:** It is not necessary for presentations to be word perfect. This is a 'deep-end' activity that enables you to discover what speaking skills students need to work on in the future.

Answers

Answers depend on the students.

C Present and answer questions about a material

- Presentations could be given in pairs or individually in front of the whole class, or in smaller groups. Note down positive aspects and any problem areas to address in feedback afterwards.

Answers

Answers depend on the students.

- Now have the students do the Workbook exercises for Unit 1, Lesson 4.

Workbook answers

Reading: note-taking

A Read the model notes

Answers depend on the students.

Language: abbreviations

A Write the abbreviations for the words

1 impossible = imp.

2 synthetic = synth.

3 greater than = >

4 compare with = comp. w.

5 very = v.

6 Pascals = Pa

7 degrees = °

8 temperature = temp

B Write sentences to summarise the talk

Answers depend on the students.

Lesson 5 — Design specifications (1)

Objectives

To practise discussing design features and their rationale.

To transfer information from textual data to diagrams.

To use new vocabulary in context.

Language

Vocabulary connected with technical drawings and considerations in airliner design.

Speaking and vocabulary

A Discuss information given by drawings

- Use the pictures to introduce the idea of airliner design.

- Point out the rubric above the pictures. Students could suggest which of the three designs they prefer.

- Set for pairwork. As you monitor, note some of the design features the students mention.

Answers

Answers depend on the students.

B Discuss design constraints

- Point out some of the design features the students mentioned in Exercise A as a way into this exercise.

- Students clarify vocabulary difficulties in the list 1–8. They should do this in pairs and/or using their dictionaries.

- Make sure that students can pronounce the items 1–8 correctly, as they will be used extensively in the following discussion.

- Run the discussion as a pyramid. Firstly, individuals work on numbering the items. They then compare their ranking with a partner and negotiate an agreed ranking. Groups of four then discuss the relative merits of their rankings to arrive at a final agreement.

- **Note:** It does not matter if there is no agreement at the end – the objective is to discuss design features and their rationale.

Answers

Answers depend on the students.

Reading and vocabulary

A Check vocabulary

- Point out to students that they are going to read about types of technical drawing which give different types of information.

- Students silently read the text once, without dictionaries.

- There are 14 bold items in the text. Students may know some of them already, but if not, this task is potentially time-consuming. To save time, ask pairs of students to divide the items between them, look them up where necessary – without making notes – and then explain them to each other.

- Feed back with the whole group to ensure that students have the correct meanings. Then ask students to record the items. Point out that it is important to note the parts of speech, and encourage students to copy the phonemic symbols.

- **Note:** You could ask students to copy down definitions which you have prepared on an OHT to save time. However, here and elsewhere in the lessons, effective use of the dictionary is a skill which is also being built. Remind students, if necessary, that they should have a technical dictionary of their own.

Answers

orthographic projection: a two-dimensional graphic representation of an object in which the projecting lines are at right angles to the plane of the projection

two-dimensional: having the dimensions of height and width only

three-dimensional: having, or seeming to have, the dimension of depth as well as width and height

visualise: to form a mental image of

pictorial drawings: show an object like you would see in a photograph

perspective: a technique of depicting volumes and spatial relationships on a flat surface

technique: the manner and ability with which someone like an artist employs the technical skills of a particular art or field of endeavour

geometry: the branch of mathematics that deals with the deduction of the properties, measurement and relationships of points, lines, angles and figures in space from their defining conditions by means of certain assumed properties of space

optics: the branch of physical science that deals with the properties and phenomena of both visible and invisible light, and with vision

impression: the first and immediate effect of an experience or perception upon the mind; sensation

point of view: a specified or stated manner of consideration or appraisal

two-point perspective: a mathematical system for representing three-dimensional objects and space on a two-dimensional surface by means of intersecting lines that are drawn vertically and horizontally and that radiate from two points

eye-level: vanishing points on a horizontal line

three-point perspective: a mathematical system for representing three-dimensional objects and space on a two-dimensional surface by means of intersecting lines that are drawn vertically and horizontally and that radiate from three points

B Read and label the pictures

- Students first read the text again right through to experience the contextualised use of the vocabulary items. Point out combinations such as *give the impression*, *produce drawings*.

- Pairs or individuals then label the pictures *1–3* and peer-check their answers.

Answers

picture 1	two-point perspective
picture 2	three-point perspective
picture 3	one-point perspective (this isn't in the text, but students may be able to guess the answer to picture 3)

C Discuss advantages of each type of drawing

- Again focus on correct pronunciation of the target vocabulary items: this discussion uses them and reinforces the concepts in the text.

- **Note:** As an extra task, you might ask students to sketch, and describe the features of, a forward-looking design of their own. It does not matter if these are fanciful and unscientific.

Answers

Answers depend on the students.

- Now have the students do the Workbook exercises for Unit 1, Lesson 5.

Workbook answers

Reading: review

A Choose the correct words

geometry

perspective

two-point

vanishing

horizontal

pictorial

orthographic

Language: nouns and adjectives connected with geometry

A Draw geometrical figures

1 rhombus (*n*)

2 right-angled triangle (*n*)

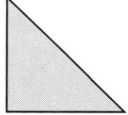

3 semi-circle (*n*)

4 octagon (*n*)

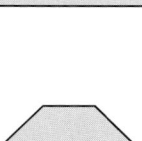

5 cube (*n*)

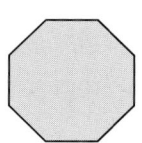

6 sphere (*n*)

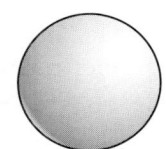

7 cylinder (*n*)

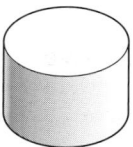

8 cone (*n*)

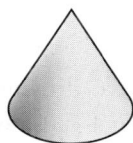

9 pyramid (*n*)

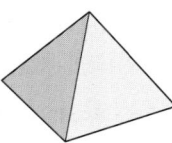

10 helix (*n*)

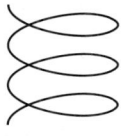

B Adjective forms

1 rhomboidal

2 triangular

3 semicircular

4 octagonal

5 cubic

6 spherical

7 cylindrical

8 conical

9 pyramidal

10 helical

Objectives

To practise speculating about the future.

To extend vocabulary in aircraft specifications.

To revise question forms.

Language

will + *certainly/definitely/probably* for speculating about the future.

Vocabulary and question forms for aircraft specifications.

Mathematical and question forms for aircraft specifications.

Mathematical symbols, units of measurement.

Vocabulary

A Discuss what the picture shows

- Set the task for pairwork. Stress that there are no correct answers required from this discussion.

- Encourage students to use *might, may, could* and *must* to speculate during their discussion. You may prefer to leave this until feedback with the whole group after their discussions. Do not spend too long focusing on this area of grammar. The target language for the lesson is *will*, but *might*, etc., may serve as a useful contrast in the Language and speaking section.

Answers

Answers depend on the students.

B Underline surprising or interesting information

- This exercise introduces the mathematical symbols and units to be focused on in the subsequent exercises.

- Encourage students to clarify for each other any vocabulary difficulties. Feed back using the whiteboard and ensure that the students are pronouncing the vocabulary correctly. They will need it in the rest of the lesson.

- Point out the abbreviations *no.* and *max.*, which mean *number* and *maximum*, respectively.

- Students do the exercise individually and then feed back to a partner explaining why they are surprised or interested by the things they marked.

Answers

Answers depend on the students.

C Write the units

- Check the spelling and pronunciation of these. Stress to students that they must be able to say and understand them accurately.

- Point out the spelling alternatives, e.g., *metres/meters, kilograms/kilogrammes, kilometres/kilometers*, etc., and the pronunciation alternatives, e.g., American English puts the stress on the first syllable of *kilometers*, whereas British English places the stress on the second syllable.

Answers

1 metres
2 kilos/kilograms
3 kilometres per hour
4 horsepower
5 cubic metres

D Write the mathematical symbols

- Again, ensure correct pronunciation.
- Point out the alternative(s). In *3*, it is *greater than*; in *5*, it is *times* for calculations, but *by* is more common and versatile; and in *6*, it is *three over four*, etc.

Answers

1 plus or minus

2 per cent

3 less than, more/greater than

4 divided by

5 multiplied/times by

6 a quarter, a half, three quarters

E Units and symbols used in students' fields

- Students may not have experience of technical fields. Encourage them to think of the science and mathematics they have studied at school, e.g., *mathematical calculations*, *dimensions*, *temperature*, *volume*, etc.
- Pairs should explain what the units and symbols refer to.

Answers

Answers depend on the students.

F Calculations

- Pairs or threes can try to do these calculations. Some use of L1 is likely during discussion, but feedback, either to the group or other groups, should be in English. Help students to prepare this.
- There may be alternative formulae proposed. Encourage students to explain and defend their proposals. The idea is to get students using the terms from the previous exercise.

Answers

1 πr^2: pi times r squared [r = radius of the circle]

2 $4\pi r^2$: (four times pi times r squared [r = radius of the sphere]

3 *6 mph* miles per hour = *10 kph* (kilometers per hour), therefore, *30 mph = 50 kph*

4 x^3: length times width times height

Language and speaking

A Question formation

- Set for pairwork and check grammatical accuracy.
- Students will probably need vocabulary such as *long, wide, carry, far*. Model some questions using the language.
- You may wish to divide the 16 questions among the group to save time if necessary.
- Common problems in the question formation, and vocabulary arising from the exercise, can be dealt with for the whole group in feedback.

Answers

Answers depend on the students.

B Ask and answer questions about other aircraft specifications

- Emphasise that students will be looking at different information sheets on different pages. They should not look at their partner's information.
- Divide the class into A and B students. Student A should look at the information sheet on page 242 and Student B should look at the information sheet on page 243. Demonstrate the information-gap activity with a strong student, if necessary, taking it in turns to ask questions about the different aircraft features (recycle questions from the previous exercise).
- Set the exercise for pairwork and monitor for difficulties.
- Conduct whole-class feedback, eliciting the main differences between the two aircraft.

Answers

Answers depend on the students.

C Discuss future air travel

- Refer students to the Language Box. The use of *will* to talk about the future is likely to be familiar to them. Highlight the use and position of the adverbs. Students may also propose *going to* and *-ing* forms to refer to the future. Decide whether you want to go into these as well, but make sure the discussion practises *will* + adverb.

- Set for small groups.

- Feed back to the other groups.

Answers

Answers depend on the students.

- Now have the students do the Workbook exercises for Unit 1, Lesson 6.

Workbook answers

Writing: form-filling

A Sketch the airliner of the future

Answers depend on the students.

B Complete the form

Answers depend on the students.

Language: predictions

A Complete the sentences with predictions

Answers depend on the students.

Lesson 7   Design rationale

Objectives

To practise discussing the function of airframe components.

To transfer items from word lists to diagrams.

To use new vocabulary in context.

Language

Vocabulary for major components of a jet fighter.

Expressions for reason and purpose, e.g., *X is for ~ing*.

Compound nouns around airframe design.

Vocabulary

A Wordsearch

- Clarify that the words can run in any direction in a straight line.

- Set the exercise for pairwork. Students can refer to the word lists at the back of their book for help.

Answers

across: engine, radar, display, tank, foreplane

down: missile

diagonally upper l to bottom r: airbrake

B Complete the diagram

1 Students can work in pairs or groups to discuss the options here. They should not worry about the fact that some terms are unfamiliar to them. The idea is to guess where the more familiar words from the wordsearch could go.

Answers

a	engine	**d**	radar
b	airbrake	**e**	foreplane
c	display	**f**	tank

2 Direct students to the word lists at the back of the book or to their dictionaries. As usual, check pronunciation, as these core words will be required again in discussions later (and throughout the course).

Answers

Answers depend on the students.

C Add labels to the diagram

- No lines are given in the picture to indicate where these items are.

- Encourage students to try to deduce enough of the meaning of the items from elements such as *flight, fuel, tank, medium*, etc., to formulate an initial guess. They can work in pairs or groups to do this. Point out to them that the 'shape' of a word – the way it is built using affixes, etc., – is revealing with regard to its meaning (a lot of work is done on the course on this point).

- Dictionaries may then be used to confirm students' ideas.

- Feed back in pairs. Then feed back with the whole group, e.g., using a drawing on the whiteboard.

Answers

The *retractable flight refuelling probe* is the right-angled tube outside the cockpit on the pilot's right.

The *external fuel tank* is the long smooth clear object under the wing, next to the wheel.

The *medium-range active missile* is one of the missiles under the wing.

D Make compound nouns

- Point out that in the examples in Exercise C, the main noun – *probe, tank, missile* – is in the final position in the string. This is usual in English.

- Set for group or pairwork. Students should focus on aircraft design rather than simply using *large/small/new/old/latest*, etc. There are many possible solutions. Students should aim to make one correct compound per item. Allow compounds such as *aircraft*. Make sure that students are aware of any incorrect combinations they propose. Outstanding queries can be checked online, e.g., **www.wikipedia.org**.

- Feed back to partners.

Suggested answers

a	starboard, port wing: wing surface, section
b	long, short fuselage; fuselage section, length, design
c	tail section, surfaces
d	wing, control surface; surface area
e	aviation fuel; fuel tank
f	single, jet, turboprop, radial, rotary, piston engine; engine housing, block
g	port/starboard wing
h	maiden flight; flight, path, test
i	carbon, tempered, stainless steel; steel parts, components
j	new, latest, special design; design rationale, features, phase, team
k	new, specialised materials; materials design, development
l	electronic, main component; component design, parts
m	high/low, production, manufacturing costs
n	full, large, low capacity

o   air brake, supply, conditioning, traffic, force, crew; aircraft, airport, airframe

Speaking and writing

A Speculate on the design rationale

- This can be done as a whole-group activity, with you eliciting ideas. Alternatively, use small groups.

- Emphasise that students are not expected to know about the rationale for the airframe design. This task reuses the language of speculation, as well as encouraging students to think analytically.

Suggested answers

1   Short-range missiles might be smaller, and therefore lighter, so they go on the weaker outer section of the wing. / Perhaps the air flows better around the aircraft if the larger medium-range missiles are kept close to the fuselage. / When they are fired, medium-range missiles could put a lot of strain on the wing, so they go on the stronger inboard section.

2   They could be for steering if they are moveable. / They might be for better stability at high/low speed.

B Describe reasons and purpose of design features

- Refer students to the Language Box. They should read it silently and then discuss it with a partner.

- Demonstrate, perhaps using a whiteboard conversation-build, several examples of this exchange.

- Students work with new partners/small groups.

- Ask students to write down some examples of the rationales they have discussed. Emphasise grammatical accuracy.

- Refer students to the many examples on the Internet of future airframe design development, such as **www.silentaircraft.org**.

These will be of immediate interest to them and of long-term use in language and conceptual terms.

Answers

Answers depend on the students.

- Now have the students do the Workbook exercises for Unit 1, Lesson 7.

Workbook answers

Language: compound (two-word) nouns

A Complete from memory

turbofan <u>engine</u>   head-up <u>display</u>

short-range <u>missile</u>   ejector <u>seat</u>

fuel <u>tank</u>   spine <u>airbrake</u>

pulse doppler <u>radar</u>   port <u>foreplane</u>

B Make compounds or phrases

Suggested answers

1   brake, plane, port

2   attendant, plan, path

3   turbofan, piston, jet

4   short-range, long-range

5   specialised, new

Reading and writing: note-taking

A Label the diagram

1   vertical stabiliser   3   airbrake

2   turbofan jet engines   4   wing spars

B Make brief notes

Answers depend on the students.

Objectives

To practise referring to illustrations to predict text content, and reading for specific data.

To discuss broader issues of design and development.

Language

Vocabulary for issues surrounding design projects.

Review of the form and concept of passives with present and past reference.

Vocabulary

A Suggest reasons why a design project might fail

- Elicit ideas from the group. This stage should be brief.

Answers

Answers depend on the students.

B Categorise reasons why projects are abandoned

1 This task contains, potentially, a lot of new vocabulary. Encourage dictionary and pairwork to scaffold it.

2 Check and correct pronunciation, highlighting word stress and word-group stress as important to meaning in English.

3 Students feed back in pairs or small groups. Point out that the politics and economics categories overlap to some extent, and that different solutions are permissible as long as students can justify their opinions.

4 Elicit ideas from the group.

Sample answers

politics: negative public opinion; changing commercial or military needs

design: accidents during trials; delays in production

economics: need for investment in new infrastructure; withdrawal of funding; lack of interest from potential buyers; innovative production techniques are expensive; high manufacturing costs; competition from other companies

Reading

A Predict and check text content

1 Refer students to the Skills Boxes. Have students read them silently. Check understanding.

Answers

Answers depend on the students.

- Direct students to look at the text, the title, layout and picture, but not to read it. Elicit ideas about the content and origin of the text, e.g., onto the whiteboard.

2 Students read silently. Feed back as a whole class to show the accuracy of their predictions.

Answers

Answers depend on the students.

B Correct the information

- Set for individual work and pair-checking. Provide answers, e.g., on OHP.

- As a follow-up, elicit or ask pairs to discuss why this particular project was abandoned, with reference to the factors in Exercise A.

Answers

1 The American government put $18 million into the project.

2 The total cost of the project was $25 million.

3 The plane spent 33 years in storage. / The plane was sent to a museum after Hughes died.

4 The plane only flew once.

5 Hughes designed the plane in 1942.

Language

A Discussion of the use of passive forms

• Technical English uses a lot of passive verb structures. These are dealt with through the course, although it is expected that students will have come across them before. In this task, the form and function of passives is initially reviewed.

• Students discuss the answer to the question, allowing them to confirm and review their knowledge. Then elicit their views.

Answer

It is more important to highlight the problems than the team who can't solve them.

B Find examples in the text

• This is a scanning exercise – students do not need to reread the text in detail. Encourage them to scan for forms of *be* and or *~ed* verb forms.

• During feedback, point out to students the context of the target items to raise awareness of their grammatical function and collocations, such as *scrap something for financial reasons*.

Answers

first paragraph: [sometimes] the project is scrapped for financial reasons; after millions of dollars have been spent; by the time a prototype has been produced

second paragraph: most of the aircraft was built of alternative materials

final paragraph: many years were spent on research and development; the plane was sent to a museum

• Students may know of other aviation white elephants. Ask them to look at:

http://www.defencetalk.com/f-35-looking-more-like-white-elephant-31347

• Students can give their evaluation of the problems affecting different stages of this project.

• Now have the students do the Workbook exercises for Unit 1, Lesson 8.

Workbook answers

Language: problems with design products

A Match the words to make phrases

1	d	5	g
2	e	6	c
3	f	7	a
4	b		

Language: the passive voice

A Rewrite the sentences

1 Factories have produced brake components in Bristol for several years.

2 The designers first built a prototype in 1926.

3 The government and the sponsors abandoned the project after only two years.

4 People will discuss the exact reasons for the failure for many years.

B True or false

1	T	3	F	5	F
2	T	4	T		

Lesson 9 — Lost classics

Reading

A Study the drawings

- Point out that it is not important that students' suggestions are factually correct; they should be based on the information provided and reasonably explained.

- Elicit some ideas from the group with reference to the pictures only.

- Then get students to read individually and either note or circle significant information in the specifications.

- Finally, pairs should compare and justify their ideas.

Sample answers

cost of development

no significant design advantage over rivals

lack of client confidence in novel designs

Brabazon:

- unusual shape of the design may have been difficult to produce, e.g., long and narrow

- small passenger capacity

- large overall dimensions: special hangar?

- large overall dimensions: expensive to build

- eight engines: high-maintenance, mechanical problems multiplied

Custer:

- need for specialist maintenance

- limited range and ceiling

- low speed

Vocabulary

A Estimate the quantities

- Refer the group to the aircraft specifications and the fact that the figures are given in metric and imperial units.

- Choose one value and elicit the equivalence, e.g., 1 ft expressed in metric. Students will probably need a calculator.

- Set for pairwork. Remind students that the answers are estimates.

- Feed back on the board.

- Point out that it is important that students learn these equivalences by heart.

Answers

1	305	**4**	121
2	1,102	**5**	1,076
3	497	**6**	15

Language

A Underline the time expressions

- Set for individual work and pairwork checking.

- Feed back onto the board/OHP.

Answers

1 By the end of 1969

2 Until the 1970s

3 no longer

4 Finally/a year later

5 After (three years' work)

6 While

7 then/for the next 12 months

8 Initially

9 From that moment on

10 since the beginning

B Look at the Language Box

- Ask students to read the Language Box and discuss it with a partner.

- Clarify any outstanding problems.

- Set the exercise for pairwork. Monitor the group and help as necessary.

- Feed back on the board. Go over common patterns, e.g., *while* is generally used with continuous tenses (and the verb *to be*); *since* is often used with perfectives.

Answers

1 PS	5 PS	8 PS
2 PP	6 PC	9 PS
3 PS	7 PS	10 PP
4 PS		

- Now have the students do the Workbook exercises for Unit 1, Lesson 9.

Workbook answers

Language: units of measurement

A Write units of measurement

Suggested answers

Length, width and height: centimetres (cm), metres (m), feet (ft), inches (ins), millimetres (mm), yards (yds)

Area: square metres (m^2), square centimetres (cm^2), cubic centimetres (cm^3)

Volume/capacity: cubic centimetres (cc), litres (l)

Weight: grams (g), kilograms (kg), milligrams (mg), pounds (lb)

Power: kilowatts (kW), milliwatts (mw)

Speed: miles per hour (mph), kilometres per hour (kph)

Writing: comparing two things

A Write sentences

1 Bristol Brabazon, Custer

2 Bristol Brabazon, Custer

3 Bristol Brabazon, Custer

4 Bristol Brabazon, engine

5 Answer depends on the students.

6 Answer depends on the students.

B Write a short report

Answers depend on the students.

Lesson 10 Design: Review

Objectives

To review the language and concepts covered in Unit 1.

To practise describing aircraft orally and in writing.

Language

Vocabulary from the unit to describe aircraft.

Phrases for describing an unsuccessful aircraft project.

Introduction

- Utilise the visuals. Elicit the names of the aircraft, why they were special and what can students remember about them.

Answers

The aircraft are the Bristol Brabazon and Custer Channel Wing (Lesson 9).

Vocabulary

A Review vocabulary for describing aircraft

- Point out that the words in the box are all words students have seen in Unit 1.

- Get pairs to discuss what they mean. Feed back orally.

- Set the task for individual work and pairwork checking. Feed back orally.

Answers

1 It was the largest plane ever built in Britain and had a <u>wingspan</u> of 70 metres.

2 It was designed to carry a maximum of just 100 <u>passengers</u>.

3 There was a lot of vibration, which made the <u>cabin</u> noisy.

4 It could fly at very low <u>speeds</u> because of its unusual design.

5 The great <u>hangar</u> that was built for it is now used for the Airbus A380.

6 It never became a commercial aircraft, although several <u>prototypes</u> were built.

7 It flew several times at air shows but never went into full <u>production</u>.

8 Its eight powerful engines allowed it to lift and <u>take off</u> quite easily.

B Identify planes from descriptions

- Set for individual work and pairwork checking. Feed back orally.

Answers

1	B	5	B
2	B	6	C
3	C	7	B
4	C	8	B

Writing

A Write a brief history of one of the projects

- Ask pairs or small groups to decide which of the options they prefer. Alternatively, you can decide, based on time/facilities available. The Internet research option provides a more genuine task in terms of information, and develops good learning strategies if students do at least part of the task outside class time. Stress that they should avoid copying the Internet or other sources verbatim. Writing Exercise D can then be done in class.

- Useful websites may include:

Custer Channel Wing

http://www.custerchannelwing.com

http://www.maam.org/aircraft/ccw5.html

Bristol Brabazon

http://www.century-of-flight.net/Aviation%20history/coming%20of%20age/Bristol%20%20Brabazon.htm

http://www.aviationarchive.org.uk/stories/pages.php?enum=GE121&pnum=0&maxp=9

B Complete a table with useful phrases

- Clarify that the phrases in the exercise will be useful for the writing activity.

- Set the task for individual work and pairwork checking.

- Feed back onto the board.

Answers

need for a new aircraft	design brief	research and development	reasons for failure	specifications
• There was no existing aircraft which	• These requirements were very demanding • It had to be able to	• A prototype was completed • Traditional materials were not suitable • Much time was spent on	• Unfortunately • It had become clear that	• length; • weight (empty)

C Find similar useful phrases in previous lessons

- Point out or elicit that Lesson 8 contains other phrases that students may find useful for the writing task.

- Set for individual work and pairwork checking.

- Feed back onto the board.

Suggested answers

(from Lessons 6 and 8):

need for a new aircraft	design brief	research and development	reasons for failure	specifications
–	• It was required to • … were in short supply	• Many years were spent on • This cost a lot of money • He completed a prototype	• scrapped for financial reasons • They no longer needed it • The aircraft was too expensive	• No. of crew • range, etc.

D Write a text

- Set the task for individual work. Monitor closely and assist students who need help.

- When they are ready, students should exchange texts and read each other's work. They should comment on the positive and negative aspects of each partner's work, and ask questions as a follow-up.

Answers

Answers depend on the students.

Speaking

A Describe an aircraft using vocabulary from the unit

- Again, this task requires reference to the concepts in Unit 1 lessons, e.g., layout and materials.

- In pairs, students choose an aircraft and, in turns, describe it to a partner, focusing on: the materials used to build it, the structures and components that are visible in the photos, and any unusual or distinguishing design features. Encourage students to speculate about its size, capacity, etc.

- The partner who is listening to the description should ask follow-up questions and guess which aircraft is being described.

Answers

Answers depend on the students.

- If you have time, you could ask students to choose and research another aircraft for homework, and present the information they find in the next lesson.
- Now have the students do the Workbook exercises for Unit 1, Lesson 10.

Workbook answers

Language: the passive voice

A Rewrite the sentences – passive to active

1 Willard Custer conceived the idea of a channel wing plane.

2 Custer hoped that the channels in the wings would enable the plane to take off quickly.

3 The high rotation speed of the plane's propellers caused a lot of vibration.

4 The American National Air and Space Museum now owns the original CCW-1 plane.

B Rewrite the sentences – active to passive

1 A design study for a long-range bomber was done in 1937 by the Bristol Aeroplane Company.

2 At the end of the Second World War, the company was asked by the government to produce two prototypes.

3 Modern jet engines were developed by designers in the 1940s.

4 The project was abandoned by the company by 1953.

Reading: specifications

A Circle the numbers

eight:	lines 12 and 13
18:	line 17
17'2":	line 13
1947:	line 8
R-4360:	line 12
11,430 sq ft:	line 9

B Write what the numbers refer to

eight:	number of engines/number of propellers
18:	normal crew
17'2":	propeller diameter
1947:	first and only flight
R-4360:	engine model (Pratt & Whitney)
11,430 sq ft:	wingspan wing area

C Write questions

Sample answers

1 How fast is the HK1 seaplane's top speed?

2 How long is the HK1 seaplane?

3 How heavy is the HK1 seaplane when empty?

4 How big are the propellers?

5 How far can it travel with 12,500 gallons of fuel?

Unit test

A test for this unit is available at:

http://www.garneteducation.com/reps/
documents/1244/SDT-u1-test.pdf

Contact your local Garnet Education
representative for information about how to
access these resources.

Objectives

To raise awareness of the difference between noun and verbs in SVO and imperative sentences.

To practise giving descriptions of different tools.

Language

Vocabulary for the appearance and purpose of tools.

Comparatives.

Preposition + verb~*ing*, e.g., *It's designed for gripping.*

Introduction

- Write the introductory text on the board, including some spelling mistakes. Students should correct you, either by consulting each other about the finished text or as you write.

- Elicit ideas from the group about a) jobs in a workshop that can only be done by hand and b) the names of some hand tools.

Vocabulary

A1 Compare noun and verb forms

- Write on the board a few verbs and nouns that are well known to students.

  Students should decide which are verbs and which are nouns, and what the difference is between a noun and a verb. Point out that the noun and verb forms of some words are the same, e.g., *oil*, and that other words have suffixes when they change their part of speech, e.g., *tighten*.

- Refer students to Language Box 1 and remind them of SVO patterns in the sentences.

- Use the first item in the list as an example and elicit from the class the fact that it is a verb (it can also be an adjective). Set the task for individual work, then pairwork checking. Feed back orally, stressing the idea of action in a verb, including the verb *be*.

- Students will need to use dictionaries. If you wish, you can divide the list between two or more students and get them to feed back to each other. Feed back as a class orally at the end, noting words that could be put in both groups.

Answers

nouns: jaws, teeth, oil (can also be used as verb, but less common), blade, head, edge, handle (verb form less common)

verbs: clean, tighten, calibrate, bend (also noun, but less common), cut (noun form less common), sharpen, hit, put, cover (noun form less common), grip

A2 Match words with pictures of tools and use them in sentences

- Elicit or present the names of the different tools and write them on the board.

- Set the task for pairwork. Feed back by eliciting suggestions and writing them on the board.

Answers

a tighten; head; handle

b jaws; handle; grip; tighten

c teeth; handle; bend; grip

d head; hit

e calibrate; jaws

f teeth; blade; edge; handle; cut

A3 Make imperative sentences

- Choose one tool and elicit an instruction using some of the words that it is matched with, e.g., *Tighten the jaws of the vice slowly and carefully.*

Suggested answers

Put the vice away.

Use the screwdriver to tighten the screws.

Bring the pliers here.

Don't hit your fingers with the hammer.

Be careful with the saw.

B Purposes of tools – dictionary work

- Again, students can work individually and feed back in pairs, or in larger groups. Feed back orally as a class.

- Elicit other ideas from the group, e.g., *joining*, *breaking*, *cooling*, *heating*. It doesn't matter if the students do not express themselves using the exact word. If you wish, you can teach any new vocabulary to them as it arises.

Answers

Answers depend on the students.

C Match each of the uses with one of the tools

- Set for pairwork. Feed back orally.

Suggested answers

measuring: callipers

gripping: vice, pliers

bending: pliers

cutting: saw

pounding: hammer

punching: screwdriver

screwing: screwdriver

D Pronunciation

- Drill the pronunciation of the vocabulary.

Speaking

A Golden rules for hand tools

- Refer students to Language Box 2. As usual, they should read it individually and then discuss it in pairs to make sure they have understood. Use the examples to highlight the forms. Ask concept-check questions and clarify where necessary.

- Go through one example together as a class, e.g., *Saws are designed for cutting wood or metal. Keep them in good condition by cleaning, oiling, straightening and/or replacing the blade regularly. Keep them in a dry place to stop them from rusting.*

- Set the task for pairwork. Feed back orally. Accept any reasonable suggestions.

Answers

Answers depend on the students.

B Read a description of a hand tool

- Refer the students to the text in the table. Elicit that the description is of a saw.

Answer

A saw.

C Complete two more tables with descriptions of tools

- Check that students understand all the categories in the table, particularly *power* and *safety*.

- Remind them to use similar language structures to the ones given in the table in Exercise B. They should recycle some of the verbs from the lesson.

- Pairs can show their tables to another pair for them to identify the tools.

Answers

Answers depend on the students.

- Now have the students do the Workbook exercises for Unit 2, Lesson 1.

Workbook answers

Language: hand tools

A Match the tools to their uses

1	e	**6**	i
2	d	**7**	b
3	h	**8**	f
4	g	**9**	c
5	a		

Writing: descriptions of tools

A Complete the boxes

Answers depend on the students.

Lesson 2 — Blade manufacture

Objectives

To raise awareness of collocation in vocabulary learning.

To practise scanning.

To practise hearing words in a stream of speech.

Language

Vocabulary for metal manufacturing.

Note forms for information in a chart.

Contrast and comparison: *but*, *and*, *whereas*; comparatives.

Vocabulary and reading

A Match word combinations

- From Lesson 1, elicit *hand* and *power* as possible combinations with *tool*. Point out that this is a useful way of remembering and recycling vocabulary. Elicit any other examples that students may know.

- Do the first one in the exercise on the board as an example.

- Set for individual work and pairwork checking. Monitor and assist as necessary.

- Do not confirm or correct answers yet – students will read to check. Ensure that you praise combinations that are likely, even though they are not those actually in this text.

Answers

No answers at this point.

B Scan the text for word combinations from Exercise A

- Set a time limit. Students should scan quickly for combinations from Exercise A.

- Feed back onto the board.

- Give students time to read the text more carefully and elicit which swords (Damascene or Japanese) the combinations on the board refer to, e.g., Damascene swords were of *superb quality*.

Answers

superb – quality (line 2)

original – shape (lines 4–5)

long – history (line 9)

national – treasure (line 12)

complex – process (line 13)

rough – shape (line 14)

specialist – polisher (line 16-17)

razor – sharp (line 18-19)

C Gap-fill

- Ask students to read the whole text.

- Elicit the fact that the six words in the box are verbs.

- Point out or elicit the use of the passive voice and remind students that the passive voice is often used to describe manufacturing processes. Elicit the form of the verb required after *was/were* and after *by* and *after*.

- Do the first one with the whole class.

- Set the task for individual work and pairwork checking.

- Feed back orally.

Answers

It is believed that Damascus craftsmen started by <u>hammering/beating</u> a batch of low-carbon wrought iron into thin sheets. These were then <u>tied</u> tightly into bundles. Next, high-carbon cast iron was <u>heated</u> until it was molten, and then the bundles of wrought iron were thrown in. The air spaces between the sheets were <u>filled</u> with the molten metal, which had the effect of welding the bundles into a solid lump of metal. This was then <u>hammered/beaten</u> into the rough shape of a sword while it was still hot. Finally, after <u>cooling</u>, the blade was filed, ground and polished.

D Check pronunciation

- Drill and/or ask students to look up the pronunciation in their dictionary.

Answers

hammering/beating

tied

heated

filled

hammered/beaten

cooling

Listening

A 🔘 Listen for key words (CD1 Track 3)

- Direct attention to the Skills Box. Stress that students are not listening to understand the whole text at the moment, but to develop the skill of hearing items in a stream of speech. They should listen for infinitive, ~*ing* and ~*ed* forms.

- Play the recording once.

- Pairs compare answers. Feed back orally. Students will have the chance to see the instances of each word later.

Answers

cool:	1
beat:	2
tie:	0
fill:	0
heat:	6 (including one noun use and one *re*~ prefix)
hammer:	1

B 🔘 Complete the information about the process (CD1 Track 3)

- Clarify any vocabulary problems. Pairs discuss any ideas they have for the gaps.

- Play the recording once. Pairs compare answers.

- Feed back onto the board.

- Refer students to the tapescript and play the recording again as they follow the text. Students can complete any information still missing.

- Point out the uses of the words in Exercise A.

Answers

1 beating

2 hammers; air (bubbles)

3 rigid; cutting edge

C Compare manufacturing processes

- Elicit one of two differences between the two processes, highlighting language for comparison and contrast, e.g., *Japanese swords are more rigid/less flexible than Damascene swords.*

- Set for pairwork

- Feed back orally. Write a few examples on the board.

Answers

Answers depend on the students.

- Now have the students do the Workbook exercises for Unit 2, Lesson 2.

Workbook answers

Grammar: passive sentences

A Match the sentence halves

1	c	4	f
2	d	5	b
3	a	6	e

B Correct order of sentences

1	2d	4	5b
2	1c	5	4f
3	6e	6	3a

Pronunciation: ~*ed* endings

A Put the words into the correct column

/d/	/t/	/ɪd/
hammered	polished	fitted
tied	finished	adjusted
sharpened	gripped	heated
measured	produced	tested

Writing: expanding notes

A Write a short description of the manufacturing process

Answers depend on the students.

Objectives

To work on the use of common collocations.

To practise scanning a text for numbers and quantities.

To practise transferring information between formats: from a text into a table.

Language

To give a short talk using the new information in the reading text.

Vocabulary for talking about computer-controlled machines.

Speaking

A Brainstorm machine tools

- Set for pairwork, but do not spend too long on this.

Answers

Answers depend on the students.

B Discussion

- Encourage students to make educated guesses at the first and third questions.
- Feed back orally.

Answers

1 CNC machines are those that are automatically operated by commands that are received by their processing units.

2 Answer depends on the students.

3 Answer depends on the students.

Vocabulary

A Focus on collocations

- Refer students to the Vocabulary Box.
- Set for individual work and pairwork checking. Feed back orally.
- You may wish to get students to write sentences of their own using the collocations.
- Ask students if they can think of any more collocations with these four words. Accept suggestions which may be from outside the technical field as a demonstration of the point.

Answers

production:	mass production, speed of production
equipment:	computerised equipment
machines:	hand-operated machines
main:	main parts; main reasons

Reading and speaking

A Scan numbers in the text

- Set for individual work and pairwork checking. Set a time limit and stress that students should not try to 'read' the text.

Answers

5 (five):	lines 11 and 17
3 (three):	line 15
70:	line 6
1,000 (a thousand):	line 21

B Write what the quantities refer to

- Use the same procedure as for Exercise A. Feed back orally.

Answers

5: the number of axes used by a five-axis milling machine

3: the number of axes used by standard milling machines

70: (microns) the tolerance in the machining of some parts of the Eurofighter

1,000: the number of holes automatically drilled in the front of a Eurofighter fuselage

C Transfer the information

- Work through the first item with the group. Set for individual work and pairwork checking. Monitor and assist.

- Feed back onto the board.

Answers

	basic function	problem in the past	advantage of this system
ICY	to allow parts made for one aircraft to be fitted to another	manufacturing tolerances – margins of error – were wide	production is kept moving
FAM	to machine surfaces	cutting limited by work on only three axes	highly complex shapes can be produced quickly
drilling machine	to make holes in (load-bearing) panels	holes were drilled by hand	greater speed and accuracy

D Practise giving a talk

- Divide students into groups (As, Bs, and Cs). Each should rehearse a short talk about one of the machines (using the table as a prompt). The other two students should listen, ask questions and suggest improvements.

- If the class finds this difficult, the talk can be limited to three or four sentences that simply expand the information in the table. Monitor and assist. Stress that students should listen carefully to each other.

- If you have a small group, students can give their talks to the whole class after they have rehearsed them.

Answers

Answers depend on the students.

- Now have the students do the Workbook exercises for Unit 2, Lesson 3.

Workbook answers

Vocabulary: word combinations

A Make compounds/common phrases with a dictionary

Suggested answers

production: food production, production line, mass production

equipment: camping equipment, sports equipment, electrical equipment

tools: power tools, garden tools, cutting tools

process: consultation process, slow process, painful process

aircraft: fighter aircraft, aircraft carrier, light aircraft

Reading: comparing tools

A True or false

1	T	**3**	F
2	F	**4**	T

B Underline examples of comparatives

smaller (line 7)

lighter (line 7)

easier (line 7)

more torque (line 8)

higher (line 8)

more quickly and effectively (lines 9–10)

less expensive (line 12)

longer (line 16)

Writing: advantages of tools

A Write an advertisement

Answers depend on the students.

Lesson 4   Two different drills

Objectives

To practise listening for key information.

To raise awareness of the mechanics and function of sentence stress.

To practise using new vocabulary and structural patterns in different contexts.

Language

Use of grammar to reconstruct sentences from a text.

Contrast and similarity: *whereas; in the same way as*, etc.

Imperatives: *Make sure …; Never …*

Speaking

A Discussion

- Set for pairwork. As usual, encourage students to make educated guesses. Elicit one or two ideas to start. Point out that students' ideas need not be complex, e.g., *one is used by a dentist while the other is used in industry; one is much larger than the other*, etc.

Answers

Answers depend on the students.

Listening

A Speculate about the different drills

- Set for pairwork.
- Feed back orally, but do not confirm or correct suggestions yet. Students will listen to check their ideas. Encourage (reasoned) disagreement among the group.

B 🎧 Listen and check (CD1 Track 4)

- Play the recording once. Students feed back in pairs.

Answers

1	1	6	1
2	1	7	2
3	BOTH	8	BOTH
4	2	9	2
5	2		

C Make full sentences

- Set for individual work and pairwork checking.
- Encourage students to form meaningful, grammatically correct sentences, but emphasise that it is not necessary to reproduce the speaker's words exactly.

Answers

No answers at this point.

D 🎧 Listen and compare with the sentences on the recording (CD1 Track 4)

- Play the recording again, or look at the tapescript to compare the original sentences with students' versions.

Answers

1 The drill bit can be moved on more than one axis.

2 The handle is usually made of plastic.

3 Power is transmitted to the drill bit by the gears.

4 Its wide heavy base helps to prevent unwanted movement.

5 The pillar supports the worktable.

6 The drill is powered by compressed air.

7 It's unsuitable for drilling teeth.

8 The drill bit can be accurately positioned.

9 The position of the bit can be controlled by the handwheel.

E Sentence stress

- Refer students to the Skills Box. Clarify if necessary what is meant by stress, using sentences from the listening as examples.
- Students work in pairs to mark their sentences. They can mark the tapescript if they prefer.
- Feed back onto the board. Draw attention to the fact that the new or key information words are stressed.

Answers

1 The <u>drill bit</u> can be moved on <u>more</u> than one <u>axis</u>.

2 The <u>handle</u> is usually made of <u>plastic</u>.

3 <u>Power</u> is transmitted to the <u>drill bit</u> by the <u>gears</u>.

4 Its wide heavy <u>base</u> helps to prevent unwanted <u>movement</u>.

5 The <u>pillar</u> supports the <u>work</u>table.

6 The drill is <u>powered</u> by <u>compressed air</u>.

7 It's <u>unsuitable</u> for drilling <u>teeth</u>.

8 The <u>drill bit</u> can be <u>accurately</u> positioned.

9 The <u>position</u> of the bit can be controlled by the <u>handwheel</u>.

Vocabulary

A Dictionary work

- Students check in dictionaries. Feed back orally.

Answers

1 align
2 workpiece
3 turning tool
4 adjust
5 spanner

B Example sentences

- It is useful to encourage students to write personalised sentences to help them remember new vocabulary. Go through the first few as a group.
- Encourage students to think of more than one context where the vocabulary might occur.
- Monitor and assist.
- Weaker students can copy the dictionary examples if you wish.

Answers

Answers depend on the students.

Language

A Match halves of imperative sentences

- Get students to read through the sentence halves first. Clarify any outstanding vocabulary problems.
- Set the task for individual work and then pairwork checking.
- Feed back onto an OHP or the board.
- Point out the use of imperatives in the sentence heads. There are also examples of prepositions with –*ing* forms: *before starting*; *while operating*, etc.

Answers

1 Don't adjust the machine *while the machine is operating.*
2 Never clean up *with your hands. Use a brush.*
3 All loose clothing should be kept *away from turning tools.*
4 Make sure that cutting tools are correctly aligned *before starting the machine.*
5 Remove all chuck keys and spanners *before operating the machine.*
6 Always wear eye protection *while operating a drilling machine.*
7 Never put tools or equipment *on the drilling table.*
8 Slow down the feed as the drill breaks through the work *to avoid damaged tools and workpieces.*

B Write imperative sentences about another tool

- Set the task for individual or pairwork. Students should choose a tool that they are familiar with.
- Put prompts on the board: *Don't ..., Never ..., Always ..., All ... should be ..., Make sure that ...*

Answers

Answers depend on the students.

- Now have the students do the Workbook exercises for Unit 2, Lesson 4.

Workbook answers

Language: tool parts

A Complete the sentences

1 This machine does not move around on the workbench because it has a heavy <u>base</u>.
2 You need to check the alignment of those <u>teeth</u> on the saw blade.
3 Modern milling machines use a range of different-shaped <u>cutters</u>.

4 The <u>thread</u> on that screw is damaged.

5 A hammer usually has a wooden <u>handle</u>.

6 Dentists' drills use a small <u>bit</u> that can be accurately positioned.

7 This knife has a very sharp <u>blade</u>.

8 Tighten the <u>jaws</u> of the spanner so that you can loosen the nut.

Writing: comparing and contrasting

A Write sentences comparing two drills

Answers depend on the students.

Lesson 5   The history of the lathe

> ## Objectives
>
> To practise transferring information: from text to drawing; from text to flow chart.
>
> To practise skimming.
>
> To practise describing how a machine operates.
>
> ## Language
>
> Vocabulary for lathes.
>
> Operation of a machine: *It is powered by …; the X drives a Y*, etc.

Speaking

A Discuss the importance of modern technology

- Put the items on the board and clarify any vocabulary or pronunciation problems.
- Set the task for pairwork. If students have trouble coming up with ideas, write prompts on the board, e.g., *speed/ease of communication/better health/labour-saving*.

Answers

Answers depend on the students.

B Brainstorm important inventions

- Elicit other important inventions from the class. If you have time, have a class vote ranking the five most important inventions.

Answers

Answers depend on the students.

Reading and vocabulary

A Introduction to the lathe

- Set for brief pairwork. Point out that it is the same technology in all four pictures.

Answers

No answers at this point.

B Read the introduction

- Ask students to read the paragraph quickly.
- Feed back orally. It does not matter if students don't know the word *lathe* yet.

Answer

A lathe is depicted in all four pictures.

C Name modern 'turned' products

- You do not need to spend long on this. Students should be aware that household items, such as bowls, pens and bottles, may be turned, as well as components of most machines, e.g., rods, screws, washers.

Suggested answers

Bowls, pens, bottles, rods, screws, washers, etc.

D Read for gist and detailed information

- Encourage students to label as much as they can on the pictures, working in pairs or small groups without dictionaries. Then tell them to use dictionaries to finish the labelling.

Answers

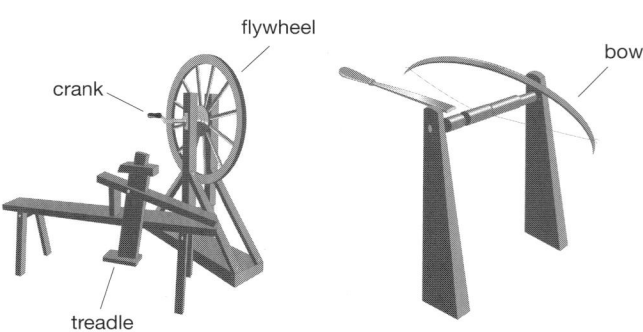

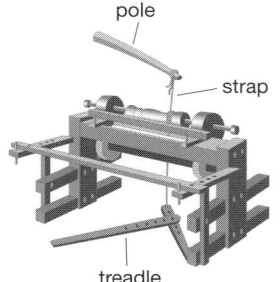

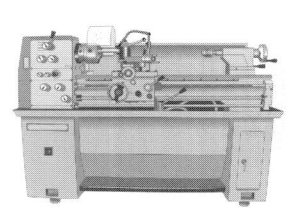

Language

A Put the vocabulary into the chart

- Get students to read the chart.
- Remind students of the similar chart they dealt with in Lesson 2. Point out that this chart moves from 1–3 in chronological order.
- Set the task for individual work and pairwork checking.

Answers

1 Early lathes: <u>reciprocal</u> variety: worked by spinning <u>material</u> between two centres and used: **a)** <u>pole</u> **b)** <u>bow</u> **c)** <u>strap</u>

2 The <u>treadle</u> <u>wheel</u> lathe: worked by foot-propelled continuous <u>rotation</u> and included **a)** <u>treadle</u> **b)** <u>crank</u> **c)** <u>flywheel</u>

3 <u>More</u> <u>complex</u> machines: partly responsible for the <u>Industrial</u> <u>Revolution</u>

B Describe the operation of a lathe

- Clarify any vocabulary problems.
- Elicit what part of speech comes after each of the expressions.
- Build up on the board notes around each expression.
- Set the task for pairwork.

- Feed back onto the board.
- **Note:** The fourth illustration, a modern lathe, does not have any of the parts labelled in bold.

Suggested answers

A basic lathe is a machine for working with wood. It consists of two centres where the material to be worked is held and rotated. It is powered by a strap, bow or pole. The strap, bow or pole turns the wood. A treadle lathe is powered by the operator's foot. The treadle is connected to a crank, which in turn powers a flywheel. The momentum of the flywheel provides continuous rotation.

- Now have the students do the Workbook exercises for Unit 2, Lesson 5.

Workbook answers

Writing: charts

A Research about a piece of machinery

Answers depend on the students.

Pronunciation: vowel sounds

A Find and circle words with different vowel sounds

1 sharpen   5 weight

2 damage   6 process

3 heat   7 knob

4 machine

B Find words with different stress patterns

Suggested answers

Oo	handle
oO	machine
Ooo	callipers
oOo	position

Lesson 6 — Modern lathes

Objectives

To discuss features and functions of screws and bolts.

To read definitions and extend vocabulary for parts of a centre lathe.

To develop the skill of listening to key words and information.

Language

Vocabulary for parts of a modern centre lathe, e.g., *headstock, apron.*

Revision of *wh~* questions.

Vocabulary for location and movement, e.g., *clockwise, rotation, between, in front of.*

Reading and speaking

A Discuss true/false statements about bolts and screws

- If you wish, elicit or clarify the difference between a bolt and screw, i.e., a screw is designed to be turned or screwed into a small hole and can be removed and reused; a bolt normally passes through a hole of larger diameter than its thread, and is held in place by a nut or similar device. It is not designed to be reused frequently.

- Set for discussion in pairs.

- Feed back ideas, but do not confirm yet.

Answers

No answers at this point.

B Scan the text to check answers

- Set a time limit. Students should scan quickly to check their answers.

- Feed back, focusing on the use of quantifiers and frequency adverbs, e.g., *all/most/the vast majority/occasionally.* Go over any new vocabulary, such as: *thread, standardisation.*

Answers

1	F	4	T
2	T	5	F
3	F		

C Discuss standardisation of screws and bolts

- Set for discussion in small groups. Emphasise that there is no single, correct answer, and encourage students to speculate and give their opinions. Monitor and prompt where necessary.

- Conduct brief feedback on students' ideas. Emphasise that virtually all modern industries, e.g., machinery, vehicles, rely on mass-produced fasteners.

Answers

No answers at this point.

Listening

A ◉ **Global comprehension: decide which questions are answered (CD1 Track 5)**

- Play the recording. Ask pairs to discuss which questions were answered.

- Feed back. Check understanding of the text by asking students to paraphrase/summarise the speaker's answers to the questions.

Answers

1 Because industry is international.

2 [not answered]

3 During the 1940s (because of the Second World War).

4 The lathe.

B **Complete the text with words from the recording**

- Set the task for individual work and pairwork checking. Refer students to the Glossary to check the meaning of unknown words.

- Feed back orally, but do not confirm or correct answers yet. Students will listen to check their ideas.

Answers

No answers at this point.

C ◉ **Intensive listening: check answers to Exercise B (CD1 Track 6)**

- Play the recording once or twice as necessary.

- Feed back orally. Check comprehension and pronunciation of the new vocabulary, referring to the diagram of a lathe for clarification. Elicit where the *lead screw* and *gearbox* are on the diagram.

Answers

The lead screw is a long, threaded rod that carries a tool along the axis of a rotating workpiece. It ensures that the workpiece moves at a constant, even speed so that threads can be cut into it. The relationship between the longitudinal speed of the tool and the rotational speed of the workpiece can be varied by means of a gearbox.

Vocabulary

A **Read descriptions of parts of a lathe and label the diagram**

- Set for individual work and pair-checking. Encourage students to label the items they already know first, e.g., *the headstock*.

- Feed back as a group. Ask questions about the different parts, eliciting where they are positioned and/or how they move, e.g., *How does the saddle move? (back and forth between the headstock and tailstock) Where is the apron? (at the front of the saddle)*

Answers

1 bed (f)

2 lead screw (g)

3 headstock (a)

4 tailstock (e)

5 spindle speed selector (b)

6 saddle (d)

7 spindle (c)

B **Ask and answer questions about parts of a lathe**

- Elicit questions about the parts of the lathe using prompts, e.g., *What is …? Where is …? How does …?*

- Students should work in pairs to ask each other about the different parts of the lathe, using the questions.

Answers

Answers depend on the students.

- If you have time, find out if anyone in the class has used, or knows how to use, a lathe. Elicit how students think it works. Encourage them to find out more about lathe operation by looking at:

  http://www.americanmachinetools.com/ how_to_use_a_lathe.htm

- Now have the students do the Workbook exercises for Unit 2, Lesson 6.

Workbook answers

Language: verbs for machine operations

A Fill in the gaps with the correct verbs

1 Turn/Switch
2 Press/Push/Pull
3 Set
4 turning
5 Turn
6 Pull

Language: finding errors

A Correct the grammar in the sentences

1 Switch the electricity supply on at the mains.
2 Before starting the machine, ensure the levers are disengaged.
3 Always stop the machine before changing gear.
4 The speed can be varied as required.
5 Without screws, modern industry would not be possible.

Writing: sequencing instructions

A Write instructions for a machine

Answers depend on the students.

Objectives

To practise reading for main ideas.

To raise awareness of pronoun referencing in text.

To raise awareness of common abbreviations in notes.

Language

Pronoun references within a text.

Connective devices: *so*, *however*, etc.

- Lead in by reviewing vocabulary for the properties of materials, such as *conductivity*, *ductility*, *strength*, *density*, etc.

Reading

A Choose a title

- Put the three titles on the board and elicit a few ideas for the content of a text with each title.
- Set the task for individual work and pairwork checking. Ensure that students understand that they should not worry about understanding every word of the text.
- Feed back orally.

Answer

Title 2 most accurately reflects the content of the text.

- Elicit the main properties of aluminium and clarify problem vocabulary, such as: *versatile*, *oxidisation* and *alloy*.

B Decide which notes best represent the information

- Ask students to reread the text individually, and then look at the notes in 1 and 2.
- Pairs compare their ideas.

Answers

Although 2 is perhaps more accessible to a less skilful reader, for the purposes of this course, the notes in 1 are better since they are more condensed. They use note forms that students will work on using themselves.

C Pronoun references

- Firstly, get students individually to circle the pronouns in the text, and check in pairs.
- Then pairs discuss what the references are.

Answers

this (line 2): the metal, aluminium

this (line 7): the fact that aluminium weighs less than steel

it (line 10): aluminium

its (line 16): aluminium

these (line 20): alloys with copper, manganese, chromium and zinc

this (line 21): corrosion

it (line 26): aluminium

Vocabulary and writing

A Abbreviations

- Elicit the first item (*e.g.*).
- Set the task for pairwork. Feed back onto the board. Point out that these are abbreviations that the students should make a habit of using as soon as possible.

Answers

e.g.	for example
v. imp	very important
+	and
Al	aluminium
elec.	electricity
g/m^3	grams per cubic metre

B Focus on connectors

- Refer students to the first text. Ask them to underline the expressions in the box. Elicit answers onto the board and clarify the use of each item.

- Set the task for pairwork. Feed back orally.

Answers

1 <u>Because</u> gold does not rust easily, it is often used in jewellery and dentistry.

2 Asbestos is a good insulator. <u>However</u>, the dust is highly toxic.

3 There are two reasons: firstly, water is cheaper than chemical options; <u>secondly</u>, it is easier to store.

4 <u>In addition to</u> the high cost of raw materials, other problems include the long development process.

5 Common cutting tools need to be rust-free, <u>so</u> they are often made of stainless steel.

- Now have the students do the Workbook exercises for Unit 2, Lesson 7.

Workbook answers

Reading: text organisation

A Put the paragraphs into the correct order

Another property of titanium is that ... (3rd)

Firstly, titanium is important as an alloying agent ... (1st)

In addition to its use in industry, titanium pigment accounts for ... (2nd)

Titanium is a lustrous, white metal. It has a low density ... (4th)

B Decide which products contain titanium

Answers depend on the students.

Language: comparisons

A Write a paragraph comparing titanium and aluminium

Answers depend on the students.

Objectives

To raise awareness of adjective and noun function; adjective-noun order.

To apply new vocabulary in context.

To develop the skill of listening for specific information.

Language

Some adjective and noun endings: ~*able*; ~*ness*.

Vocabulary for material properties: *malleable*, *dense*, etc.

Vocabulary for working sheet metal: *scriber*, *burr*, *bend*, etc.

Expressions for asking for and giving spoken instructions: *How do you ...?*

Expressions for recommendations: *You mustn't ...*

Vocabulary

A Word-building: nouns and adjectives

- Begin by eliciting a few simple adjectives and nouns. Students do not need to know the words *noun* and *adjective* yet. Establish the order adjective-noun in English.

- Refer students to the Language Box. Point out the terms *noun* and *adjective*.

- Remind students of Unit 1, Lesson 3, where they looked at the properties of solid materials such as metal. Write the table on the board and work through *soft* and *lightness* as a group.

- Set the task for pairwork with dictionaries. Feed back onto the board.

- Check the meaning of the words in the table.

- Drill pronunciation. Take the opportunity to point out that the dictionary gives the pronunciation, for students who may be unaware of this.

- Point out the adjective and noun endings, which are common – and consistent – in English. Elicit further examples with some of the endings, e.g., *tolerance*.

Answers

adjective	noun
soft	softness
malleable	malleability
ductile	ductility
light	lightness
resistant	resistance
conductive	conductivity
dense	density
durable	durability
flexible	flexibility
tough	toughness

Speaking and writing

A Discuss the properties of aluminium

- Set the task. Point out that students should use the nouns and adjectives from the previous exercise, e.g., *Aluminium is light, so it is suitable for drink cans.*

- Feed back orally.

Answers

Answers depend on the students.

B Write sentences about the properties of aluminium

- Set for individual work. Encourage students to structure their sentences as follows: *Aluminium is suitable for … because it is …*

- Monitor closely and assist. Pairs compare answers.

- Elicit a few answers and put them on the board.

Answers

Answers depend on the students.

Listening

A Pre-teach vocabulary

- Elicit, explain or simply ask students to look up the words in the dictionary.

- Feed back orally.

- Drill pronunciation.

- Explain that these words are all used in a conversation between two engineering students and their instructor that students are going to listen to.

Answers

bend: to force an object into a different form

shears: scissors of large size

scriber: a tool for scribing things such as wood

mark out: to mark the boundary of something by making lines

edge: a line at which two surfaces of a solid object meet

smooth: free from projections or unevenness of a surface; not rough

handle: a part that is designed to be held or operated with the hand

hole: an opening through something

radius: a straight line extending from the centre of a circle or sphere to the circumference or surface

burr: a small, hand-held, power-driven milling cutter

B Listen and identify instructions using new vocabulary (CD1 Track 7)

- If you wish, clarify the fact that Alclad refers to *a process where a skin of pure aluminium is applied to both sides of a metal to protect it from corrosion.* With stronger classes, you may wish to wait for students to hear the definition of the process on the recording.

- Set for individual work and pairwork feedback. Elicit other information that students learned about the process.

Answers

He tells them to mark out and drill the sheets.

C Listen for specific information: decide whether statements are true or false (CD1 Track 7)

- Give students time to read through the statements and check problem vocabulary, e.g., *corrosion, scriber*.

- Set for pairwork. Feed back orally, but do not confirm answers yet.

- Play the recording again so that students can check their answers.

- Feed back on any problems.

Answers

1 True
2 False (only Alcladded sheets)
3 False (four)
4 True
5 True
6 True

D 🔘 Listen, read and pick out useful expressions (CD1 Track 7)

- Students follow the tapescript as they listen to develop a sense of the relationship between sound and written form, and to hear an extended piece of spoken language under relatively little pressure.

- Ask students to underline useful expressions and monitor as they do this.

- Elicit more examples of useful language, ready for the speaking activity. Check students are aware of the correct verb form to use with each one.

Suggested answers

Asking about processes:

What's the point of that?

Does this stuff have a special name?

So how exactly do you …?

What about …?

Are there any special things to remember?

How do you know if …?

Giving instructions:

You must/mustn't …

You should be able to …

Make sure you …

You don't have to …

The basic rule is that you mustn't …

The important thing is to …

Speaking

A Write and rehearse a conversation

- Set for pairwork. Stress that students should use as many of the expressions from the previous exercise as they can.

- With weaker groups, you could build an example dialogue with the class on the board.

- Monitor and help with pronunciation.

- If possible, pairs should rehearse their conversation for another pair, who should in turn suggest improvements in the usual way.

- Get pairs to perform their conversation to the group or a new pair.

- If you have time, ask students to look again at the vocabulary in Listening A. They should make two lists: *verbs* and *nouns*.

Answers

Answers depend on the students.

- Now have the students do the Workbook exercises for Unit 2, Lesson 8.

Workbook answers

Language: modal verbs

A Complete the sentences with a suitable modal verb

1	MUST	5	can
2	should	6	should not
3	cannot	7	must
4	should	8	must not

Writing: instructions for working with Alclad

A Write sentences using the modal verbs

Answers depend on the students.

Objectives

To transfer information from text to drawing.

To develop the skills of: using existing knowledge to interpret and predict information; reading for main idea; deducing meaning.

Language

Vocabulary for fasteners, e.g., *bolt, screw*.

Affixation, e.g., *airworthy, non-standard*.

Speaking

A   Distinguish between fasteners: screw, bolt, rivet, pin

- Check that students agree on what each drawing represents. Ask them to agree among themselves on a translation for each fastener in their L1.

- Elicit which ones have threads (for screwing) and which are hammered or pushed. Point out that bolts and screws are very similar to one another, but that bolts are often used with washers and their heads distinguish them (screwdrivers are used to tighten screws, whereas spanners are used on bolts).

Answers

1	screw	**3**	pin
2	bolt	**4**	rivet

B   Discuss everyday uses of fasteners

- Ask students to point out screws, pins, rivets and bolts in the classroom.

- Elicit other uses for each, e.g., screws and pins in furniture (wooden objects) and masonry, bolts in engines, bolts and rivets in other machinery and metal sheets.

Answers

Answers depend on the students.

Reading

A   Read definitions and identify different riveting processes

- Ask students to study the drawings and discuss them in pairs.

- As feedback, elicit a few ideas.

- Ask students to briefly look at the two short texts under the pictures and to agree in pairs which process is shown in the pictures.

- Check students understand the differences between the two processes by asking concept-check questions, e.g., *Which one uses a pin?*

Answers

The pictures show standard riveting.

B   Label diagrams with key terms

- Set the labelling task for individual work and pairwork checking.

Answers

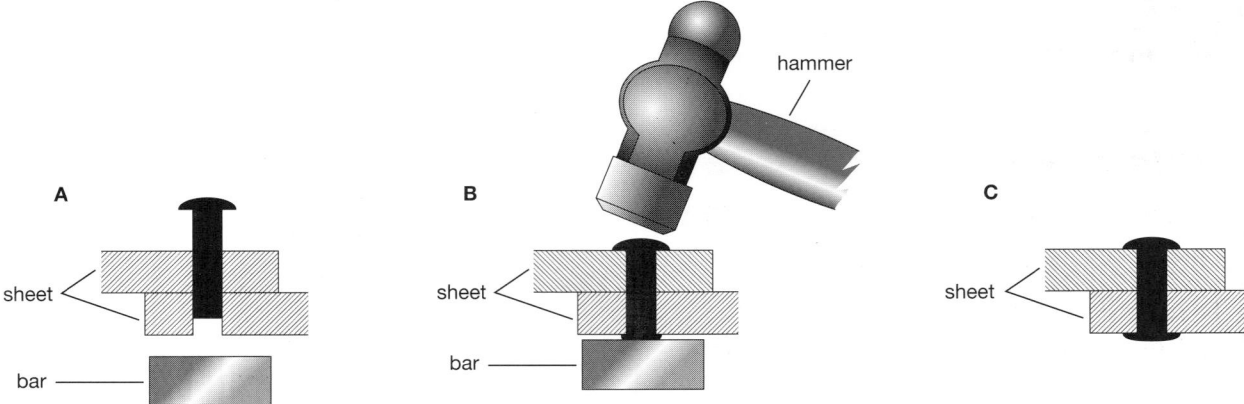

A

sheet

bar

B

hammer

sheet

bar

C

sheet

C Identify and order topic sentences in a text

- Set the task for individual work and pairwork checking. Ensure that students are aware that they do not need to understand every word of the texts.

- Feed back orally.

Answers

a 2

b 3

c 1

Vocabulary

A Focus on new vocabulary in the text

- Set for individual work and pairwork checking. Students should be able to guess new terms from the individual words or parts of words.

- Feed back onto the board. Elicit the prefix (*non-*) and suffix (*worthy*) and see if students know any other words that use them.

Answers

1 airworthy (an airworthy plane)

2 internationally recognised standards

3 shear strength

4 resistance to vibration

5 non-structural

- Now have the students do the Workbook exercises for Unit 2, Lesson 9.

Workbook answers

Language: prepositions

A Underline examples of prepositions

Answers depend on the students.

B Complete the sentences

1 Protective shields are fitted <u>over</u> delicate electrical components.

2 Then the pin is moved <u>back</u> to its initial position by a small spring.

3 This is a view of the fuselage <u>from</u> the rear.

4 During testing, the meter sits <u>on</u> top of the assembly.

5 If it presses <u>against</u> this pressure plate, the system cuts out.

6 Plastic plugs are put <u>into</u> the valve openings to keep them clean.

Writing: describing applications of fastenings

A Choose two fastenings from the box and describe/explain

Answers depend on the students.

Objectives

To review the engineering concepts covered in Unit 2.

To practise note-taking and raise awareness of note-taking conventions, e.g., abbreviations.

Language

would + infinitive for hypothetical situations.

Review of the language covered in this unit.

Speaking and listening

A Discuss the questions

- Discuss the photo. Elicit the fact that this is a kit-plane.

- Refer students to the Language Box. Elicit ideas for the first question from the group. Highlight *would* + infinitive.

- Set the task for pairwork. Feed back orally.

Answers

Answers depend on the students.

B Listen to the interview (CD1 Track 8)

- Point out that students do not need to write grammatical sentences. This is an introduction to note-taking. If students need help with scaffolding their notes, write the following headings on the board:

  1 *Why?*
  2 *Where?*
  3 *How much?*
  4 *How long?*
  5 *What tools and facilities?*

- Play the recording.

- Get pairs to compare answers.

- Feed back some answers onto the board and mention some note-taking conventions, e.g., abbreviations, omitting articles. This will be focused on later in the course.

Suggested answers

1 **Why?** cheaper than ready-made, enjoys a challenge

2 **Where?** his garage (24 ft x 9 ft x 9 ft), wings attached outside

3 **How much?** kit: just under £14,000, tools: £200

4 **How long?** just over 2 yrs (approx. 800 hrs.)

5 **What tools and facilities?** good set of hand tools + tin snips, new elec. drill, hand rivet gun

Vocabulary

A Group words by number of syllables

- Use the examples of *engine* and *micrometer*, and other vocabulary well known to students, until they have understood the idea of syllables.

- Set the task for pairwork. Feed back onto the board.

Answers

one syllable: grip, drill, lathe

two syllables: alloy, rivet, hammer, transmit, special

three syllables: technical, interested, elements

four syllables: experience

B Word stress

- Refer students to the Skills Box.

- Again, use familiar vocabulary to review the notion of word stress.

- Set the task for pairwork. Tell students to say the words aloud to each other in order to hear the stress.

- Feed back onto the board. Point out the tendency in English to stress the first syllable, especially of two-syllable nouns.

- Point out also that a stressed syllable contains a core vowel, which may be short or long, but is clearly audible.

Answers

First syllable stressed: *alloy, rivet, hammer, special, technical, interested, elements*

Second syllable stressed: *experience, transmit*

Language

A Sentence correction

- Set for individual work and pairwork checking. Encourage students to look back at the language points they have covered in this unit.

Answers

1 Tin snips are usually used for cutting sheet metal.

2 It's better to use blind riveting when you can't reach both sides of the job.

3 Correct

4 Some people say that hand tools last longer than powered tools.

5 Correct

- Now have the students do the Workbook exercises for Unit 2, Lesson 10.

Workbook answers

Language: questions with *would*

A Make questions

1 Why would you like to build a plane?

2 Where would you build it?

3 How much would it cost?

4 How long would it take?

5 What tools would you need?

B Write an answer for each question

Answers depend on the students.

Reading and writing: safety tips

A Think about how to avoid problems

Answers depend on the students.

B Write safety tips for using tools in a workshop

Answers depend on the students.

Unit test

A test for this unit is available at:

http://www.garneteducation.com/reps/documents/1245/SDT-u2-test.pdf

Contact your local Garnet Education representative for information about how to access these resources.

Objectives

To practise using prepositions of location.

To practise discussion/comparison of characteristics of airframes.

Language

Vocabulary for major structural parts of an aircraft, e.g., *wings/tail/power plant*.

Prepositions for describing parts of the airframe: e.g., *the back of the plane, in the cockpit*.

Introduction

- Exploit the visuals. Elicit what kind of aircraft these are, and how old they might be.

Speaking and vocabulary

A Discuss similarities and differences between the aircraft

- Elicit a few ideas.
- Challenge students to work in pairs to be the first to come up with five or ten similarities or differences.
- Feed back orally.

Answers

Answers depend on the students.

B Match definitions to parts of the airframe

- Get students to read through the words, and in pairs discuss which they think they know, without reference to definitions or the Glossary.
- Then they read definitions *1–10* to check.
- Refer students to the Glossary. Set for individual work and pairwork checking.
- Feed back orally.

Answers

1	wings	**6**	struts
2	power plant	**7**	flaps
3	landing gear	**8**	fin
4	fuselage	**9**	propeller
5	skin	**10**	rudder

C Raise awareness of word stress

- Drill the pronunciation of the vocabulary from Exercise B.
- Write the first item on the board. Elicit and mark the stress.
- Set the task for pairwork. Feed back onto the board. Drill again.

Answers

(**Note:** Words with one syllable do not have a marked stress.)

fuselage	**pow**er plant
wings	propeller
landing gear	skin
rudder	fin
struts	flaps

Language

A Complete gapped sentences with prepositions

- **Note:** The exercise aims to use a limited group of phrases to introduce students to the idea that they should memorise a preposition in combination as part of a predictable phrase.
- Refer students to Language Box 1. Clarify as necessary, using examples from the lesson.

- Set the task for individual work and pairwork checking.

- Ask students to find as many of the answers as they can in the definitions in Speaking and Vocabulary Exercise B.

- Feed back onto the board. Point out the use of *the* in the phrases.

Answers

1 They are rods attached to the wings.

2 It's a hinged vertical surface at the back of the plane.

3 It supports the aircraft on the ground.

4 These blades are on the front of the power plant.

5 The seats are in the cockpit.

B Sentence writing to reinforce use of prepositions

- The objective of the activity is to reinforce and personalise the use of prepositions. Elicit a few suggestions, e.g., *The flaps are on the wings. The tail is at the back of the plane.*

- As students write their sentences, go around and monitor and make suggestions.

Answers

Answers depend on the students.

Speaking

A Expand prompts to make fuller descriptions

- Use Language Box 2 to clarify the use of 's, e.g., *The main body's covering surface.*

- Elicit a few phrases or uses of *of*, e.g., *the back of the room, the majority of the class.* Point out that it is a very high-frequency word in English.

- Ask students to cover the left-hand page. They will use it to check answers later.

- Go through the example sentence in the exercise on the board. Point out the prepositional phrase: *on the wings.*

- Set for pairwork. Monitor and assist.

- If they are unable to expand some of the prompts, students can check against the definitions in Speaking and vocabulary Exercise B.

- If you have time, test the students informally on the prepositional phrases. You may wish to work on relative pronouns *which/that*, or on articles if you have time.

Answers

1 These are *the* large surfaces *that/which* project horizontally *from the* main body *of the* plane.

2 This *is the* equipment *that/which* supports *the* aircraft *on the* ground.

3 This *is the* main longitudinal body *of the* plane.

4 These *are the* strong rods and bars *that/which* are attached to the wings.

5 These *are the* small, hinged control surfaces *on the* wings.

6 This *is the* fixed vertical surface *at the* back.

7 These *are the* pitched blades *at the* front *of the* power plant.

8 This *is the* hinged vertical surface *at the* back *of the* plane.

- Now have the students do the Workbook exercises for Unit 3, Lesson 1.

Workbook answers

Language: relative pronouns

A Write sentences including *who, which* or *that*

The project manager is the man <u>who</u> takes responsibility for the whole project.

Mr Johnston is the man <u>who</u> designed the prototype.

There are a lot of people <u>who</u> would like to build their own aircraft.

A blind rivet is a fastening <u>which</u> can be fitted from one side of the work.

A strut is a bar <u>which</u> is attached to the wing to support it.

Writing: definitions

A Complete the definitions

Aluminium is a metal that is used a lot in aircraft construction.

Lathes are machines which are used to cut screw threads.

CNC machine tools are tools that are controlled by computers.

B Write definitions

Answers depend on the students.

C Compare your definitions

Answers depend on the students.

Language: prepositions

A Complete the gaps with *on*, *in* or *at*

1 There are numbers printed <u>on</u> each skin panel to aid assembly.

2 If you're not sure, the procedure is <u>in</u> the manual.

3 Look at the diagram <u>on</u> page 78.

4 Holes are drilled <u>at</u> intervals of 12 cm.

5 Fuel is stored <u>in</u> tanks <u>in</u> the fuselage and wings.

6 <u>At</u> each end of the hangar, there are large access doors.

7 Before it flies, each aircraft is tested <u>on</u> the ground. Further tests are done <u>in</u> flight.

8 Most small components are turned <u>on</u> a lathe.

9 Computers can easily show designs <u>in</u> 3D.

10 Many important tests are done <u>in</u> the wind tunnel.

11 The only surviving example is kept <u>at</u> an RAF base <u>in</u> Scotland.

12 It is most effective <u>at</u> high speeds and <u>at</u> high altitude.

B Copy ten phrases

Answers depend on the students.

Lesson 2 — Force

Objectives

To develop scanning skills.

To introduce and discuss five types of mechanical stress.

Language

Flight manoeuvres: *bank, roll, hover*, etc.

Directions: *anticlockwise, forwards*, etc.

Noun/verb forms: *compress/compression*.

Speaking

A Introduce vocabulary for flight manoeuvres

- Briefly elicit ideas of what each machine or insect in the illustrations is and how it moves.

- Demonstrate, using drawings on the board or objects, or by miming the words in the box.

- Set the discussion for pairwork. Feed back orally.

Answers

Answers depend on the students.

Reading

A Scan the text for information about movement

- Set for individual work and pairwork checking.

- Stress that the students do not need to study and understand every word. Set a time limit.

- As feedback, elicit ideas from the group.

Suggested answers

The text mentions: *lift, hover, go backwards, up, down, sideways, accelerate, decelerate, roll, spin, rotate (clockwise and anticlockwise).*

B Circle words in the text

- Write the first few letters of a word on the board, e.g., *bac* … Demonstrate to students that they can 'read' the whole words even though they can see only the beginnings. This is an important part of scanning for specific items, as jumping between the beginnings of words increases speed.

- Set the task. Encourage students to work quickly.

- Feed back orally. Ask students to refer to line numbers in the text.

Answers

anticlockwise:	line 12
clockwise:	line 11
direction:	line 9, line 10
forces:	line 2, line 12
gravity:	line 3
lift:	line 4
roll or spin:	line 7
speed:	line 9
stresses:	line 13
the ground:	line 5

C Find words that collocate in the text

- Point out the example and get students to find it in the text.

- Set for individual work and then pairwork checking.

- Feed back onto an OHP of the text if possible. Point out that words are often used together in familiar combinations (collocation) and that it is useful to notice and remember them.

Answers

anticlockwise	turn anticlockwise (line 12)
clockwise	the propeller is rotating clockwise (line 11)
direction	change direction (line 9)
	there is an equal force in the opposite direction (line 10)
forces	a variety of different forces (line 2)
	opposition of forces (line 12)
gravity	the force of gravity (line 3)
	to counteract gravity (line 3)
lift	a way to provide lift (line 4)
roll or spin	be able to roll or spin (line 7)
speed	change direction and speed (line 9)
stresses	produces stresses (lines 12–13)
the ground	hovered above the ground (line 5)

- Clarify any vocabulary problems in these expressions. There is no need to go into all the vocabulary in the text.

D Match definitions of stresses with diagrams

- Set the task for individual work with dictionaries. Monitor and assist.
- Pairs compare their answers.
- Feed back orally.

Answers

a	2	d	5
b	4	e	3
c	1		

Vocabulary

A Use the vocabulary for stresses in context

- Elicit verb forms for the five types of stress: *torque, compress, tense, shear, bend*.

- Do the first one as a group.
- **Note:** In this sentence alone, the word is used in a non-physical sense.
- Set the task for individual work and pairwork checking. Stress that the gaps may require either a noun or a verb form.

Answers

1	compression	**4**	tension
2	Torsion	**5**	shear
3	bend		

B Memorise word units

- Students can test themselves on how many of the expressions containing the words in Reading Exercise B they can remember with the text covered.
- Alternatively, you could work on the expressions from Exercise D: *a result of, the opposite of, one of, the combination of*, which will help students become familiar with the concepts in the five types of stress.

Answers

Students should memorise the following words:

anticlockwise	lift
clockwise	roll or spin
direction	speed
forces	stresses
gravity	the ground

Speaking

A Examples of stresses

- Elicit a few examples.
- Set for pairwork. Students should make a brief note of their ideas. It does not matter if students cannot think of two examples for each force or if they can't think of any engineering examples.

Examples

compression: a hammer on a nail; a person's spine when they stand; the air in a tyre

torsion: wringing out a cloth; driving in a screw; twisting an object (string, scrap paper) in the hands

bending: the sole of a shoe as you walk; a wardrobe rail with heavy clothes hanging on it; a horizontal beam

tension: clothes hanging from pegs; a violin string; an inflated balloon; a rope pulling a car or an animal

shear: a picture hanging on a nail; a geological fault/landslide; bolts securing a metal plate onto a vertical surface

B Describe forces

- Ask a stronger student to describe one of his examples to the class. They must say which of the five forces (stresses) is operating.
- Set the task for pairwork.
- As feedback, elicit a few further examples from the group.

Answers

Answers depend on the students.

- If you have time, elicit and build up on the board definitions of the types of stress in Reading Exercise D.
- Students can look up mechanical forces on the Internet at:
  www.wikipedia.org
  (detailed technical information)
- Now have the students do the Workbook exercises for Unit 3, Lesson 2.

Workbook answers

Language: spelling

A Correct the misspelled words

1	correct	**10**	opposition
2	contract	**11**	structure
3	correct	**12**	correct
4	hover	**13**	anticlockwise
5	sideways	**14**	correct
6	accelerate	**15**	compression
7	correct	**16**	correct
8	correct	**17**	correct
9	exert	**18**	radius

B Put each word into a combination

Answers depend on the students.

Writing: descriptions of movement

A Write sentences describing diagrams

Suggested answers

The helicopter is taking off vertically.

The eagle is doing a shallow dive.

The kite is rotating and moving in a circular direction in the sky.

The hummingbird is moving forward horizontally and hovering by a flower.

Lesson 3   Manoeuvres

Objectives

To develop skills in listening for general meaning; for identifying items in a stream of speech.

To apply the language of flight manoeuvres to a context.

To raise awareness of and practise sentence stress.

Language

Expressions for flight: *gain height, climb away*.

Speaking

A Vocabulary revision

- Elicit vocabulary for flight manoeuvres, using the drawings/mimes you used in Lesson 2.

Suggested answers

take-off	lift
climb	forwards
dive	backwards
bank	rotate
hover	vertical
roll	horizontal
fall	

B Discuss bird flight

- Set for pairwork. Feed back orally.

Answers

Answers depend on the students.

Vocabulary and listening

A Listen for gist (CD1 Track 9)

- Set for individual work and pairwork checking.
- Play the recording once.
- Feed back orally.

Answers

1 Outside, near a lake.

2 An eagle (in a tree).

3 The guide.

4 The eagle is going to start hunting/(try to) catch the fish.

B Listen for specific information (CD1 Track 10)

- Drill the pronunciation of the items *1–9*. Students will need to recognise them.
- Point out that the items may appear in the text in a different form, e.g., with ~*ing* or ~*ed* or a preposition between the two parts of a phrase, etc.
- Play the recording once. Students tick which of the items they hear and then compare answers. Feed back orally.
- Play the recording again. Students number the items in order. Check in pairs and feed back orally.

Answers

1 4: … he's taken off from the branch

2 8: … he's climbing away

3 9: … it's diving straight down towards him

4 3: Is she trying to pull him up higher?

5 1: … if she can gain enough height

C Summarise the movements of the eagle

- In pairs or threes, students should agree on what happened during the sequence.

- Elicit the sequence onto the board using a simple drawing and the language of manoeuvres from the text and the previous lessons.

Suggested answers

He's taken off from the branch; he's climbing away; it's diving straight down towards him; she's trying to pull him up higher; she's gaining height, etc.

D Listen and delete the extra word (CD1 Track 9)

- Ensure that students understand the task. If necessary, read out the first line from the tapescript as an example.

- **Note:** There is no extra word in Line 6.

- Play the recording without stopping it. Students compare answers in pairs.

- Play the recording again.

- Feed back as a group onto an OHP of the text if possible.

Answers

line 1:	just	line 9:	My
line 2:	really	line 10:	his
line 3:	back	line 11:	around
line 4:	side	line 12:	surface
line 5:	there	line 13:	very
line 6:	–	line 14:	probably
line 7:	only	line 15:	indeed
line 8:	now		

E Write exactly what was said (CD1 Track 10)

- Demonstrate the task using, e.g., *doing,* in Mr Lewis' question. When students hear it, they must shout *Stop.* Rewind and replay the question *What's he doing?* Students write it down.

- Elicit the question onto the board.

- Set for individual work.

- This is a demanding exercise. You will need to play the relevant sentences more than once.

- **Note:** Students will hear the words in the order listed.

Answers

No answers at this point.

F Edit the sentences

- Pairs compare answers.

- Students must agree on a version which is grammatically correct.

- Feed back onto the board.

Answers

1 Guide: He'll dive at quite a *shallow angle* so the fish won't see him coming …

2 Mr Lewis: It's another eagle – it's diving *straight down* towards him.

3 Guide: It's a female … they are usually bigger and *heavier.*

4 Guide: If she can gain enough height, she will try a *special manoeuvre* … yes … watch.

5 Mr Lewis: She's climbing really *fast* … she must have some power in those wings.

Speaking

A Mark the stress (CD1 Track 10)

- Use the recording to elicit and mark the stress on the sentences and drill them.

1 Guide: He'll **dive** at quite a *shallow angle* so the **fish** won't see him coming …

2 Mr Lewis: It's another **eagle** – it's diving *straight down* towards him.

3 Guide: It's a **female** … they are usually **bigger** and *heavier.*

4 Guide: If she can gain enough **height**, she will try a *special manoeuvre* … yes … watch.

5 Mr Lewis: She's **climbing** really *fast* … she must have some **power** in those **wings**.

B Mark the tapescript

- Refer students to the Skills Box. Use the sentences you have marked for stress to point out that the stressed words – the 'content words' – are more prominent because they are spoken more loudly and at a higher pitch than the neighbouring words in the sentence. The stressed vowel in these words is clearly and fully articulated.

- Set for pairwork. Students should work on a few lines of the tapescript only. Monitor and assist.

- As feedback, ask some students to read out their section.

- Students and teachers can find accessible discussion of bird versus powered flight at:

  http://www.baesystemseducationprogramme .com/games-and-multimedia/interactive-learning/

- There are many websites on the structure of birds and the mechanics of bird flight, e.g.:

  http://www.garden-birds.co.uk/information /flying.htm

- Now have the students do the Workbook exercises for Unit 3, Lesson 3.

Workbook answers

Vocabulary: manoeuvres

A Discuss similarities

Suggested answers

1 dive = fly downwards quickly, often to attack; descend = fly down slowly, possibly to land; hover = stay in the same place in the air

2 bank = turn left or right in the air; roll = rotate completely while travelling forward; rotate = turn around an axis

3 take off = leave the ground and fly up; climb away = gain height quickly after taking off; gain height = general term

4 turn around = turn 180 degrees to face the direction from which you came; move sideways = move horizontally left or right; reverse direction = travel backwards or go back in the direction from which you came

5 accelerate = increase speed; speed up = increase speed; lift = carry up

B Add additional words

Answers depend on the students.

Pronunciation: long vowels

A Put these words into the right column

/eɪ/	/əʊ/	/ɑɪ/	/aʊ/
wait	coat	file	amount
made	go	eye	clouds
safe	load	by	sound
weigh	lower	line	loud
aim	throw	light	ground

B Look at the table

1 /eɪ/   4 /eɪ/

2 /əʊ/ or /ɔː/   5 /aʊ/

3 /əʊ/

C Practise saying the words aloud

Answers depend on the students.

Language: present continuous v present simple

A Choose the correct form of the verb

1 They <u>are not taking</u> any new orders at the moment.

2 He usually <u>works</u> in our office, but he is <u>replacing</u> someone at headquarters today.

3 The R&D department <u>are developing</u> an improved adhesive, which we hope will be ready next year.

4 I think the project <u>is going</u> very well.

5 These alloys <u>contain</u> zinc, manganese, chromium or copper.

6 We <u>have</u> everything we <u>need</u>. We <u>are just waiting</u> for instructions to start.

7 Sections of the aircraft <u>are arriving</u> as sub-assemblies, and we just <u>put</u> them together here.

8 The managers <u>are looking at</u> the possibility of a new system for next year.

B Find examples of the present continuous

CD1 Track 9: coming

CD1 Track 10: climbing, feeding, turning, coming, doing, having, dragging, beginning, diving, doing, flapping, trying, rolling, falling, having, falling

C Write four sentences

Answers depend on the students.

Lesson 4   Skeletons

Objectives

To put into context the language of airframe structure and flight demands.

To develop the skill of extracting important information from a text.

To work on editing for written accuracy.

Language

Pronunciation of the diphthongs /eɪ/, /ɑɪ/ and /aʊ/.

Reading

A Discuss the skeletons

- Exploit the visuals.

- Elicit one or two ideas. Students who have done Internet research for Lesson 3 will be able to discuss the questions in more detail.

Suggested answers

1 function

Bird
   To allow energy-efficient flight using feathered wings/tail. A way of finding food and travelling.

Flying machine
   To fly using the power of the single human passenger, who operates the flapping wings with his arms and legs.

Jet fighter
   To provide a high-speed weapons/ reconnaissance platform. The skeleton houses a jet engine to power the plane.

2 construction

Bird
   Very light; hollow bones; large breastbone to anchor flight muscles; long neck for improved vision/preening.

Flying machine
   Wood, metal, fabric, leather, ropes – materials available at the time. Much too heavy to fly effectively.

Jet fighter
   Modern metal alloys and plastics. Aerodynamic; fixed wings and tail. Heavy.

B Skimming for gist

- Set for individual work and pairwork checking.

Answer

The text is about birds.

Reading and writing

A Underline the most important information

- Work through the first three lines with the group. Establish that the important points are that birds' skeletons need to be light and strong.

- Set for pairwork. Monitor and assist, but encourage students not to spend time looking up vocabulary.

- Feed back onto an OHP if possible.

Answers

Answers depend on the students.

B Complete the summary

- Again, do the first part with the group.

- Set for individual work. Pairs compare answers.

- Feed back onto the board.

Model answer

Birds' skeletons:
  need to be *light* and *strong*

Special characteristics:
  skeleton light in relation to size

  lightweight beaks and reduced or non-existent bones

  bones of limbs are hollow, with struts/other bones more rigid than those of mammals

  large breast bone which *anchors huge muscles*

C Check spelling and grammar

- Before students edit each other's work, they should read Skills Box 1. Point out that this is both a language exercise and a closed reading task.

- Individuals should use a pencil to suggest corrections. Feed back in pairs.

- Monitor and assist.

Answers

Answers depend on the students.

Pronunciation

A Raise awareness of syllables, vowel sounds and word stress

- Remind students of the importance of word stress.

- Draw attention to the highlighted words in the text and clarify that students need to identify which have *two* and which have *three* syllables.

- Set Exercise 1 for individual work and pair-checking.

- Students should identify the vowel sounds Exercise 2 in pairs.

- Conduct feedback, drilling the words as necessary. Point out that the weak form /ə/ is often used for the unstressed syllable(s).

Answers

2 syllables: w*i*ngspan (stressed syllable vowel /ɪ/ as in b*i*g), h*o*llow (stressed vowel /ɒ/ as in h*o*t), un*i*que (stressed vowel /iː/ as in h*ea*t) m*u*scles (stressed vowel /ʌ/ as in c*u*p)

3 syllables: sk*e*leton (stressed syllable vowel /e/ as in pen), int*e*rnal (stressed vowel /ɜː/ as in b*i*rd), rel*a*tion (stressed vowel /eɪ/ as in d*a*y), ext*e*nsions (stressed /e/ as in pen)

B Practise distinguishing between vowel sounds: diphthongs

- Point out that the /eɪ/ sound in rel*a*tion is a long sound made up of /e/ and /ɪ/. Refer students to Skills Box 2. Drill the three sounds.

- Use a few vocabulary items that are well known to students to elicit which one has which diphthong, e.g., *plane*, *fly* and *flower*.

- Write the words from the box in *2* on the board. Pairs should briefly decide how they are pronounced, without using the Course Book or a dictionary.

- Drill the pronunciation of the words.

- Pairs complete the table in *2*. Feed back in pairs.

/aɪ/	/eɪ/	/aʊ/
height	gain	around
flight	straight	power
dive	tail	down
climb	away	ground
	weight	

- Repeat the words so that students can check their ideas. Then drill the pronunciation again.

Answers

- If you have time, ask students to write and then read to each other sentences containing the words in the Pronunciation exercise, focusing on the diphthong sounds.

  OR build up on the board the summary of the writing text without reference to notes.

  OR get students to find words in the Glossary or in their dictionaries which contain one of the three sounds.

- For use in class or as research, as well as the websites in Lesson 3 on bird skeleton structure, the Internet has many sites about Leonard da Vinci's flying machine, e.g.:

  http://www.lairweb.org.nz/leonardo/ornithopters.html

  http://www.suberic.net/~goodmanj/davinci/

- Now have students do the Workbook exercises for Unit 3, Lesson 4.

Workbook answers

Reading and writing: editing

A Underline the errors in the text

B Rewrite the text correctly.

Bird's skeletons are extremely light in relation to there size. Frigate bird, for example, which flies long ditsances over tropical oceans, has a wingspan of over two feet, but it's skeleton is lighter than its feather's. Birds also have lightwait beaks instead of heavy teeth and jawbones. Som eother bones are very small, or – like the taill-bones – non-existent. The bones of a birds main limbs are hollow with and special

struts inside them, to strengthen them.

(handwritten corrections: Birds' / extremely / their / The frigate / distances / a / its / feathers / lightweight / Some other / tail-bones / bird's)

Pronunciation: the sound /ɜː/

A Circle words which have the same vowel sound as *bird*

first	third
world	earn
turn	learn
term	word
hurt	burst

B Make a note of the spellings

Answers depend on the students.

C Practise saying the words aloud

Answers depend on the students.

Language: superlatives

A Compare the words

Answers depend on the students.

Objectives

To practise transferring data to aircraft drawings.

To practise scanning.

To give a short talk on the conceptual content of the lesson.

Language

Airframe parts: *longerons*; *skin*; *bulkhead*.

Airframe construction: *These run laterally through the wing and act as the main support for the wing structure.*

Speaking

A Discuss the advantages and disadvantages

- Exploit the visuals.

- Elicit what the illustration shows and what kind of drawing this is. Revise the words *skeleton (frame)*, *skin*, *panels*.

- Ensure that students understand the difference between the two types of airframe design. (The illustrated semi-monocoque is a combination of the two: an internal frame and a load-bearing outer skin.)

- Elicit one or two ideas.

- Ask pairs to try to think of more.

- Feed back orally.

Answers

1 *pure monocoque designs:* give increased space inside the aircraft; are very strong; BUT need thick skins, which are heavy and may deform.

2 *trussed airframe designs:* fabric covering cheap and easily applied; BUT early wooden/steel beams were heavy; fabric skin prone to damage.

- **Note:** Most modern car and aircraft designs are semi-monocoque, with a thin aluminium skin on a light aluminium frame.

Reading

A Underline the vocabulary

- Set for individual work and pairwork checking. Remind students of the technique of skipping between the beginnings of words when scanning.

- Feed back onto an OHP if possible, or a simple drawing on the board.

Answers

aircraft	(lines 1 and 4)
skin	(lines 1 and 5)
longerons	(lines 9, 10 and 13)
frames	(lines 3, 4, 6 and 14)
bulkhead	(lines 7 and 10)
fuselage	(lines 4, 6, 11 and 12)

B Label the drawing

- Set the task for individual work and pairwork checking. Monitor and assist.

- Feed back onto an OHP if possible; if not, orally.

Answers

a	longerons	**c**	bulkheads
b	frames	**d**	stringers

Vocabulary

A Put the words into four columns

- Get students to look at the drawing of the wing and read the description individually. Pairs check that they have understood. Clarify vocabulary problems.

- Set Exercise A for pairwork. Feed back onto the board.

Answers

airframe part	verb	direction	purpose
spar	run	laterally through the wing	act as the main support

B Add more words to the lists

- Work through the first item with the group.

- Set the task for pairwork. Students can use dictionaries if they want to. Monitor and assist.

- Feed back onto the board.

Answers

airframe part	verb	direction	purpose
skin (panels)	are attached to	run the length of	to allow access
internal framework	extend	run (right) round	maintain the shape
frames	provide	lengthwise	act as both X and Y
lateral members	transfer	from X to X	the X provide
fuselage	are located at	longitudinally	to provide X
bulkheads			
mounting plates			
longerons			
stringers			

Writing and speaking

A Read a text and make notes for a talk

- Form pairs of As and Bs. Student A should use the Additional material on Course Book page 244 and Student B should use the Additional material on Course Book page 245. Ensure that students understand they must use the information to prepare a short three- to four-minute talk. They should not try to write out the whole talk.

- Monitor and assist. Encourage use of the language from the Vocabulary section.

Answers

Model notes for Student A: Monocoque airframe

- '*monocoque*' French = 'single shell'

- skin = structural element. Bears shear + bend

- v. light and low-volume skin – handle high load

- transfers load, heavy component (frame – extra) $\longrightarrow$ light (skin) = save weight + space inside

- e.g., cardboard box; honeycomb for military aircraft skin; F1 car body + chassis

- modern carbon fibre esp. good for m'coq:

  1 can design for strength in spec. directions + flex. in others

  2 v. light, v. strong

  3 easier to work than steel/Al.

Model notes for Student B: Trussed airframe

- load-bearing internal structure w/ separate skin

- internal frame: wood/steel tubes

- skin: fabric/sheet metal/plywood

- strong in all directions BUT heavy; uses space inside plane

- easy to make by hand BUT diff. and expensive for mass-production

- old-fashioned, pre-1930s
- late 1920s Al cheap $\longrightarrow$

  1 big Al frame – (strong) + thin Al skin riveted (light)

  OR

  2 small frame + thick skin

  OR

  3 comb. light Al frame + stressed Al skin = strong + v. light

B Give a short talk

- Explain that students will give their talk to a member of the other group. Get them to rehearse together first. Partners should suggest improvements to certain areas, e.g., grammar, manner of delivery and content.

- Mix Students A and B. Stress that listeners must pay close attention because they must ask questions at the end. It does not matter if the speaker does not know the answer – he can say so or speculate.

- **Note:** You may wish to ask students to do some research before presenting their talk in the next lesson.

- After students have given their talks, ask them whether their notes helped them. Let them compare their notes with the model notes below.

Answers

Answers depend on the students.

- For descriptions and relative advantages of different types of chassis design in the car industry:

  http://www.autozine.org/technical_school/ chassis/tech_chassis.htm (Note: English not perfect)

- For a discussion of airframe design:

  http://wikipedia.org

- Now have the students do the Workbook exercises for Unit 3, Lesson 5.

Workbook answers

Language: definitions

A Complete the words below. They are all important parts of an aircraft.

1	wings	**5**	tank
2	fuselage	**6**	rudder
3	rib	**7**	landing gear
4	struts	**8**	propeller

B Write short definitions

Answers depend on the students.

Language: parts of speech

A Identify the parts of speech

light:	adjective
strength:	noun
strengthen:	verb

B Complete the sentences

1 Longerons are structures that provide *stiffness* in the forward fuselage.

2 Frames are members which run the *length* of the fuselage.

3 Stringers are strips of wood or metal which run *along* between the frames.

4 Spars are strips which run *laterally* through the wing.

5 Bulkheads are frames which act as *protection* shields.

Writing

A Write a description

Answers depend on the students.

Objectives

To make students aware of 'signpost' aids to meaning when listening.

To discuss the stages of production of a modern jet fighter.

To use vocabulary for the production process in context.

Language

Present simple passive structures: *flight tests are carried out.*

Expressions used to 'signpost' a talk.

Speaking and listening

A   Discuss the stages of production

- Exploit the visuals. Elicit ideas as to what each picture shows. If students don't have the vocabulary to discuss the pictures, prompt with key words such as *fit, assembly, tests, electrical cables, spray paint.*

- Set the ordering task for pairwork. Feed back orally.

Suggested answers

1. 3: fitting assemblies together (welding)
2. 2: testing electrical cables
3. 5: installing control systems (with screwdriver)
4. 1: flight tests
5. 4: spray painting

B   Listen and number production stages (CD1 Track 11)

- Get students to read the list of stages *1–9*. Elicit whether any of the students' ideas

from Exercise A are in the list. Point out, using Skills Box 1, that the picture has already helped them to 'know' the content of the listening text.

- Set for individual work and pair-checking.
- Play the recording once.
- Do not confirm or correct answers yet.

Answers

No answers at this point.

C   Discuss answers

- Pairs compare answers.
- Elicit similarities/differences.
- Confirm the correct order.

Answers

1. 2 (main assemblies)
2. 7 (wings)
3. 9 (electronic tests)
4. 5 (hydraulics tests)
5. 1 (engines)
6. 4 (flight control systems)
7. 6 (pre-flight tests)
8. 8 (flight tests)
9. 3 (painting)

Vocabulary

A   Sequence signposting language (CD1 Track 11)

- Refer students to Skills Box 2.

- Get pairs to look at the list *1–7* to see if they can find any errors immediately. Do not confirm ideas yet.

- Play the recording. Pairs compare answers.

- **Note:** Remind students to use these and other signposts in talks that they prepare during the course.

Answers

1 ~~In~~ *At* the first stage
2 ~~Nextly~~ *Next*
3 At ~~these points~~ *this point*
4 At *the* third stage
5 Following ~~on~~ satisfactory completion
6 Following ~~to~~ this
7 Once ~~this has~~ *these have* been carried out

Language

A Complete passive sentences

- Elicit the past participle of each of the verbs onto the board. Clarify vocabulary problems.

- Go through the first sentence as a group.

- Set the task for individual work and pairwork checking. Feed back onto the board. Do not give students the answers at this point.

Answers

No answers at this point.

B 🔘 Listen and check answers (CD1 Track 11)

- Ask students to tick off the items as they hear them (they are not in order).

- Pairs compare answers.

- Alternatively, you could ask students to stop you when they hear the verb structures.

Answers

1 The three main assemblies *are fitted/spliced* together.

2 The aircraft *is painted* in air force colours.

3 Pre-flight tests *are carried out*.

4 The wings *are attached* to the fuselage.

5 Navigation systems *are installed*.

6 Electrical cables *are tested*.

C Ask and answer questions using the passive voice

- Ask students to work in pairs. One partner should cover the picture; the other, the questions.

- Set for pairwork.

- Elicit the correct form for all the questions. Use two strong students to demonstrate the exercise.

- Students then work with a new partner and take a new role.

- **Note:** This exercise is an introduction to the notion of reduced passive structures used in maintenance documentation. Students will study it in more detail later in the course.

Answers

Answers depend on the students.

- If there is time, ask students to listen again to the recording as they follow the tapescript. Remind them of the signposting expressions.

- Images of the painting process are available at:

  http://www.airbus.com

- The YouTube website contains videos on delivery, assembly and testing.

  Parts delivery and assembly

  http://flightglobal.com/blogs/flightblogger/atom.xml/

  Tests

  Jet engine test

  http://www.youtube.com/watch?v=Kvoyj0bwlzw

  Brake test

  http://www.youtube.com/watch?v=m1dv_y_3EK0

Ingestion test

http://www.youtube.com/watch?v=rSaf
RuLB0c0

A380 Blade off test

http://www.youtube.com/watch?v=j973
645y5AA

- Now have the students do the Workbook exercises for Unit 3, Lesson 6.

Workbook answers

Language: words that go together

A Write nouns or noun phrases that go with the verbs

Suggested answers

1 carry out ... a plan, tests, etc.,

2 install ... navigation systems, electrical cables, etc.,

3 attach ... wings, windows, etc.,

4 perform ... tests, an operation, etc.

B The ~ed endings of regular past tense verbs

1 Divide the verbs into three groups

/d/	/t/	/ɪd/
installed	spliced	fitted
used	attached	painted
performed		tested
carried		

2 One syllable and two syllable words

one syllable	two syllable
fit	install
splice	perform
paint	attach
use	carry
test	

Writing: punctuation

A Complete the listening text

The final production assembly line for the German version of the Eurofighter is located in the South German town of Manching. It is here that, as well as equipping the main fuselage, the engineers and technicians put together more than 300 pieces of equipment sub-assemblies and assemblies to produce the finished aircraft.

At the first stage, the three main assemblies are fitted or spliced together. These are the centre fuselage, the rear fuselage and the cockpit. Next, the flight control surfaces, such as the wings, flaps and fins, are attached to the fuselage. At this point, the fighter starts to look like a real plane. At the third stage, all the electrical cables are thoroughly tested and then the aircraft is moved on to Station 4, where the mechanical, electrical and hydraulic systems are subjected to rigorous testing with detached computerised equipment. Following satisfactory completion of these system tests, the aircraft is ready to have its engines and weapons systems fitted.

Following this, the flight control and navigation systems are installed and the plane is now ready for its pre-flight tests. Once these have been carried out, the aircraft is tested in flight before finally being moved to the paint shop to be painted in the colours of the German air force.

Objectives

To practise scanning.

To practise checking of data in data transfer.

Language

Speculation: *I doubt if …*; *I'd say …*; *must*.

Quantity: *over, thousands of, twice as many as*.

Speaking

- Students should read the introduction to the topic. Elicit the meaning of *in situ* (a Latin phrase meaning 'in the place'. Equipment on board aircraft must be tested 'in situ' or 'in place' to confirm everything functions properly as a system).

- Point out the passives *were built* and *are manufactured*.

A Discuss the photograph

- Elicit one or two ideas. Remind students of the vocabulary from Lesson 6 (*attach*, *fit*, etc.). Set the task for pairwork. Feed back orally.

Answers

Answers depend on the students.

B Discuss the questions

- Refer to the Skills Box. Clarify the language as usual.

- Get students to read questions *1–3*; clarify vocabulary. Emphasise that they should use the language in the box when guessing the answers to the questions.

- Set the task for pairwork, and remind students to use the expressions in the Skills Box. Monitor and assist. Feed back orally and clarify that students will discover the correct answers when they read the text on the next page.

Answers

Answers depend on the students.

Reading

A Scan text to find numbers

- Set for individual work. Specify the time limit. Point out that some of the numbers are in figures and others are in words.

- Ask students how many they circled. Pairs compare answers.

- Write the numbers from the table on the board.

- Point out that some items in the table are not correct and students must change them accordingly. Go through the first item as a group. Set the task for individual work and pairwork checking. Monitor and assist.

- Feed back onto the board. Check pronunciation of high numbers and knowledge of measurements and abbreviations.

- Elicit whether students guessed the answers to the questions in Speaking B correctly.

Answers

a	number of homes that could fit inside the factory	200	2,000
b	hangar door width	900 m	90 m
c	parts in a 777 liner	300,000	3,000,000
d	bridge cranes	25+	CORRECT
e	build time for a 747	±150 days	±120 days
f	shifts	3	CORRECT
g	employees	2,000	25,000
h	maximum crane load	40 tons	CORRECT
i	height of hangar doors	90 ft	ROUGHLY CORRECT

Speaking

A Discuss the Boeing plant

- Set the task for small-group work.
- Feed back orally.

Answers

Answers depend on the students.

- If you have time, read out the text as students follow. Make deliberate errors which students must spot, OR work further with vocabulary, e.g., *long/wide/high – length/width/height* or *production line, shifts, just-in-time*, etc.

- More photographs and information about the Boeing assembly line can be seen at:

  http://www.boeing.com/commercial/facilities/

- Now have the students do the Workbook exercises for Unit 3, Lesson 7.

Workbook answers

Language: measurements

A Calculate the measurements

1	27.43 m	5	11.05 lb
2	85.30 ft	6	38.10 cm
3	96.56 km	7	439.94 gal
4	36.29 tonnes	8	25.75 km

B Write out the calculations

Answers depend on the students.

C Practise saying the sentences aloud

Answers depend on the students.

Writing

A Circle all the capital letters

See text on Course Book page 55.

B Mark the statements T or F

1	T	4	F	7	T
2	F	5	T	8	F
3	T	6	F		

C Put capital letters where they are needed

The Boeing MD-11, the world's only modern large, wide-cabin trijet, was produced in Long

Beach, California (USA), at the Douglas Products Division of Boeing Commercial Airplanes. It was launched on December 30th, 1986. Assembly of the first unit began March 9th, 1988, and the first flight was on January 10th, 1990. Three engines – General Electric CF6-80C2 and Pratt & Whitney 4460 and 4462 – powered the MD-11, providing maximum efficiency in their thrust class.

Language: passives

A Put the words into the correct order to make sentences

1 Before the 1930s, aircraft structures <u>were built</u> using a trussed frame.

2 After that, the sub-assembly <u>is transported</u> to the construction plant.

3 Brake components <u>have been produced</u> at a plant in Bristol since 2002.

4 Aircraft are not in operational service until <u>they are armed</u>.

5 Problems <u>can be avoided</u> by rigorous testing.

6 The design <u>is admired</u> all over the world.

B Underline the passive form in each sentence

See above.

C Decide which sentences express stronger probability

Answers depend on the students.

 **Lesson 8**   Robots

Objectives

To develop awareness of key aids in reading for general understanding.

To extend vocabulary for manufacturing and new technology.

Language

Prediction: *I don't think X will …*; *will definitely*; *we might have problems.*

Speaking

A Discuss predictions about future uses of robots

- Elicit students' opinions about the first item. Put some useful language on the board, e.g., *I don't think X will …*; *will definitely …*; *probably won't …*, etc.

- Clarify vocabulary problems, e.g., *sophisticated.*

- Ensure the students understand the scale 1–5. Set the task for individual work.

Answers

Answers depend on the students.

B Discuss your answers

- Discuss the first one or two as a class, eliciting reasons for students' opinions. Pairs compare and discuss opinions.

- As feedback, elicit a few ideas.

Answers

Answers depend on the students.

Reading

A Skim the text for gist and answer the questions

- Stress that students need not understand all the words in the text.

- Set for individual work and pairwork checking. Specify a time limit.

Answers

1 Snake-arm robots are robotic assembly and maintenance machines which can reach into spaces that are currently inaccessible to machines.

2 Students may wish to modify 4, 5 and 6; 4 and 6 are evidenced by the text; by implication, more and more manual jobs will probably be assigned to machines (5).

B Raise awareness of reading sub-skills: skimming

- Refer to the Skills Box. Have a general discussion about how these things might help understanding of the text.

- Get students to find examples of each in the text.

Answers

There are examples of all of the skimming elements in the Skills Box in the text.

Speaking

A Work in small groups

- Ideally, there should be an equal number of industry/home/business groups, but this is not essential.

- Groups should clarify why their chosen area would be the most useful of the three for research. Discussion should be in English as far as possible, as preparation for the next stage. Monitor and assist.

- Get students to consider current challenges; possible solutions; advantages that new systems and technology would bring.

- Stress that all members of the group will give the talk. The groups should rehearse carefully, as usual, and suggest improvements to each other.

B Form new groups

- Students should use the language of speculation, prediction and signposting learnt so far in the course.

- As usual, those listening must ask questions at the end of each presentation.

- New groups must decide on which of the fields now seems the most appropriate for research. It does not matter if they can't agree.

C Present decisions

- As feedback on the task, each group should nominate a member to briefly summarise points made and the final choice of the group.

Answers

Answers depend on the students.

Language

A Sentence completion using future tenses

- Work through the Language Box. Stress that *will* in English is not inseparably pegged to future meaning, but that it means 100% commitment to the proposition.

- Do the first sentence on the board.

- Set the task for pairwork. Monitor and assist.

- Feed back onto the board.

Suggested answers

1 The road traffic problem will get worse before it gets better.

2 If the world climate gets warmer, there might be/we may have serious problems.

3 In the future, everybody could live for 120 years.

4 Space tourism will become popular during this century.

5 Poverty won't be eradicated in the 21st century.

6 I doubt if the population of the world will pass 10 billion.

7 In my work, I may/might have to fly.

B Discuss statements from the previous exercise

- This exercise repeats Speaking Exercise 2 to some extent. Students should now be able to use a broader range of expressions and vocabulary in order to complete it.

- Set for pairwork. Feed back orally.

Answers

Answers depend on the students.

- Students can visit Internet sites dealing with snake-arm robotics, such as:

  http://www.ocrobotics.com/

  http://www.cnn.com/2011/11/18/tech/innovation/snake-arm-robot-nuclear/

- Videos of emerging snake-arm research is available at:

  http://www.snakerobot.com/

  http://www.ece.clemson.edu/crb/octor/multimedia.htm

- Now have the students do the Workbook exercises for Unit 3, Lesson 8.

Workbook answers

Language: the future

A Correct the grammatical mistake in each sentence

1 One of these days, he will definitely <u>buy</u> a new one.

2 Take plenty of water with you because it might <u>be</u> hot outside.

3 We're not sure yet, but there <u>might be</u> a possibility of getting a bigger budget.

4 No. There's a flaw in the plans: it <u>won't</u> work.

5 I don't think we <u>will have</u> any problem finishing by the deadline.

6 What about welding it? You never know, it <u>might work</u>.

7 He's not here today, so I <u>can't</u> speak to him until tomorrow.

8 I <u>might</u> get a chance to look at it later – it depends.

Language: expressions for probability

A Put the words into the gaps

1 doubt if   **5** probably

2 are bound   **6** must

3 I'd say   **7** perhaps

4 might

Writing: opinions about the future

A Complete the dialogue

Answers depend on the students.

Objectives

To listen to talks on specific roles in production and design.

To develop the use of abbreviations for note-taking.

To practise asking and answering questions about engineering jobs.

Language

Vocabulary for roles and skills in design and production: *technician*, *analysis*, *modifications*.

Question forms.

Speaking and reading

A Read the job advertisements quickly

- Set the task for individual work. Students should read without dictionaries and check in pairs.

- Feed back onto the board any information that students have, without referring back to the text.

Answers

1 **duties:** complete modifications on the aircraft and fabricate parts, etc.,

  skills: structural/sheet metal background; experience of all aspects of aircraft structures; riveting experience

2 **duties:** produce manufacturing and design data to enable manufacture and assembly of clients' products

  skills: experience in sheet metal weatherproof and non-weatherproof enclosure experience, structural steel detail design, awareness of prevention of galvanic corrosion, complex CNC machined

component, understanding of need to record modifications to drawings and documents

3 **duties:** perform stress analysis in the form of finite element (FE) or fatigue and damage tolerance; work closely with customers; deliver to time, cost and quality

  skills: degree with a minimum of 1 year's industry experience of stress engineering; good team player; able to work without much assistance; PC-literate; able to use Mathcad and Microsoft and Microsoft Office (Word, Excel)

4 **duties:** working on a jet trainer project; be the point of contact for queries; reviewing flight tests, steering tests, systems tests and mission systems

  skills: requirements experience; mathematical, navigation, steering and weapons-aiming background

B Read the advertisements more closely

- Set again for individual work. Ask students to underline or make brief notes on each job.

- Monitor and assist. Point out that students must also note disadvantages of their preferred vacancy.

Answers

Answers depend on the students.

C Discuss your choice in pairs

- Set for pairwork. Feed back orally.

Answers

Answers depend on the students.

Listening

A 🔘 Listen and choose suitable jobs (Number 12)

- Work through the Skills Box.
- Ask students to find examples of abbreviations in the course so far.
- Ask them what makes English hard to understand, and what might make their English hard to understand.
- Set the task for individual work and pairwork checking.

Answers

Answers depend on the students.

B 🔘 Listen again and take notes (CD1 Track 12)

- Ensure that students understand the four categories: *qualifications, background, relevant experience, reasons for wanting the job.*
- Play the recording. Pause between speakers.

Answers

No answers at this point.

C Compare notes

- Set for pairwork. Monitor. Feed back good use of abbreviations onto the board.

Answers

	1	2	3	4
qualifications	BSc Maths & Physics, MSc Metallurgy	BSc Engineering	apprenticeship	BSc Maths & Physics, MSc Avionics & Control Systems
background	stress testing	drawing/painting, UK-based	RAF overseas	family Italian, Maths
experience	small lab, London: tight deadlines, big aerospace clients	CAD, UK-based	helicopters/Hercules, skin fabrication, machinist	PhD, Uni teaching, short contr. for RAF
reasons	OK to start @ bottom, can travel	short contract	bored-change, wants to work in sheet metal again, Middle East, poss. of saving money	wants career, see project thru, stay in S Wales

Speaking

A Practise asking and answering job interview questions

- Elicit examples of one question for each job onto the board.
- Set for individual work. Monitor closely and assist.
- Set the pairwork task. Stress that interviewees should give as full an answer to questions as possible. Interviewers should make brief notes. Again encourage the use of note forms.

- As feedback, ask how successful the interviews were.

Answers

Answers depend on the students.

- If you have time, exploit the texts further for vocabulary OR play the recording again while students follow the text.
- Now have the students do the Workbook exercises for Unit 3, Lesson 9.

Workbook answers

Language: prepositions

A Complete the sentences with the correct preposition

1 of
2 to
3 from
4 with
5 for
6 of
7 by
8 from/to

B Underline the prepositional phrases

1 of maintaining single-engine aircraft
2 to it
3 from drawings
4 with other departments
5 –
6 of Cessnas
7 by a week's leave
8 from five to ten a month

C Memorise the prepositional phrases

Answers depend on the students.

Pronunciation: contractions

A Practise saying the contractions

1 I've /aɪv/
2 I'll /aɪl/
3 I've /aɪv/
4 I'm /aɪm/
5 I'd /aɪd/

B Write the full form of the contracted words

1 I have
2 I will
3 I have
4 I am
5 I would

Writing: job applications

A Write a job application letter

Answers depend on the students.

Lesson 10 — Speaking clearly

Objectives

To use new vocabulary in context.

To produce a short written text using concepts and language from the unit.

Language

Focus on correction and review of language covered in Unit 3.

Introduction

- Exploit the visual. Elicit uses of the items pictured.
- Get students to read the text above the pictures and discuss in pairs how the things pictured can be useful for safe, efficient aircraft production and maintenance.

Vocabulary

A Parts of speech

- Review the idea of *noun*, *verb* and *adjective*.
- Set the task for individual work and pairwork checking.

Answers

1 b (both)

2 n (noun)

3 n (noun)

4 v (verb)

5 n (noun)

6 b (both)

7 n (noun)

8 b (both, but usually a verb)

B Write items in the table

- Do the first item as a group.

- Set for pairwork. Feed back onto the board, allowing students to argue the case for putting words in particular columns, e.g., *shout* could be placed in the spoken and written communication column because writing an e-mail in capital letters is commonly referred to as *shouting*. Elicit more words for each category, e.g., *note*, *distortion*, *chat*.

Answers

connected with written communication	connected with spoken communication	connected with written and spoken communication
log	intercom	instructions
leaflet	repeat	terminology
	shout	order

Reading and speaking

A Divide into five groups

- Assign a paragraph to each group. Students should read silently, then discuss vocabulary problems. Monitor and assist.

B Prepare a talk

- Students prepare their talk in the usual way. Encourage use of vocabulary from this and the previous lesson. As usual, monitor and assist.

- Students must rehearse their talk and advise each other on improvements as part of the process.

Answers

Answers depend on the students.

C Give your talk

- Rearrange the groups.

- Students deliver their talks. Listeners take notes and ask questions in the usual way. They will need their notes in the next stage.

Answers

Answers depend on the students.

Writing

A Prepare a leaflet or notice

- Students work in pairs or small group. Discuss what the finished product will be. Students could write/draw a classroom poster or large notice with each of the five headings on it, or they could each write or word-process a paragraph to be compiled into a leaflet or printed handout.

- Their text should be concise and cover all the information given to them by their colleagues' talks. Monitor and assist. Encourage editing and correction.

- Pass the work to other groups for comment.

- Students reread the original text to see if they included all the information.

- Work on errors you have noted, especially with relation to the language covered in Unit 3.

Answers

Answers depend on the students.

- If you have time, work on specific vocabulary from the text. Refer students to the text at the beginning of the lesson: '*Working in aircraft production …*' Ask them to find three verbs, ten nouns and five adjectives.

- Now have the students do the Workbook exercises for Unit 3, Lesson 10.

Workbook answers

Language: obligation and advice

A Choose the best alternative

1 Always ask
2 Never pretend
3 It is a good idea
4 It is especially important
5 You shouldn't
6 must be
7 If you have to

Writing: giving advice

A Write sentences giving advice

Answers depend on the students.

Reading: reading for specific information

A Answer the questions from the information in the text.

1 The extra weight of mounting the wing on a single pivot increases fuel consumption and manufacturing and maintenance costs.

2 They can make existing designs more versatile, and make specialised aircraft capable of many kinds of flight manoeuvre. They can extend, expand, bend and contract on command, changing into the best shape for the kind of flight required.

3 Any gaps that appear as you open up a joint in a wing will cause drag. If an aircraft is travelling at supersonic speeds, the stresses this drag creates could rip a wing apart.

4 Lockheed Martin are developing an origami-like folding wing. NextGen Aeronautics envisage a 'batwing' comprising a series of 'sliding skins' that disconnect, fan out to create new wing shapes and then connect together again. Raytheon is working on a telescopic wing that moves in and out of the fuselage. Boeing, NASA and the US Air Force are developing an 'active aeroelastic wing'.

Unit test

A test for this unit is available at:

http://www.garneteducation.com/reps/documents/1246/SDT-u3-test.pdf

Contact your local Garnet Education representative for information about how to access these resources.

The power of water

Objectives

To practise scanning and marking a text for relevant sections.

To practise describing hydrology.

Language

Language of water supply, e.g., *pump*, *supply*.

Two-word combinations (compound nouns), e.g., *water supply*, *flow rate*.

Pronunciation: stress in two-word phrases.

Reading and speaking

A Discuss these questions

- Discuss water supply and elicit key concepts.
- Elicit as many instances of everyday uses of water as students can think of.
- Set the task for pairwork. Feed back orally.
- Drill key vocabulary, e.g., *reservoir*, *supply*, etc.

Suggested answers

1 bathing, drinking, cooling machinery, e.g., car radiators, washing clothes, cleaning, watering parks/gardens, fountains, fish tanks, flushing toilets, producing energy in hydroelectric projects, irrigation for agriculture

2 in pipes from rivers/treatment plants to factories and cities; in tankers by road; using animals to carry it from wells in more remote areas; using channels for irrigation

3 desert countries have little fresh water – desalination is expensive and requires energy; drinking water needs to be clean; used water needs to be taken away; modern cities use a large amount of water; users may be situated in upland areas; river levels may be seasonal, causing shortage at certain times of year; pipelines are expensive and need maintenance – also may run through environments where, e.g., extreme heat requires special materials, are vulnerable to illegal siphoning and armed attack

B Scan the text

- Ask students to read quickly and mark sections of the text *1*, *2* or *3* in the margin according to which of the questions in Vocabulary and speaking they refer to.
- Pairs compare answers. Feed back onto an OHP if possible.

Answers

1 bathing, drinking, beauty (fountains), power

2 by natural water pressure from reservoirs

3 river water level low at certain times of year; flow rate needed to be controlled by managing gradients and size of pipe/channel; water had to be raised into the reservoirs

- **Note:** There is an extensive and accessible illustrated description of the golden age of water management technology in the Islamic World at:

  http://www.saudiaramcoworld.com/issue/200603/the.art.and.science.of.water.htm

C Discuss the similarities and differences in meaning

- Some of this vocabulary will have been dealt with in Vocabulary and speaking above. Students can use dictionaries to look up the vocabulary.
- Feed back orally.

Suggested answers

1 A reservoir is man-made for water supply; a pool can be natural, and is used by people for decoration/bathing. Both contain a body of water.

2 A pipe is closed, a channel open; both carry a flow of water.

3 Dimensions = size; capacity = maximum amount held; both refer to measurement.

4 Level = how high or low; gradient = how steep; both refer to lines.

5 Sophisticated = perhaps nuances of modernity and cleverness of design; complex = nuances of difficulty to understand. Both mean intricate.

6 Suction = draw water up/out; pump = cyclical draw + push water.

Vocabulary

A Match the two-word combinations

- Ask students to cover the text.
- Do the first one as a group.
- Set the task for individual work and pairwork checking.
- Feed back onto the board. Clarify the meaning of the combinations and reiterate that students should think of these larger chunks of language as vocabulary items.

Answers

1 b (water – pressure)
2 d (constant – supply)
3 e (kinetic – energy)
4 g (water – management)
5 c (flow – rate)
6 a (reciprocating – piston)
7 f (inlet/outlet – valve)

B Complete sentences using the vocabulary

- Ask students to read through all the sentences first. Elicit answers to a sentence that students can immediately complete (perhaps 2 or 6).

- Set the task for individual work and pairwork checking. Feed back orally.

Answers

1 *Water pressure* is normally measured in psi (pounds per square inch).

2 The *flow rate* of a liquid is determined by the dimensions of the channels and the pipes it runs through.

3 *Inlet valves and outlet valves* are used to control the flow of water in and out of pumps and tanks.

4 Reservoirs can ensure a *constant supply* of water even when the water level is very low.

5 A *reciprocating piston* sucks water into the cylinder of the water pump.

6 *Kinetic energy* is the energy that something possesses because of its motion.

7 Some ancient civilisations used sophisticated systems of *water management*.

C Pronunciation: stress patterns in compound words

- Refer students to the Skills Box. Point out that they should mark the stress on two-word combinations in their vocabulary records as they do on single words, and memorise the stress pattern as part of the language item.

- Drill the phrases.

- Set pairs to read aloud sentences in Exercise B above with the correct stress.

Answers

Answers depend on the students.

- **www.alhambra.org** gives brief news updates on the 2007–2009 restoration of the water system of the Alhambra in Spain.

- Now have the students do the Workbook exercises for Unit 4, Lesson 1.

Workbook answers

Language: past simple

A Write sentences

1 The water ran down a slope.

2 The rate of water flow depended on the gradient of the slope and the dimensions of the pipe.

3 The water came from a source in the mountains.

4 Arab engineers built the Alhambra Palace.

5 The pressure from the water powered the Alhambra fountains.

6 The reservoirs had a sufficient capacity to provide a constant supply.

7 Al-Jazari contributed a lot to hydraulic technology.

8 He designed pumps which used the power of naturally flowing water.

9 A piston pushed water through outlet valves.

10 Inlet valves closed through a pumping stroke.

Vocabulary: water supply

A Complete the sentences using vocabulary from Lesson 1

1 supply
2 reservoirs
3 level

4 pressure
5 pipes
6 pump

Writing: giving information

A Write about the water supply in your area

Answers depend on the students.

Lesson 2   Pushing water

Objectives

To practise listening to a description of a process and labelling a diagram.

To increase awareness of sentence stress and strong/weak vowel pronunciation.

To practise explaining how a system works from notes.

Language

Language to discuss and describe hydraulic ram operation: *delivery pipe, cycle*, etc.

Speaking

A Discuss different types of ram

- Clarify that both the animal and the mechanical device are called a *ram*. Elicit ideas from the group about similarities and differences.

Suggested answers

Animal/mechanical ram; large/small; one drinks water, the other doesn't; one is used in industry, the other in farming; (possibly, both are found in farming – livestock/irrigation); both push things in front of them.

B Activate schemata: discuss true/false statements about hydraulics

- Set the task for pairwork Students may not know about hydraulics; tell them to use their general knowledge.

- Feed back orally.

- Students could look up *hydraulic ram pump/ram/hydraulic(s)* in a good dictionary.

Answers

All the statements are true.

Vocabulary and listening

A Review key vocabulary

- Write the words from the box on the board.

- Elicit any meanings the students remember.

- Drill the pronunciation of the items in the box. Students will need to recognise and produce them later in the lesson.

- Refer students to the Course Book illustrations. Set the task for pairwork.

- Feed back orally.

Suggested answers

1 increase, upwards

2 –

3 cycle

4 spring, momentum

5 –

B Preparation for listening: study the diagram

- Refer students to the diagram and the rubric. Elicit what the shaded area and the arrows show. Point out that *clack* is the noise that the valve makes.

- Set the task for pairwork. Students should label the sections they are sure about.

- As feedback, elicit a few ideas. Do not confirm or correct ideas yet. Students will listen to check.

Answers

No answers at this point.

C Listen for confirmation (CD1 Track 13)

- Set for individual work and pairwork checking. Students should write in the letters for any labels that they were unsure of.

- Check understanding of the text by asking students to explain how the pump works.

Answers

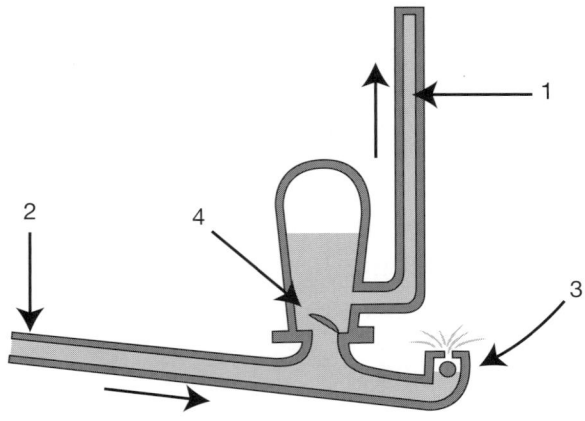

D Listen and put the correct words in place (CD1 Track 14)

- Allow time for students to read the text silently. Elicit whether/where they found words missing.

- Work through the first line orally with the group.

- Play the recording. Pairs compare answers.

- Feed back onto an OHP if possible, or by checking with the tapescript.

Answers

So when this happens to **the** water, there is this backward pressure in the pump, and this **is** enough to open the delivery valve here in the centre of the diagram **and** push water upwards through the outlet pipe. Of course, then the pressure inside the pump goes down **again**, and as it returns to normal, **our** spring-loaded valve at the bottom – it's sometimes called a 'clack' valve **because** it makes a clacking sound like someone hitting **two** pieces of metal together ... actually, sometimes it doesn't have a spring, just a weight – **anyway,** this valve opens again, and of course the – what happens to the delivery valve up here?

Speaking

- It has been shown by research that vowel quality, and especially vowel length, is key to comprehension in spoken English. Students must learn to give stressed vowels their full length, in contrast to those in unstressed words.

A Focus on stressed words and weak sounds

- Go through the Skills Box with the group.
- Write extracts *1–5* on the board.
- Read out the first extract and elicit which words are stressed.
- In pairs, students mark the stressed words in *2–5*. Elicit ideas, then say the sentences yourself for students to check.
- Use sentence *1* to elicit the /ə/ sounds. Again, pairs mark the rest of the sentences and check as you say them.

Answers

1 *and* the <u>whole</u> <u>cycle</u> <u>starts</u> *a*<u>gain</u>

2 *it* <u>flows</u> down *a* <u>pipe</u> *on a* <u>slope</u>

3 <u>Yes</u>, *and* <u>when</u> *the* <u>water</u> hits *the* <u>ram</u>,

4 *and* it <u>doesn't</u> <u>need</u> *an* <u>electric</u> <u>motor</u> *or an* <u>engine</u>.

5 *the* <u>energy</u> *is* <u>changed</u> *into* <u>pressure</u>

B Practise saying the extracts using appropriate stress and weak forms

- Set for pairwork. Feed back orally. Drill the correct pronunciation.

Answers

See above.

C Practise saying a longer extract with appropriate stress patterns

- Set the task for pairwork. Students can focus on a small section of the text. Monitor and assist.
- As feedback, ask some stronger students to say their part of the text.

Suggested answers

So **when** this **happens** to the **water, there** is this **backwards pressure** in the **pump,** and **this** is **enough** to **open** the **delivery valve here** in the **centre** of the **diagram** and **push water upwards** through the **outlet pipe. Of course,** then the **pressure inside** the **pump** goes **down** again, and as it **returns** to **normal,** our **spring-loaded valve** at the **bottom** – it's sometimes called a **'clack' valve** because **it** makes a **clacking** sound like **someone** hitting **two** pieces of **metal** together ... **actually, sometimes** it doesn't have a **spring,** just a **weight** – anyway, this **valve opens** again, and **of course** the – **what** happens to the **delivery valve** up **here?**

- **Note:** The *delivery pipe* is also known as an *outlet pipe*.
- Many illustrated descriptions of hydraulic rams can be found on the Internet, e.g.:

  http://www2.warwick.ac.uk/fac/sci/eng/ research/structures/dtu/pubs/tr/lift/rptr14/ tr14.pdf

Writing and speaking

A Prepare notes on how a hydraulic ram works

- Set for pairwork. Point out that students should use note forms, not whole sentences. They should include abbreviations as necessary.
- The introduction of factual 'errors' needs to be carefully prepared. As usual, students must rehearse the talk with their partner and make improvements based on their partner's feedback. Encourage the use of weak forms of *and, a, an, the*.

Answers

No answers at this point.

B Practise giving and correcting factual information

- Set for new pairs. Stress that listeners must pay attention to detail, but should wait until the end of their partner's talk to point out the errors, e.g., 'You said …, but …'
- If this is too challenging for students, give the talk yourself and elicit corrections from the students.

Model answer

Electricity is the source of energy for this machine.

The water flows through the pipe towards the spring-loaded valve at the bottom and pushes it closed.

Then, because the moving water has nowhere to go, there is a sudden *decrease* in pressure. The water rebounds, creating backwards pressure in the pump. This is enough to open the delivery valve and push water *downwards* through the outlet pipe.

The pressure inside the pump goes *up* again.

- Now have the students do the Workbook exercises for Unit 4, Lesson 2.

Workbook answers

Language: verbs

A Fill in the gaps with the words from the box

1	fill	7	forces
2	move	8	falls
3	lift	9	closes
4	flows	10	reopens
5	causes	11	starts
6	opens		

Language: articles

A Correct where necessary

1 *The* new factory makes plastic pipes for houses.

2 In *the* human body, blood flows more quickly away from the heart.

3 Correct

4 This system takes water from *the* sea and turns it into drinking water.

5 You can see *pumps* here on *the* diagram.

6 A valve is *a* device which controls the flow of air through *a* pipe or *a* tube.

7 Alhambra Palace fountains use *the* natural pressure of water.

8 Correct

9 *Irrigation* has been used for centuries by *farmers* all over the world.

10 All *moving* bodies have momentum.

Speaking

A Discuss two diagrams using key terminology

- Refer students to the diagrams. Elicit briefly what is shown. Clarify the word *fulcrum* in the left-hand diagram and check understanding of the units of measurement in the right-hand diagram.

- Ask students to read the short text above the diagrams and check their understanding of it.

- Set the task for pairwork. Feed back orally, using an OHP or drawing on the board.

Suggested answers

1 In the first diagram, a long lever (the crowbar) is moved a long distance, using a small force, on one side of a fulcrum. On the other side of the fulcrum, this produces a large force moving over a small distance (enough to move the rock). The mechanical advantage is in the creation of a large force from a small one. The force is transmitted immediately from hand to rock using the lever.

2 In the second diagram, the left-hand piston moves 4 inches with a force of 100 lb. This pushes the right-hand piston only 3 inches, but with a force of 900 lb. Again, the mechanical advantage is in the creation of a large force from a small one. The force is transmitted immediately from one piston to the other using the (green) liquid.

- There are some simple illustrated examples of mechanical advantage at:

  http://library.thinkquest.org/CR0120120/ Mechanical%20Advantage.html

  and video at:

  http://www.youtube.com/watch?v=yf AdmRJDKIc

Reading and vocabulary

A Read the text and choose a title for each paragraph

- Ask students to read all five headings. Clarify vocabulary problems.

- Point out that the three paragraphs begin at lines 1, 8 and 15.

Answers

Paragraph 1: d (Everyday applications of hydraulics)

Paragraph 2: e (Fluid properties)

Paragraph 3: b (Hydraulics in the aircraft industry)

B Complete the gaps

- Set the task for pairwork using dictionaries.

- Monitor and ensure that students look up the correct form of the word.

- Students should reread the text silently without writing anything, then in pairs agree on the words for the gaps.

- Monitor and assist. Feed back orally.

Answers

1	line 1	ram
2	line 2	engineering
3	line 4	increasing
4	line 7	atmospheric
5	line 8	properties
6	line 11	conditions
7	line 13	fluid
8	line 16	loading
9	line 19	electrical
10	line 21	surfaces

Language

A Identify and underline adverbs in sentences

- Students read the Language Box silently. Give or elicit examples using adverbs which do not end in ~ly.

- Set the task for individual work and pairwork checking. Feed back onto the board. Elicit which ones are adverbs of degree (*hardly, considerably, only*), which are adverbs of frequency (*normally, usually*) and which are adverbs of manner (*effectively, hydraulically*).

Answers

1	hardly	5	usually
2	considerably	6	hydraulically
3	effectively	7	only
4	normally		

B Circle the verbs that the adverbs refer to

- Again students work individually, then compare answers. Feed back onto the board.

Answers

1 there is hardly any aspect of modern life

2 the effectiveness of the system can be increased considerably

3 the brakes work effectively with fluid at atmospheric pressure

4 water is not normally used

5 which is usually some kind of oil

6 the flaps, air brake and landing gear are hydraulically operated

7 some planes are only fitted with mechanical linkages

C Focus on use of adverbs in the text

- By underlining the phrases, students will review the context and the conceptual content of the text, as well as practise scanning.

Answers

Students should find:

obviously (line 12)

very (line 13)

Conversely (line 13)

- As an extension to the Speaking exercise, ask students to work in pairs.

- Students take turns to tell each other something they remember about one of the three aspects of hydraulics described in the text: *Hydraulics in the aircraft industry, Everyday applications of hydraulics or Fluid properties*.

- Students then look back at the text and find a better way to phrase what they said. They rehearse it and tell a second partner the same information.

- If you have time, do further work on passive constructions, e.g., those in Language Exercise A OR do further work on some of the word endings for parts of speech in Reading and vocabulary Exercise B: noun/adjective ~*ing*; adjective ~*al*/~*ic*; noun ~*ion*; plural ~*s*/-*es*.

- Now have the students do the Workbook exercises for Unit 4, Lesson 3.

Workbook answers

Language: adverbs

A **Separate the adverbs into three groups**

time/frequency adverbs

soon	now
simultaneously	later
usually	often

adverbs of place/movement

longitudinally	vertically
laterally	around

adverbs of manner

simply	tightly
clearly	carefully
clockwise	fast

B **Choose five adverbs and write sentences with them**

Answers depend on the students.

C **Check the pronunciation of the adverbs**

soon	/suːn/
simultaneously	/sɪmlteɪniːəsliː/
usually	/juːʒuːəliː/
now	/naʊ/
later	/leɪtə(r)/
often	/ɒfn/
longitudinally	/lɒŋgɪtjuːdɪnliː/
laterally	/lætərəliː/
vertically	/vɜːrtkliː/
around	/əraʊnd/
simply	/sɪmpliː/
clearly	/kliːə(r)liː/
clockwise	/klɒkwaɪz/
tightly	/taɪtliː/
carefully	/keəfliː/
fast	/faːst/

D **The correct/appropriate adjective or adverb form**

1 ~~specifically~~/specific
2 ~~simultaneous~~/simultaneously
3 dangerously/~~dangerous~~
4 ~~easy~~/easily
5 ~~probably~~/probable
6 ~~quick~~/quickly
7 automatically/~~automatic~~ flexibly/flexible
8 ~~exactly~~/exact
9 lateral/~~laterally~~
10 ~~High~~/highly ~~good~~/well-

Pronunciation: vowel sounds

A **The odd one out**

1 park
2 say
3 fine
4 hole
5 turn

Writing: making notes

A **Use notes to write a short text**

Answers depend on the students.

B **Find examples in the notes**

Answers depend on the students.

Objectives

To discuss the operation of a specific hydraulic system.

To develop skimming skills in a technically dense text.

Language

Vocabulary for hydraulics: *select*, *actuate*.

Speaking and reading

A Discuss the schematics

- As an introduction, elicit from students how they think a plane slows down and stops: the pilot lowers the engine power and extends *flaps* (see Unit 3, Lesson 1); the landing gear touches the ground, causing friction; a parachute opens in some cases; air brakes are opened.

- Ask students to look at the schematics and discuss the questions in pairs.

- Feed back orally. Do not confirm or correct students' ideas yet.

Answers

No answers at this point.

B Read and check your ideas

- Set for individual work and pairwork checking. Stress that students will not understand a lot of the terminology in the text.

- Feed back and confirm/correct ideas.

Suggested answers

The left-hand schematic shows an extended air brake.

The right-hand schematic shows the brake in closed and open positions.

The brake slows down the aircraft for landing.

A hydraulic piston pushes the brake open and then pulls it closed again.

The pilot controls this operation using a switch/selector.

Vocabulary

A Check the meaning of the words

- Set the task. Point out that *control* is both a noun and a verb.

Answers

extend	to stretch out; draw out to the full length
retract	to draw back or in
actuate	to put into action; start a process
control	to exercise restraint or direction over
select	to choose in preference to another or others
indicate	to show, as by measuring or recording

B Complete the sentences

- Set for individual work and pairwork checking.

- Feed back orally. Clarify outstanding vocabulary problems, e.g., differentiate between commonly confused concepts such as *fluid* and *liquid* and opposites such as *extend* and *retract*.

Answers

1 The main control surfaces of an aircraft are operated by *actuators*.

2 The air brake system is hydraulically *operated*.

3 It *extends* and retracts the air brake.

4 A hydraulic power system supplies hydraulic *fluid* to the system.

5 There are *selector* switches and valves in the cockpit.

6 These switches energise the selector valve *solenoid*.

Reading

A Underline the phrases from Vocabulary Exercise B.

- Set the task for individual work. Pairs compare answers. Point out that some of the word combinations occur in different forms in the reading text.

- Check understanding of how the system works by asking checking questions, e.g., *Where is the hydraulic fluid when the air brake is extended? What about when the air brake is retracted?*

Suggested answers

operated	(line 1)
extending	(line 2)
fluid	(line 2)
actuator	(line 5)
indicator	(line 5)
solenoid	(line 6)
selector	(line 6)
piston	(line 8)
extend	(line 9)
selector	(line 11)
solenoid	(line 12)
fluid	(line 12)

B Find and correct the spelling mistakes

- Set for individual work and pairwork checking.

- Feed back onto the board.

- Elicit or use the Glossary to check the meaning of key terms such as: *actuator, selector valve, piston, chamber, solenoid.*

Answers

piston head ~~chambre~~ chamber

~~hydramechinical~~ hydromechanical lock valve

air brake ~~selecter~~ selector valve

air ~~break~~ brake in/out

position ~~mikroswitch~~ microswitch

air brake ~~actueator~~ actuator

- Point out to students that they have seen the extend/retract use of hydraulic pistons in earth-moving equipment, such as bulldozers, where the cylinders are clearly visible on the arms and shovels of the machinery, e.g., the View Specifications pdf brochures at:

  http://deutschland.cat.com/machines/ hydraulic-excavators

- If you have time, drill sentences from Vocabulary Exercise B. Get students to memorise some of them.

- Now have the students do the Workbook exercises for Unit 4, Lesson 4.

Workbook answers

Vocabulary: engineering verbs

A Complete these engineering verbs

1 act<u>ua</u>te

2 c<u>o</u>ntr<u>o</u>l

3 ind<u>i</u>cate

4 fl<u>ow</u>

5 <u>e</u>xtend

6 r<u>e</u>tract

7 rel<u>a</u>y

8 <u>e</u>nergise

B Use/change the verbs to complete the sentences

1 actuated	5 energised
2 flows	6 extended
3 relayed	7 retracted
4 indicate	8 control

Language: suffixes

A Decide which words are nouns

1 pressure	6 selector
2 extension	7 direction
3 actuator	8 manager
4 energy	9 operation
5 deflection	

B Meaning of ~ise/~ize, ~ion, ~or/~er

~ise/~ize: to become, make or make like

~ion: the action or state of

~or/~er: a person or thing that

C Words with ~ise/~ize, ~ion, ~or/~er

Answers depend on the students.

D Write sentences

Answers depend on the students.

Lesson 5 — Happy landing

Objectives

To practise listening to check information.

To practise comparing different systems.

To develop skills in taking notes: reduced sentences.

Language

Vocabulary for manoeuvres and movement, e.g., *twist*, *impact*.

Omission of *a(n)/the*; *be* and *I* in notes.

Introduction

- Elicit where on a car hydraulics are used, and how the system works.
- Ask students where on an aircraft the same principle might apply.
- Refer students to the header in Lesson 5.

Answers

On a car, hydraulics are used, in conjunction with large springs, in the suspension. The wheels are linked to the chassis by hydraulic cylinders. The pressurised fluid in the hydraulic cylinder acts as a buffer, absorbing impact smoothly when the wheels pass over irregularities in the road surface, and returning the spring smoothly to its normal position. The principle is used in aircraft landing gear.

Speaking

A Discuss the pictures: review movement verbs

- Remind students of the description of the eagle's movement in Unit 3, Lesson 3. Elicit and drill the pronunciation of the verbs *twist, fall, turn, land, extend*.

- Point out that the pictures 1–5 are in the correct order. Go through the first one as a group.
- Set the task for pairwork.
- Feed back orally.

B Match pictures with description

- Ask students to read sentences *a–e*.
- Clarify vocabulary problems.
- Set the task for individual work and pairwork checking. Again, go through the first one with the group.
- Feed back orally.

Answers

1	e	**4**	d
2	c	**5**	b
3	a		

C Discuss similarities between a cat's ability to land and an aircraft landing system

- Set for pairwork. Keep the activity short if students do not have many ideas.

Suggested answers

The pilot checks that the aircraft is level as it descends; an aircraft extends landing gear, spaced evenly under the body and wings of the aircraft; the landing gear absorbs impact (using hydraulic cylinders); the aircraft lands on its two main sets of landing gear simultaneously.

An aircraft cannot twist or modify the airframe in flight to help absorb impact in the way that the cat uses its back.

Vocabulary

A Match words with similar meaning

- Put the words on the board. Elicit the first match.
- Set the task for pairwork.
- Feed back onto the board.

Answers

1 c (descent - fall)

2 e (determine - decide/judge)

3 g (flexibility - mobility)

4 b (impact - shock)

5 f (manoeuvre - turn)

6 d (precise - exact)

7 a (spread - distribute)

- Elicit differences between the matched pairs. Emphasise that some pairs are used interchangeably, whereas others may be distinct in particular contexts.

Suggested answers

descent (n) – usually implies control; *fall*, a lack of control; *fall* is both *(n)* and *(v)*.

determine – decide/judge (v) are very similar.

flexibility (n) – implies the capacity to change (e.g., shape) in general; *mobility*, the capacity to *move* from one place or position to another.

impact (n) – refers to physical contact; *shock* to the resultant force(s); *shock* is also *(v)* in UK English; both are *(v)* in US English.

manoeuvre (v) – is a general term covering a range of movements: climb/turn/dive/bank; *turn* is a less technical term and often refers to a movement to the left or right. Both are *(n)* and *(v)*.

precise – exact (adj) are similar.

spread – distribute (v) are similar; *spread* is *(vt)* and *(vi)* so things may spread by themselves, e.g., *news*, *illnesses*; distribute is *(vt)* only, so things must be distributed by someone or something, e.g., *the food was distributed by lorry*.

B Check pronunciation

- Students will need to recognise the words in the Listening section. Drill the words, correcting word stress.

Listening

A 💿 Listen for the main ideas (CD1 Track 15)

- Clarify what the pictures show: the inner ear of a cat; the skeleton of a cat; a chamber filled with pressurised fluid (hydraulic strut assembly).

- Play the recording. Pairs compare their ideas.

- Feed back orally.

Answers

The order is: inner ear of a cat; skeleton of a cat; a chamber filled with pressurised fluid.

B 💿 Check answers to true/false statements (CD1 Track 15)

- Set for pairwork discussion. Elicit, but do not confirm answers.

- Play the recording again while students check their answers. They may need to hear it twice, as the answers are not in consecutive order.

- Feed back: go over any problems or queries and elicit the basic similarities and differences between the way cats and aircraft land, e.g., cats have a lot of flexibility and change their shape/position, whereas planes manoeuvre and use their control surfaces.

Answers

1	T	7	T
2	F	8	F
3	F	9	T
4	T	10	T
5	F	11	T

6 T (although planes have less flexibility)

C Look at sketches of an aircraft landing and make notes

- Ask students to read through the Skills Box. Clarify with further examples if necessary.

- Look at the sketches together with the class, go through the example and elicit ideas for other notes.

- Set the note-taking task for individual work. Monitor and assist.

- Pairs should compare and discuss notes.

- Give additional feedback by showing the model notes on OHP or writing them on the board.

Model answer

- **A** Pilot checks artificial horizon.
- **B** Control surfaces extend to slow aircraft down.
- **C** Pilot checks altitude.
- **D** Pilot manoeuvres.
- **E** Pilot looks at ground.
- **F** Plane lands on main wheels.
- **G** Landing gear absorbs impact.

- If you have time, review the meaning and pronunciation of the words in Vocabulary Exercise A OR get students to listen again to the recording as they follow the tapescript.

- Now have the students do the Workbook exercises for Unit 4, Lesson 5.

Workbook answers

Language: verb forms

A Decide which form is used in passive sentences

The third form is used.

B Write three forms of the verbs

1 cut cut cut
2 rotate rotated rotated
3 spread spread spread
4 protect protected protected
5 make made made
6 fall fell fallen

7 build built built

8 have had had

9 develop developed developed

10 distribute distributed distributed

11 align aligned aligned

12 take off took off taken off

13 find found found

14 reduce reduced reduced

15 send sent sent

16 write written wrote

17 require required required

18 check checked checked

Writing: note form

A Write note forms of the following sentences

1 Runway oriented north-south.

2 Most internal structure protected fireproof bulkheads.

3 Disc rotated 90 degrees.

4 Speed reduced using parachute stored back fuselage.

5 Two assemblies aligned accurately.

6 Aircraft lands impact distributed between wheels.

Language: word-building

A Complete the table

noun	verb (1)
rotation	rotate
land	land
distribution	distribute
alignment	align
construction	construct
reduction	reduce
requirement	require

B Write and memorise example sentences

Answers depend on the students.

Lesson 6 — Tricycles or taildraggers?

Objectives

To practise listening skills: comparing different types of landing gear assemblies.

To practise evaluation and decision-making: suitability of landing gear types for a range of scenarios.

Language

Vocabulary for describing landing/take-off/landing gear layout: *centre of gravity, nose wheel.*

Introduction

Use a drawing on the board to elicit *bicycle* and the fact that *bi~* means *two*. Draw a *triangle* and elicit the meaning of *tri~* (*three*). Elicit the idea of *tricycle* for *a vehicle with three wheels*. Elicit or teach the verb *drag*, and the *tail* of an aircraft.

Speaking

A Introduce and discuss different types of undercarriage

- Set the task for pairwork. Clarify that *undercarriage* is another word for *landing gear*.

- Feed back orally.

Suggested answers

1 To allow safe taxiing, take-off and landing. All aircraft landing gear must be able to support the weight of the aircraft while it is stationary, while it is taking off and while it is landing. When the aircraft is landing, the gear must be able to support the downward and forward forces resulting from gravity and the aircraft's velocity.

2 The left-hand picture is a taildragger; the right-hand picture is a tricycle.

3 Students will probably only suggest one or two ideas at this stage. They will discuss this issue later in the lesson, so do not 'give away' information that they do not suggest.

taildragger: pilot cannot see the runway ahead; easier to land on rough ground; tends to spin if not kept in line; vulnerable to high winds when parked because of the upward angle of the wings; faster and more economical because the landing gear is relatively small.

tricycle: on uneven ground, nose wheel can collapse or go into a hole, causing the propeller to touch the ground; easier for the pilot to get into; aircraft easier to load because it is level with the ground.

Vocabulary

A Check correct pronunciation and label diagrams

- Drill the words, correcting stress. Ask students to underline the stressed part of each word/phrase.

Answers

<u>cen</u>tre of <u>gra</u>vity

<u>lan</u>ding gear

<u>cock</u>pit

pro<u>pel</u>ler

<u>nose</u> wheel

1

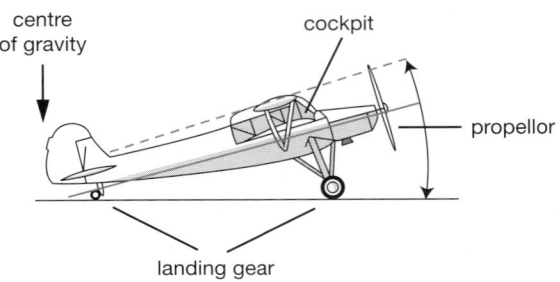

2

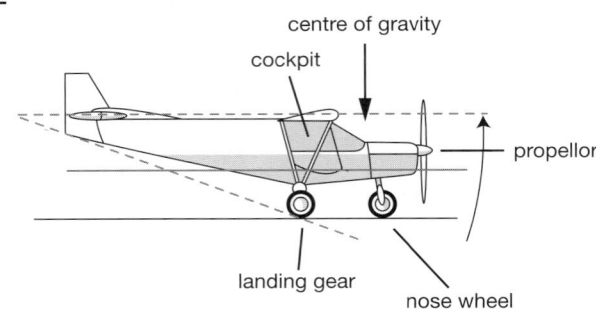

B Decide which statements apply to each design

- Get students to read silently through *1–10*. Clarify any vocabulary problems.

- Elicit ideas for item *1*.

- Set the rest of the task for pairwork. As feedback, elicit one or two more ideas, but do not confirm or correct yet.

Answers

No answers at this point.

Listening

A Listen and check ideas (CD1 Track 16)

Answers

1	A	6	A
2	A	7	B
3	B	8	A
4	B	9	A
5	B	10	B

B 🎧 **Intensive listening practice: listen and correct sentences (CD1 Track 16)**

- Ask students to read the sentences silently and then aloud in pairs.

- Play the recording. Pairs compare answers. Feed back onto the board.

- Drill the pronunciation of the sentences.

Answers

1 There's no danger of breaking ~~your~~ *the* nose wheel in a hole.

2 I'd much rather make an emergency landing in ~~yours than mine~~ *mine than yours*.

3 The centre of gravity is ~~between~~ *behind* the wheels.

4 And the cockpit is pointing up ~~towards~~ *at* the sky – you can't see.

5 You just have to pay ~~close~~ *more* attention.

6 I've done ~~a lot of~~ *a couple of* groundloops myself.

C Discuss aircraft design features

- Elicit the advantages and disadvantages of tricycle/taildragger landing gear covered so far in the lesson. See suggested answers for the Speaking section above.

- Put on the board one of the scenarios a–d below. Elicit what kind of aircraft would suit the situation. Get students to suggest as much information as you can, e.g., overall size, number of engines, seats and landing gear layout. Point out that there is no correct answer, but that they must give reasons for their suggestions.

- Set the task for pairwork. Feed back orally.

Suggested answers

1 small; single-engined; two-seat; large loading area; tricycle undercarriage for easy loading of parcels, etc.

2 traditional tricycle undercarriage; single-seat high-performance/similar to what he flew in the airforce

3 small; powerful; tricycle for speed and uneven runways

- If you have time, you may wish to use the following scenarios to add to the ones in the Course Book for more varied groupwork, e.g., individual groups study one case and present their conclusions to their colleagues:

  a for a middle-sized commercial passenger company whose aircraft stays on the tarmac all the time

  b for an inexperienced pilot who has just got his license

  c for the police, who need to fly small groups of men to places which sometimes have no hard runway

  d for a club which does not have much money for fuel

Suggested answers for additional scenarios

4 large; twin (jet)-engined; tricycle undercarriage for ground stability and ease of baggage loading

5 light; two-seat; tricycle for good visibility

6 4–8 seat; tricycle for uneven runway/fuel economy; passenger compartment not finished to commercial standards

7 small; single-engined; two-seat, tricycle for fuel economy

- If you have time, play the recording from Listening Exercise A while students read the tapescript OR elicit the word stress from the items in Vocabulary and speaking Exercise A.

- Additional reading/research on landing gear layout, including drawings and descriptions, can be done at:

  http://www.aerospaceweb.org/question/design/q0200.shtml

- Now have the students do the Workbook exercises for Unit 4, Lesson 6.

Workbook answers

Pronunciation: short vowel sounds

A Group the words according to the sound they contain

/æ/ as in *bad*	/e/ as in *ten*	/ɪ/ as in *pin*	/ɒ/ as in *stop*	/ʌ/ as in *cut*
advantage	propeller	windspeed	hot	just
happen	level	resistance	cockpit	son
gravity	momentum	fitted	coffee	enough
landing	attention	flying	velocity	stuff

B Mark the unstressed syllables

advantage propeller windspeed

happen   level   resistance cockpit

gravity   momentum fitted   coffee   enough

landing   attention flying   velocity

C Practise saying the words

See the tapescript of CD1 Track 16.

Language: *if* and *when*

A Decide what will happen in each situation

Answers depend on the students.

B Complete the sentences

1 Next week when the full plans are ready, we will know more.

2 If the impact is too heavy, it might crack.

3 If it gets hot, that component won't work.

4 A light will come on in the cockpit when the air brake extends.

5 Production can only work well if they deliver the assemblies on time.

6 When I want to think about a machine, I always draw a diagram.

Objectives

To develop skills in reading reduced notes.

To practise interpreting and comparing schematics.

Language

Components of a hydraulic actuation system: *retract chamber*; *pressurised fluid*.

Introduction

• Elicit as many parts of an airframe as students can remember.

Speaking

A Work with a partner

• Set the task for pairwork. Feed back onto a drawing on the board or an OHP.

Answers

See Unit 3.

B Discuss the meaning of the notes

• Ask students to look at the Skills Box and their notes for Listening Exercise C on page 71.
• Work through the first line with the group.
• Set the task. Feed back orally.
• Have a class discussion on the meaning of features of the notes: underlining; dash; colon; '+'; brackets; '='.

Suggested answers

For take-off, the flaps are part-extended. This creates lift without slowing the aircraft down.

For landing, the flaps are fully extended. This creates lift and air resistance, called *drag*. The effect is to reduce the aircraft's speed, but not its height.

The pilot decreases the aircraft's forward speed and controls the angle of descent.

Vocabulary

A Divide the labels into three groups

• Draw a table on the board with the three categories, *1*, *2* and *3*. Clarify the meaning of *familiar*.
• Use one student. Do you know *stop nut*? Do you know *stop*? Do you know *nut*? Elicit from the group which column that student should write *stop nut* in.
• Set the task for individual work.

Answers

Answers depend on the students.

B Discuss meaning of vocabulary with a partner

• Set the task. Encourage pairs to clarify any words they can to each other.
• Students should look up any outstanding terms in their dictionary. Monitor and assist.
• Use a drawing on the board or an OHP of diagram *a* for feedback.

Answers

Answers depend on the students.

Reading

A Compare diagrams *a*, *b* and *c*

• Point out that the diagrams show the system in different positions. Highlight the coloured sections on each diagram and elicit that they represent hydraulic fluid.

- Set the task for pairwork. Emphasise that students may find the system complex at the moment, but that they will have time to read and discuss it in detail later in the lesson.

- Feed back some ideas, but do not confirm or correct yet.

Suggested answers

In *a*, the hydraulic fluid fills the retract chamber, and the primary piston (on the left) is pulled into the unit. In picture *b*, the hydraulic fluid fills the middle – intermediate – chamber, and the primary piston is pushed partly out. In picture *c*, the hydraulic fluid has filled the piston head chamber and pushed the primary piston fully out.

- The Concord website:

  http://www.concordesst.com/gear.html

  gives details of the operation of the aircraft's landing gear.

- Now have the students do the Workbook exercises for Unit 4, Lesson 7.

Workbook answers

Language: prefixes

A What *re~* means

It means 'again'.

B Write the meaning and an example sentence for the verbs

Answers depend on the students.

C Find more examples in your dictionary

Answers depend on the students.

Writing: capital letters

A Label the diagrams using capital letters

No answers at this point.

B Add useful labels

See Course Book pages 14, 28 and 64.

Lesson 8   First flight basics

Objectives

To practise listening and filling in a table.

To practise the pronunciation of consonant clusters.

To practise explaining how a control system works.

Language

Vocabulary and phrases to describe controls and control surfaces: *pitch*, *roll*, *ailerons*.

Introduction

- Draw a simple picture of, or use an OHP, to show a hydraulic actuation system. Elicit briefly how it works.

Vocabulary and speaking

A Label axes on a diagram

- Use lines on the board to elicit the words *horizontal*, *diagonal*, *vertical*. Draw a simple graph and elicit *horizontal axis* and *vertical axis*.

- Monitor as students label the axes on the diagram.

Answers

3 vertical axis

4 lateral axis

5 longitudinal axis

B Read definitions and complete the labelling

- Set for individual work and pairwork checking. Feed back onto the board or an OHP.

Answers

1 pitch

2 roll

6 yaw

Listening and writing

A Listen for specific information: list controls mentioned in the recording (CD1 Track 17)

- Set the scene. Point out that students do not need to spell correctly, but simply to write the control systems and surfaces they hear.

- Pairs compare answers and discuss the role of each element.

Answers

joystick (control column)

elevators

flaps

ailerons

rudder

B Complete notes in the table from memory

- Refer students to Skills Box 1. If necessary, ask them to look at notes in previous lessons, e.g., Unit 4 Lesson 7.

- Set the task for pairwork. Do not confirm or correct answers yet.

C Listen and correct answers

- Play the recording again. Allow students time to amend their notes.

- New partners compare answers. Feed back onto the board.

Model answer

	where on aircraft?	determine(s) which movement?	controlled how by pilot?
elevators	on the horizontal tail surfaces	pitch-raising and lowering the nose	joystick pulled/pushed
ailerons	on the wings	rolling movement of plane (banking)	joystick moved laterally from side to side
rudder	on the vertical tail	yaw-turning the nose to the right or left	two pedals pressed down

Speaking

A Raise awareness of consonant clusters

- Work through the Skills Box with the students.

- Drill the consonant clusters in the phrases, then get students to practise them in pairs.

B Find the phrases in the tapescript

- Set for individual work and pairwork checking.

- Put the sentences on the whiteboard or OHP and elicit which are the stressed words. Drill the sentences, then get students to practise saying them in pairs.

Answers

See tapescript to CD1 Track 17.

Vocabulary and speaking

A Underline the five or six key phrases

- Elicit the first sentence.

- Set for individual work and pairwork checking. Monitor and assist. Feed back onto the whiteboard. Accept alternatives if there is consensus that the sentence is key to how the control surfaces work. In any case, keep the number of sentences to six.

Suggested answers

Now, when I pull the stick back the elevators move up.

... we can also move the stick from side to side.

... *these* control surfaces are called ailerons and they control the rolling movement of the plane.

We use it to swing or turn the nose of the plane to the right or left.

All the linkages are mechanical.

B Practise two conversations

- This exercise is designed to contextualise the language from the lesson.

- Students work in two different pairs for the conversations.

- Monitor and collect errors for review at the end of the lesson. Pay particular attention to consonant clusters and the use of language for describing controls and control surfaces.

- If you have time, ask students to rehearse and present one of their conversations to the class.

Answers

Answers depend on the students.

- Wikipedia has clear illustrations of flight controls, although the text is fairly dense.

  http://en.wikipedia.org/wiki/Flight_controls

- Now have the students do the Workbook exercises for Unit 4, Lesson 8.

Workbook answers

Language: cause and effect

A Match the two parts of the sentences

1	e	**4**	a
2	b	**5**	c
3	d		

B Complete the missing parts of the sentences

Answers depend on the students.

Pronunciation: sentence stress

A Mark the words that will be stressed

No answers at this point.

B 🔘 **Check your answer (CD1 Track 17)**

1 it's <u>sometimes</u> called the <u>joystick</u>, or <u>even</u> just the <u>stick</u>

2 it's <u>this control</u> that we <u>use</u> for <u>what's known</u> as the <u>pitch</u> of the plane

3 The <u>flaps</u> are actually <u>further in</u>, <u>near</u> the <u>fuselage</u> you see.

4 these <u>control surfaces</u> are called <u>ailerons</u> and they <u>control</u> the <u>rolling movement</u> of the plane.

5 And <u>lastly</u>, there are <u>these two pedals</u> down <u>here</u>.

Lesson 9 — From pilot to control surface

Objectives

To provide further practise in note-taking from a text and in interpreting notes to convey detailed information.

Language

Language for describing mechanical linkages: *cable*, *rod*, etc.

Introduction

- Ask students:

  the names of the controls and what they do from Lesson 8,

  what type of linkages the plane in the Listening text had (they were mechanical),

  what linkages do (they connect the pilot's controls in the cockpit to the aircraft's control surfaces).

- Write the opening sentence to the lesson on page 78 on the board, but give only the first and last letters of each word. Elicit the sentence. Students should practise this technique with sentences and individual vocabulary items.

Vocabulary

A Match the words to their meaning

- Set the task for pairwork and individual checking. Students can use their dictionary.

Answers

1	g	**6**	i
2	j	**7**	h
3	d	**8**	b
4	e	**9**	c
5	f	**10**	a

B Focus on correct pronunciation

- Drill the items, then get students in pairs to practise saying them. The students will need to use this vocabulary later in the lesson.

Answers

1	tube	/tju:b/
2	rod	/rɒd/
3	cable	/keɪbl/
4	hollow	/hɒləʊ/
5	rigid	/rɪdʒɪd/
6	tension	/tenʃn/
7	play	/pleɪ/
8	pulley	/pʊli:/
9	inaccessible	/ɪnæksesəbl/
10	stiff	/stɪf/

C Review vocabulary using diagrams

- Set for pairwork. Feed back orally. Check that students can differentiate between rods and tubes.

Answers

1	tube	3	cable
2	rod	4	pulley

Reading and speaking

A Work in groups of three

- Divide students into groups A, B and C and assign the texts. Get students to cover the two texts they are not working on.

- Get students to copy the headings *1–5* in their notebooks and clarify the meaning of these.

- Review the idea of note-taking, i.e., students should not copy large pieces of text; that they should extract key points only; that they should use abbreviations and symbols. Use a few lines of text from an earlier lesson, e.g., page 17, to build notes on the board to demonstrate this.

- Set the task for individual work and checking in groups A, B, C. Monitor closely to assist and correct.

Answers

Answers depend on the students.

B Explain the linkage system from your notes

- Students rehearse their talk as usual within their group, and partners must listen and suggest improvements to pronunciation, grammar, delivery, etc.

- Students move around to form groups of three and take turns to give their talk. Listeners, as usual, must listen and ask at least one question each at the end of each talk. The speaker may well not know the answer: if so, s/he must say so or make an informed guess.

Answers

Answers depend on the students.

Extension: writing

- Ask students to look again at the Vocabulary section on page 78. They should choose five words and write an example sentence for each. Monitor and assist.

- Ask students to repeat the sentences to each other and memorise them, to reinforce the sentence stress, new vocabulary and engineering concepts.

- The wikipedia website again has a clear general illustration of the mechanical flight control linkage system, with links to further illustrations of ailerons, elevators and rudders:

  http://en.wikipedia.org/wiki/Aircraft_flight_control_systems

- Now have the students do the Workbook exercises for Unit 4, Lesson 9.

Workbook answers

Vocabulary: movement and change

A Complete the gaps

1 pulls it forwards
2 from side to side
3 backwards and forwards
4 increase or decrease
5 to the right or left
6 up and down
7 raised and lowered
8 back
9 turn around
10 to the left/to the right

Language: *there is/are* and *it is/they are*

A Complete the sentences

1 There are/it is
2 there is
3 They will be
4 It is
5 Is there
6 it is not
7 there are
8 they are

B Find examples of *there + be*

Lines 4, 5 and 6.

Writing: spelling

A Circle the correctly spelt words

stiff
damage
insufficient
torque

adjusted
cables
control column
landing gear

B Practise writing the vocabulary words

Answers depend on the students.

Lesson 10   Manoeuvres: Review

Objectives

To review the language and concepts of Unit 4.

To discuss and describe flight manoeuvres using appropriate terminology.

To practise listening for specific items in a commentary.

Language

Manoeuvres and operation of control surfaces.

Vocabulary

A Focus on vowel sounds

- Research suggests that vowel length, even more than vowel quality, is crucial in comprehension in EIL. Where appropriate in other pronunciation work, take time to focus on this.

- Go through the Skills Box. Make sure students understand the idea of short and long vowels and diphthongs.

- Elicit the pronunciation of one or two of the words in the left-hand column, drawing attention to the vowel sound(s). Point out that words with /eɪ/ and /ɑɪ/ occur twice in each column.

- Get students to read through the lists and in pairs, discuss the pronunciation of each one. Again, elicit one or two ideas. Students should be able to pronounce all of the words.

- Drill all the words.

- Set for individual work and pairwork checking. Feed back onto the board.

Answers

pitch – which	dive – five
loop – group	bank – tank
roll – hole	break – make
turn – learn	straight – weight
climb – time	spread – head

- Elicit one or two ideas. Set the task for individual work and pairwork checking. Feed back orally.

Answers

Answers depend on the students.

- Again, elicit ideas and then set the task for pairwork. Feed back onto the board, marking the three kinds of vowel sound in three different ways. Redrill the words. Then ask pairs to practise saying the words together.

Answers

short vowel sounds: pitch, which, bank, tank, spread, head

long vowel sounds: loop, group, turn, learn

two sounds (diphthongs): roll, hole, climb, time, dive, five, break, make, straight, weight

B Review vocabulary from the unit

- Students can work in pairs to look up/discuss the meanings of the words. Monitor and assist.

Answers

Answers depend on the students.

C Write a sentence with each word

- This exercise consolidates the meaning and pronunciation of the vocabulary. Ask students to write sentences for a limited number of words if you prefer. Monitor and assist.

- In pairs, students should read their sentences to each other. Note common errors. Ask strong students to read out some sentences to the group.

Answers

Answers depend on the students.

Speaking

A Look at the sketches

- Work through the first sketch with the group. Ensure they understand that the arrows represent individual aircraft and the arrow head shows the direction of flight.

- Set the task for pairwork. Feed back orally.

Suggested answers

1 The planes climb together in a line, turn and dive and spread (out).

2 Four planes dive straight down. Then two turn, climb, dive again and go left and right; the other two turn and dive, then climb and loop.

3 Two planes fly straight. Two other planes turn around them.

4 Seven planes dive together, then climb and spread out.

5 Two planes loop down and cross at the bottom of the loop. Another plane flies through the loop.

6 Nine planes fly in a diamond shape and turn together.

Listening

A Gist listening: listen to part of a radio commentary (CD1 Track 18)

- **Note:** The Red Arrows are a famous display team which performs at many air displays in Britain and abroad every year. The team uses Hawk jets. Students may like to read about them at the team's website.

- Set the task for individual work and pairwork feedback.

- Play the recording once.

- Feed back orally or onto an OHP.

Answers

The order is 6, 5, 3, 2, 4, 1.

B Listen for specific words (CD1 Track 18)

- Ask the students to read through *1–6*. Ask the group to say the words to each other to remind them of the pronunciation: they will need to recognise the words. If necessary, drill some of the words.

- Set the task for individual work and pairwork checking. Clarify that *1–6* are in the order in which the manoeuvres are described, i.e., *1* refers to the *Diamond Bend*, *2* to the *Heart*, etc. Point out that at least one, and sometimes two of the words in each item are not used by the speaker.

- Play the recording.

- Feed back orally.

Answers

The words used are:

1 horizontally, turn together

2 climb, cross, bank, dive towards each other

3 roll in long horizontal loops, straight and flat, inside

4 climb, loop, dive down vertically

5 horizontally, close together, break apart

6 turn together, accelerating downwards, spread apart

Writing and speaking

A Describe the manoeuvres from the listening

- Students did a similar exercise in the Speaking section, but this time they should use the terms from the Listening text.

- Work through the first illustration together.

- Set the task for pairwork. Monitor and assist. Feed back orally.

Answers

Answers depend on the students.

B Write notes to describe the pilot's actions

- Review briefly the operation of the control surfaces in the previous Lessons 7–9.

- Point out that students should focus on a single manoeuvre from illustrations 1–6, and that they should write note forms.

- Put the example on the board. Elicit the full operation.

- Set the task for individual work. Monitor and assist.

Suggested answer

1 Opposition Loop

Aircraft climbs. Pilot moves stick fully to right: aircraft rolls 180 degrees. Pilot pulls back joystick: aircraft loops. Pilot pushes joystick slightly: aircraft flies downwards.

2 Twizzle

a) (extreme top left of diagram) Aircraft dives down vertically. Pilot pushes joystick to left: left-wing aileron moves down + right-wing aileron moves up: aircraft rolls. Pilot presses l/h rudder pedal: aircraft banks + turns left. Pilot pushes joystick forward and presses rudder pedal (aircraft turns + dives). Pilot pulls back joystick. Ailerons move up: aircraft climbs + loops.

b) (aircraft second from left at top of illustration) Aircraft dives down vertically. Pilot pulls back joystick hard + presses l/h rudder pedal. Ailerons move up: aircraft climbs + turns left. Pilot pushes joystick forward: aircraft banks left + loops.

3 Corkscrew

a) Pilot keeps joystick central. Aircraft flies horizontally.

b) Pilot presses r/h rudder pedal and pushes joystick to right. Aircraft banks + rolls slowly to right.

c) Pilot presses l/h rudder pedal + pushes joystick to left. Aircraft banks + rolls slowly to left.

4 Vixen Break

Pilot quickly pulls joystick back + presses l/h or r/h rudder pedal: aircraft banks + climbs.

5 Heart

a) Aircraft climbs. Pilot pulls back joystick hard – aircraft loops + dives down.

b) Pilot keeps controls central: aircraft flies horizontally.

6 Diamond Bend

Pilot pushes joystick to left: aircraft banks. Pilot pulls back joystick: aircraft turns slowly 180 degrees.

C Discuss other manoeuvres

- Set the task for groups of three.

- Stress that students do not need to design complex display manoeuvres – although they can if they wish. They simply need to hold a discussion using the language from the lesson, produce sketches and, in D, describe them. All students should make copies of the sketches, as they will need them in Exercise D.

- Put a simple sketch on the board as an example. Do not describe it yet.

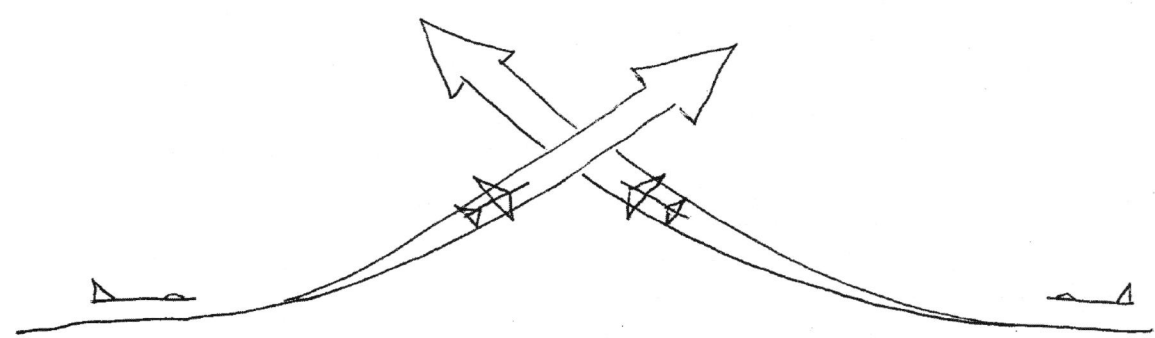

Answers

Answers depend on the students.

D Explain the manoeuvres

- Change the groups round. Set the task. Stress that all members of the new groups must speak. As usual, listeners should ask questions.
- If you have time, students can listen to the tapescript again as they follow the text.

Answers

Answers depend on the students.

- Now have the students do the Workbook exercises for Unit 4, Lesson 10.

Workbook answers

Language: word-building

A What meaning do these suffixes have?

un~   not, the opposite of

im~   not, the opposite of, to put into the condition mentioned

bi~   two, twice, double,

multi~ more than one, many

B Complete the table

noun	verb	adjective
management	*manage*	*managerial*
load	load	*loaded*
increase	increase	*increasing*
cycle	*cycle*	cyclical
technology	–	technological
atmosphere	–	*atmospheric*
inaccessibility	–	inaccessible
rigidity	–	rigid
stiff	*stiff*	stiff
electricity	–	*electrical*
selection	select	*select*
spin	*spin*	–
alignment	align	*align*
reduction	reduce	*reduced*
brake	*brake*	*brake*
–	*exceed*	excessive
resistance	*resist*	*resist*

Language: cause and effect

A Make sentences using *if* or *when*

Suggested answers

1 If you press the pedal, the rudder will move.

2 If the electricity supply is not constant, there will be problems with the machinery.

3 When you land on rough ground, the plane will tip forward.

4 When you arrive at the site, phone me.

5 If you follow the procedures, everything will be fine.

Language: *there* + *be* and *it/they* + *be*

A Delete the incorrect sentence

1 ~~There can sometimes be difficult to stay on time.~~

It can sometimes be difficult to stay on time.

2 This is the trainer aircraft. It is the plane you will be working on.

~~This is the trainer aircraft. There is the plane you will be working on.~~

3 In this country there are more and more people working in aviation.

~~In this country they are more and more people working in aviation.~~

4 ~~Next week it will be a meeting.~~

Next week there will be a meeting.

5 In each team there should be four people.

~~In each team they should be four people.~~

Reading and writing: the Red Arrows

A Think of five questions you would like to ask

Answers depend on the students.

B See whether your questions are answered in the text

Answers depend on the students.

Unit test

A test for this unit is available at:

http://www.garneteducation.com/reps/documents /1247/SDT-u4-test.pdf

Contact your local Garnet Education representative for information about how to access these resources.

UNIT 5

Lesson 1

Engine basics

Speaking and vocabulary

- Set the introductory question, *How much do you know about car engines?* for pairwork.

- Elicit a few ideas, e.g., they are internal combustion engines, there are petrol and diesel engines, etc.

- Feed back orally.

A Discuss the differences

- Set for pairwork. Feed back orally.

Answers

assembly, e.g., internal combustion engine

sub-assembly, e.g., carburettor

component, e.g., spring

- **Note:** The boundary between an *assembly* and a *sub-assembly* is not always clear, but as a rule, a *component* is a single item; a *sub-assembly* is a group of components which, with other *sub-assemblies*, goes to make up the whole unit, the *assembly*.

B Look at and start to label the exploded cutaway of an old-fashioned engine

- Set for pairwork. The vocabulary load in

this lesson is high: students will probably not know many terms for engine parts in English. Allow the use of L1 here.

- Monitor to see which parts students are able to label.

Answers

Answers depend on the students.

C Identify vocabulary for engine parts

- Set for individual work and pairwork checking, and again stress that students are not expected at this stage to know all the parts of the engine.

Answers

Answers depend on the students.

D Work in pairs to finish labelling the diagram

- Students first work with a new partner or in groups to compare ideas.

- Individuals then use dictionaries and check with a partner.

- Monitor and assist. Allow L1 again.

- You can also make use of diagrams such as those at:

  http://www.autoeducation.com/rm_preview /engine_intro.htm

  and

  http://auto.howstuffworks.com/engine2.htm

  for further information.

- Do not wait for students to find all the labels before starting feedback. Feed back orally, using the projection on the board. Go over the new vocabulary.

Answers

1	carburettor	13	pushrod
2	distributor	14	valve
3	fuel pump	15	valve springs
4	fly wheel	16	pushrod rocker (arm)
5	starter motor	17	rocker cover
6	oil pump	18	cylinder block
7	oil filter	19	spark plug
8	oil strainer	20	fan
9	crankshaft	21	fan belt
10	camshaft	22	alternator
11	timing chain	23	engine block
12	piston	24	sump

Reading

A Match terms to descriptions of parts

- Like the diagram on page 82, the table contains a large amount of information. In this exercise, students orient themselves to it.
- Set the task for pairwork. Monitor and assist.

Answers

fuel pump (3)

crankshaft (9)

pushrod rocker (arm) (16)

spark plug (19)

alternator (22)

sump (24)

B Complete table

- This is an important stage in dealing with the task. Pairs or groups should pool information.
- Students will now have a lot of information to help them complete the rest of the items. Encourage them to guess the definitions they are unsure about. Monitor again and assist.

Answers

part number	description	name
1	small chamber which mixes fuel and air to be burnt in the cylinders	carburettor
11	flexible chain or belt which connects the crankshaft and camshaft	timing chain/belt
5	motor which turns the flywheel to start the engine	starter motor
3	pump which supplies fuel to the carburettor	fuel pump
12	cylindrical component which moves up and down inside the cylinder	piston
9	the main shaft which converts linear to rotary motion	crankshaft
13	thin rod which pushes a rocker arm up	pushrod
7	replaceable filter which removes dirt and particles from the lubricating oil	oil filter
17	metal cover which protects the valves and rocker arms	rocker cover
19	small plug screwed into the cylinder which ignites the fuel mixture	spark plug
22	belt-driven AC generator powered by the engine	alternator
24	large container which stores the lubricating oil at the bottom of the engine	sump
10	small shaft with cams that drive the push rods up	camshaft
23	largest part of the engine which houses the cylinders, pistons and crankshaft	engine block
6	small pump which circulates lubricating oil	oil pump
16	small arm which rocks and moves valves up and down	rocker/rocker arm
8	permanent metal oil strainer situated in the sump	oil strainer
14	small component which allows fuel mixture to enter and gas to leave a cylinder	valve
2	switch which sends electricity to the spark plugs in turn	distributor
15	small springs attached to the rocker arms	valve springs
20	rotating blades which help to cool the engine	fan
18	top section of the engine which houses the top of the cylinders	cylinder block
4	heavy wheel which maintains the momentum of the crankshaft	flywheel
21	belt which drives the alternator and fan from the crankshaft	fan belt

Extension: speaking

Task 1

- Put students in pairs. They must take turns to ask and answer questions about the information in the table.

- Demonstrate using a strong student, e.g., *What is the camshaft?* or *What does the flywheel do?*

- Set the task. Monitor as usual.

Task 2

- Students individually choose eight new words for engine components to remember from the table. They write them in a list and exchange lists with a partner. Pairs then test each other to see if they can remember what they are with the table covered. They can look at page 82 if they wish.

- Now have the students do the Workbook exercises for Unit 5, Lesson 1.

Workbook answers

Language: adjectives

A Circle the nouns and underline the adjectives

1 nouns = engine, version
   adjectives = new, previous

2 nouns = generator
   adjectives = external

3 nouns = selector, cockpit, ignition
   adjectives = small

4 nouns = material
   adjectives = good, strong, expensive

5 nouns = systems, capacity
   adjectives = full

B Use endings to make adjectives

1	lubricating	9	specific
2	varying	10	technological
3	rotating	11	cooling
4	cylindrical	12	basic
5	replaceable	13	logical
6	flexible	14	dangerous
7	moveable	15	economic
8	electrical	16	reliable

C Make ten sentences using the adjectives

Answers depend on the students.

Language: *which/that/who*

A Write shorter versions of the sentences

1 The timing chain is a flexible belt connected to the crankshaft.

2 The springs attached to the rocker arms open and close the valves.

3 There are three kinds of linkages found in aircraft construction.

4 Everything you need is on the bench.

5 You need to watch the numbers on the right.

6 Alclad is the material used to cover the frame in this case.

Pronunciation: word stress

A Mark the stressed syllable

1	elec<u>tri</u>city	7	com<u>po</u>nent
2	ro<u>ta</u>tion	8	<u>re</u>move
3	con<u>nect</u>	9	carbu<u>rett</u>or
4	<u>gen</u>erator	10	<u>pis</u>ton
5	<u>en</u>gine	11	<u>crank</u>shaft
6	con<u>tai</u>ner	12	<u>pro</u>tect

B The way dictionaries show word stress

Answers depend on the students.

C Practise saying the words

Answers depend on the students.

Objectives

To practise describing engines and use vocabulary related to factors involved in engine design.

To develop skills in listening: to check accuracy of information; to identify specific information.

Language

Vocabulary for describing features of engine types, e.g., *reliability*, *weight*.

Vocabulary

A Review adjectives

- Elicit some factors that might affect the design of a new engine.
- In pairs, students should check the meaning of the words in the box, using dictionaries as necessary. Monitor and assist.
- Set the task for pairwork. Feed back orally.
- Drill the words in the box, as students will need to recognise them in the Listening text.

Suggested answers

size:	big, small
simplicity:	simple, complex
power:	powerful, low-powered
price:	expensive, cheap
reliability:	reliable, unreliable
weight:	heavy, light
beauty:	beautiful, ugly

Speaking

A Compare factors that affect engine design

- Elicit why, for example, simplicity is important, e.g., a simple engine has low production costs and is easy and cheap to maintain.

- Set the task for groups of three.
- Feed back orally.
- **Note:** The most important issue in aero engine design is often considered to be the power-to-weight ratio: engine designers must get as much power as possible, combined with as little weight as possible.

Answers

Answers depend on the students.

Listening

A Listen for gist and specific information: factors in engine design (CD1 Track 19)

- Ask students to cover the table at the bottom of page 84.
- Play the recording once. Pairs check answers. Feed back orally.

Answers

The order is: reliability, weight, power, size, simplicity.

- Ask students to read through the table. In pairs, they should clarify any vocabulary difficulties and share initial ideas. Monitor and assist. Stress that they should not mark anything yet.
- Set the task for individual work and pairwork checking.
- Play the recording again. Feed back orally.

Answers

size:	(a) is not true
weight:	(b) is not true
simplicity:	(b) is not true
reliability:	(a) is not true
power:	(b) is not true

Listening and speaking

A 🎧 Listen for gist and specific information: advantages and disadvantages of different engines (CD1 Track 20)

- Ask students to look at the pictures of engines on page 85. Elicit a brief description of the *horizontally opposed twin-cylinder engine*, e.g., *two cylinders, one on each side.*

- Pairs should work through the other pictures in the same way.

- Set the task for individual work and pairwork checking.

- Play the recording once.

Answers

The three engines Tom has experience on are: *4 twin-cylinder in-line, 1 radial* and *3 turbine.*

- Refer students to the table of advantages and disadvantages. Clarify any vocabulary problems.

- Set the task. Play the recording again. Pairs compare answers.

- Feed back onto the board.

Answers

advantages	disadvantages
It's streamlined *(c,d)*	It needs a lot of spare parts *(a)*
It's powerful *(a,c)*	It's expensive *(c)*
It's simple *(d)*	It isn't very powerful *(d)*
It runs smoothly *(a)*	It's complicated *(a)*

- If you have time, focus on the phrases from the tapescript for CD1 Track 20 for describing the three different engine designs and ask pairs to describe one of the engines on page 85 for their partner to identify OR, as usual, ask students to follow the tapescript as they listen to it again.

- Now have the students do the Workbook exercises for Unit 5, Lesson 2.

Workbook answers

Writing: evaluating strengths and weaknesses

A Write a passage

Answers depend on the students.

Language: comparative expressions with *the x, the y*

A What does he mean?

Suggested answer

The more power you can get out of an engine, the better means that it will be beneficial to get as much power out of the engine as possible.

B Make similar sentences

Answers depend on the students.

Objectives

To raise awareness of the technique of using topic sentences to: aid quick reading, reference explanations and examples.

To do further work on vowel quality and length.

Language

Vocabulary to explain basic theory of momentum: *velocity*, *rotation*.

Reading and speaking

A Discuss the pictures

- Set the question. Ensure that all students understand *in common*.

- Feed back orally.

Suggested answer

All three pictures show something spinning.

B Focus on topic sentences

- Refer students to the Skills Box, and allow them to go through it alone and with a partner.

C Predict content of the text: look at the topic sentences *1–4*

- Let students read the four topic sentences.

- Ensure that they understand that these are the first sentences in four different paragraphs.

- Set for pairwork. Stress that students are not looking for correct answers, but speculating, on the basis of the topic sentences alone, what kind of information might come after it.

- Feed back orally. Accept all suggestions, within reason.

Answers

Answers depend on the students.

D Match topic sentences to text

- The discussion in Exercise B should allow students to do this exercise without reading all of the paragraphs. Set a time limit.

- At the end of the time limit, ask students to cover the text on pages 86–87 and compare answers. Feed back onto the board.

Answers

No answers at this point.

E Read the text more carefully and check answers

- Set the task. Allow students to read in their own time. Feed back orally.

Answers

Paragraph 1:   4 – Have you ever watched an ice-skater on TV?

Paragraph 2:   2 – The answer is momentum.

Paragraph 3:   3 – Think of the skater in mathematical terms for a moment.

Paragraph 4:   1 – This fact was used to provide the answer to an old and difficult problem.

Speaking and writing

A Categorise words according to vowel sounds

- First drill the pronunciation of all the words in the box.

- Set the task for pairwork.

- Feed back onto the board.

- Ask students to practise saying the words, taking care to stress the correct syllable.

B Complete skeleton notes

Answers

end	stop	back	big	make	learn
momentum stress	top velocity problem	mass	spin distance radius	stable radial vibration skater rotation	turn circle

C Ask and answer questions about momentum

Answers

Answers depend on the students.

- Now have the students do the Workbook exercises for Unit 5, Lesson 3.

Workbook answers

Spelling: vowels

A Complete the vocabulary with the missing letters

1 linear motion
2 gyroscope
3 radial engine
4 angular momentum
5 the problem of vibration
6 combustion cylinders
7 perpendicular to the axis
8 arranged in a circle
9 in a specific direction
10 a heavy flywheel

B Check the spellings

See Unit 5, Lesson 3, in the Course Book.

Vocabulary and writing: writing notes

A The meaning of the abbreviations

1 e.g. = for example
2 etc. = etcetera
3 NB = nota bene (take note of something)
4 i.e. = id est (explain exactly what the previous statement means)
5 ditto = used to avoid repeating something
6 v. imp = very important
7 IAW = in accordance with

B Use the abbreviations to rewrite these sentences.

1 Several possibilities exist, e.g., rubber, plastic, and perhaps even aluminium alloy.

2 <u>NB COMPLETED FORMS MUST BE HANDED INTO THE SUPERVISOR!</u>

3 The gyroscope will continue to turn, ditto for the spinning top.

4 The smoothing out of vibration is v. imp in engine design.

5 Technological advances rely on materials: alloys, plastics, fluids, coatings, etc.

6 The project will be reviewed after the first month: i.e., on the first of November.

7 Make sure you replace the fuse IAW the way it says on the manual.

Language: expressing similarity and difference

A Fill in the gaps with the right expression

1 Flying an airliner is certainly not _the same as_ flying a trainer.

2 Small workshops are very _different from_ production lines.

3 A glider is quite _similar to_ a seabird in the way it flies.

4 Is 30 cm _the same as_ 12 in? Well, not exactly, no.

5 The new system is _similar to_ the old one, but not the same.

Objectives

To practise reading, listening and note-taking.

To practise talking about advantages and disadvantages of the rotary engine (from notes).

Language

Vocabulary for describing the firing cycle of an internal combustion engine: the power stroke begins; the fuel/air mixture is compressed.

Speaking and vocabulary

A Discuss the diagrams

- Elicit the fact that the illustrations are of a car engine. Draw attention to the labelling and review the vocabulary (most of the labelling words are in the Glossary).

- Set the task for pairwork. Encourage students to use the vocabulary in the labels and discuss the process.

- Feed back orally. Elicit ideas about what the diagrams show, but do not go into too much depth or you will pre-empt the reading task.

Answers

No answers at this point.

B Discuss meanings of key vocabulary

- Set for pair or group work. Stress that students are not expected to know all of the words. Point out that they should explain to one another/remind each other of meanings where possible.

- Feed back onto the board. Point out that students know a lot of the vocabulary already.

- Explain any unknown words or refer students to their dictionaries.

Answers

Answers depend on the students.

C Match the verbs on the right with the phrases on the left

- Elicit a few ideas.

- Set the task for pairwork. New pairs compare ideas. Monitor and assist.

Answers

the exhaust valve *opens, closes*

the ignited gas *expands*

the fuel/air mixture *is sucked in, is compressed, is admitted*

the inlet valve *opens, closes*

a vacuum is *created*

the power stroke *begins*

D Complete the labels on the first picture

- Set the task for pairwork. Feed back on the OHP.

Answers

1	spark plug	**4**	cylinder
2	valve	**5**	flywheel
3	piston		

Reading and writing

A Number the stages of an engine cycle

- Ask students to read through all the stages silently.

- Feed back by eliciting the sentences in the correct order. Check students' understanding of the process by asking questions, e.g., *What's happening to the valve in the second picture? Why? Which picture shows the ignition spark? What happens next?*

Answers

The correct order is 4, 2, 5, 3, 1.

B Read and swap information

- Exploit the three illustrations. Elicit the characterising feature of a rotary engine, i.e., that the whole cylinder block, with the propeller fixed to it, rotates around a stationary crankshaft. This is the opposite of the conventional model, in which the cylinder head is fixed, and the crankshaft rotates with the propeller fixed to it.

- Elicit possible advantages and disadvantages of such a design. Do not worry if students do not have many ideas at this stage.

- Remind students of note-taking techniques such as dashes, colons, abbreviations, brackets, and the focus on content words.

- Set the task for individual work. Monitor and assist.

Suggested answers

Text A

rotary engine – advs.

advs of circular arrngmnt + no extra weight (cyl. block acts as flywheel)

easy start by hand: momntum of cyl. block fire engine

air cooled: no radiator/fluid

Text B

rotary engine – disadvs.

only max. power

= use a lot of oil + fuel

= hard to manvr (gyroscpic effect): high rotating weight

e.g., easy turn l./hard turn rt.

C Read your notes to yourself

- Set the task. Students should check that their notes cover the information in the text. Emphasise that students should make no attempt to memorise their text.

D Swap information with a partner

- Refer students to the Skills Box. The idea of recording key information was important in the last exercise, and students will test the effectiveness of their note-taking in this exercise.

- Put Students A together first to rehearse their talk, then form AB pairs.

- Point out that listeners must take notes as their partner speaks. Emphasise that speakers should use a natural speed of delivery, and not pause to allow their partner make his notes.

Answers

Answers depend on the students.

E Check notes against the text

- Set for individual work and pairwork feedback in AB pairs.

- In their pairs, students must identify any key points missing from their notes on the talk.

- Encourage students to decide a) why these things were left out (oversight, lack of writing time, omission from the talk), and b) whether or not the missing information is important.

- If you have time, exploit the texts for useful phrases OR review vocabulary from Reading and writing Exercise A.

Answers

Answers depend on the students.

- Now have the students do the Workbook exercises for Unit 5, Lesson 4.

Workbook answers

Language: prepositions and adverbs of direction

A Draw arrows showing directions

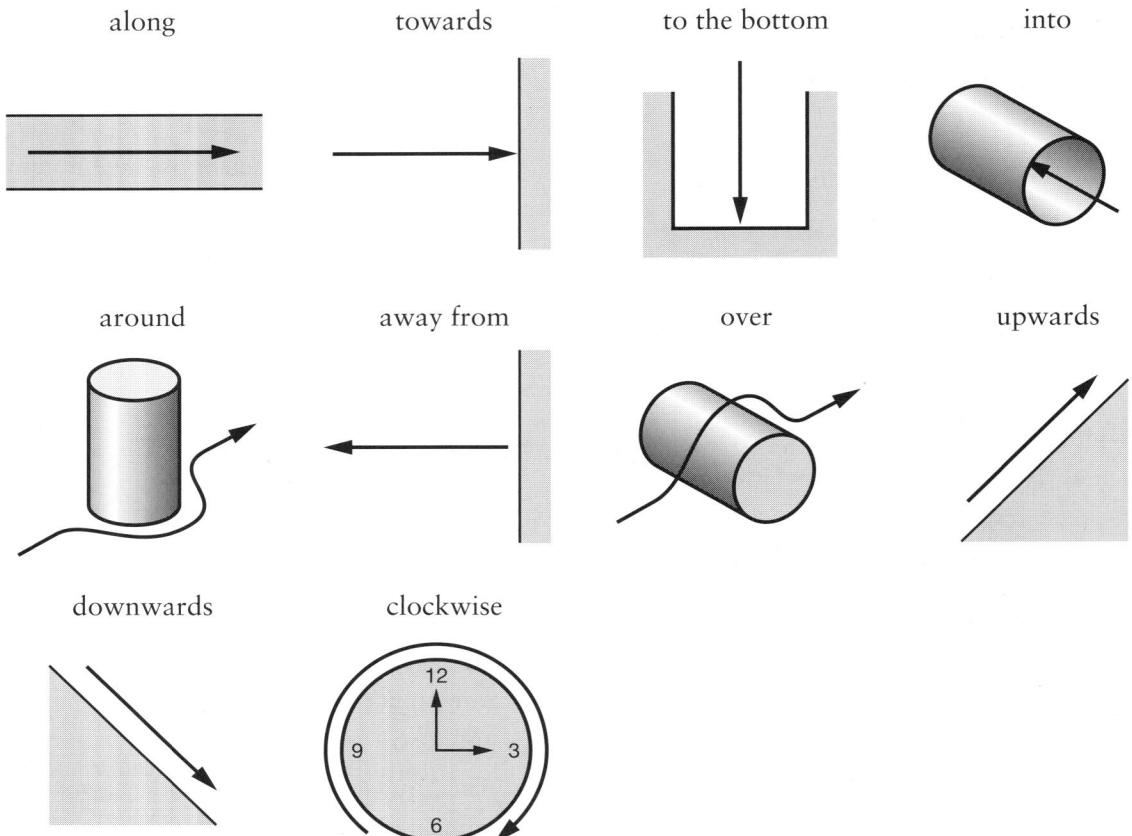

along   towards   to the bottom   into

around   away from   over   upwards

downwards   clockwise

Writing: descriptions of processes

A Fill in the gaps with phrases from the box

1 fuel/air mixture is delivered
2 vacuum is lost
3 moves upwards
4 reaches the top centre
5 the power stroke

B Prepositions and adverbs of direction

prepositions: through, in, by, downwards, before, upwards

adverbs of direction: to the centre, to the bottom

Language: word-building

A Complete these words to make nouns using ~*ure* and ~*ion*

1 temperature
2 pressure
3 compression
4 innovation
5 fixture
6 combustion
7 reduction

B Write sentences with each word from Exercise A

Answers depend on the students.

Objectives

To develop scanning skills for finding information on a specifications list.

To locate required information in a table of technical data.

Language

Vocabulary for engine specifications, e.g., *fuel system*, *weight*.

Question formation with *How/What/Is/Does?*

Speaking and vocabulary

A Look at the diagram of an engine in Lesson 1

- Refer the students to the exploded diagram on page 82. Give them time in pairs to remind themselves of the components and sub-assemblies of the engine.

- Get students to compare the diagram on page 82 with the picture on page 90. Keep this brief, as the engines are not very similar and students will not be able to identify many parts. Establish that the Merlin is an aero rather than a car engine.

- Set the task for pairs. Feed back orally.

Suggested answers

Students should be able to identify more distinctive components on the Merlin, e.g., *the flywheel*, *cylinder block*, *sump* and *spark plugs*.

- **Note:** The Merlin has a *supercharger*, i.e., a special high-speed fan which forces air into an internal combustion engine and increases its power.

Reading

A Skim the text

- Write the three options *1–3* on the board and check that students understand them.

- Set the task for individual work and pairwork checking. Feed back orally.

Answer

2 the specification of an engine

B Find items in the text as quickly as possible

- Elicit ideas as to how students will find the items in the text, e.g., by looking for one word of a group of two/three and then checking to see if the others are present; by looking for numbers and particular combinations such as *kg*.

- Set the task, giving a time limit. Ask students to underline the items. Pairs compare answers.

- Feed back onto an OHP if possible.

Answers

weight: 698 kg

dimensions of the spark plugs: 14 mm x 14 mm

fuel type: 100/130 grade aviation fuel

number of valves: 48

material for the crankcase: forged steel

C Find and circle numbers in the text

- In the workplace, students are likely to have to deal with tables containing a large amount of data. This exercise gives further practice in finding information, in this case, capital letters, mathematical symbols and numbers. Point out to students that any features that stand out from the surrounding text should be their focus. Other examples in this case are *mm*, -, *m/sec*.

- Set the task for individual work and pairwork checking.
- Feed back onto the OHP.

Answers

15.2 m/sec	= max piston speed
0.42:1	= spur reduction gear ratio
SU SUX-601	= type of variable stroke fuel injection pump
BTH CA4750	= starter
757 mm	= width

D Interpret data from the text

- Ask students to read questions *1–5*. Point out that the presence of unfamiliar vocabulary will not prevent them finding the information.
- Set the task for individual work and pairwork checking.
- Feed back orally.

Answers

1 steel
2 1,803 mm
3 75 gallons per hour
4 10.2 inches
5 no, it's automatic

Extension: writing and speaking

- Clarify the units of measurement in the specification: *lb*; *ft/min*, etc.
- Ask students to write five or six questions for other items of data in the table, beginning *How…?, Is …?, What …?, Does …?*, e.g., *What is the maximum piston speed?* or *How wide is the engine?*
- Monitor and assist.

- In pairs, students ask and answer their questions.
- Now have the students do the Workbook exercises for Unit 5, Lesson 5.

Workbook answers

Language: fixed prepositions

A Underline the fixed prepositions

replaced by	(line 2)
design of	(line 3)
addition of	(line 4)
type of	(line 4)
produced in	(line 7)
from the 1930s to the 1950s	(line 7)
compensate for	(line 9)
run at	(line 11)

Language: numbers and abbreviations

A The pronunciation and meaning of abbreviations

6:1	six to one
75 hp/cu in	seventy-five horsepower per cubic inch
60 psi	sixty pounds per square inch
5.4"	five point four inches
163 hp	one hundred and sixty-three horsepower
152 mm	one hundred and fifty-two millimetres
1,540 lb	one thousand, five hundred and forty pounds
3,000 rpm	three thousand revolutions per minute
778 kg	seven hundred and seventy-eight kilograms

15,500 ft	fifteen thousand, five hundred feet
1,679 cu in	one thousand, six hundred and seventy-nine cubic inches
1,635 hp	one thousand, six hundred and thirty-five horsepower
3,000 ft/min	three thousand feet per minute
60.7 hp/litre	sixty point seven horsepower per litre
600 km/h	six hundred kilometres per hour
15.2 m/sec	fifteen point two meters per second

14 mm x 14 mm	fourteen millimetres by fourteen millimetres
7.06:1	seven point o-six to one

B Practise saying the expressions

Answers depend on the students.

Writing: descriptions

A Answers depend on the students.

Lesson 6 — Rebuilding an engine

Objectives

To practise manipulating note forms.

To extend practice in listening for general ideas and for specific information.

Language

Vocabulary for engine overhaul, e.g., *replace*, *reassemble*, *fit*.

Simple past passive: reduced passive forms in MRO documentation.

Speaking

A Use clues in a text to predict content of a news article

- Ask students to cover the text.
- Set the task. Feed back orally, but do not confirm or correct yet.

Answers

No answers at this point.

B Read to check answers

- Set for individual work and pairwork checking. Feed back orally.

Answers

1 It is a kind of technical college.

2 The five people are students at the college.

3 They have been given permission to rebuild a Merlin engine (similar to the one in Lesson 5) which was found in a college storeroom.

Reading and writing

- Before doing this section, you may wish to do a general review of past passive forms with the group, as this task relies on their having a good grasp of it.

A Focus on the language of check sheets

- Work through the Skills Box.
- Establish that the sentences in the exercise all use the passive voice apart from Sentence 1.

- See if students can guess the meanings of any unfamiliar verbs, e.g., *strip down, mount, sandblast, attach*. Clarify where necessary.

- Set the task for individual work and pairwork checking.

- Feed back, eliciting which words have been omitted in the check sheet, e.g., *the* and *was/were*.

Answers

1 d   **4** a

2 m   **5** j

3 h

B Write full sentences for the rest of the check sheet notes

- This exercise is somewhat unrealistic in the context of MRO documentation; passives are usually reduced rather than expanded: the aim here is to help clarify the formation of reduced passives.

Suggested answers

a The engine was moved from the storeroom to the workshop.

b The supercharger was removed and stripped down.

c We took the carburettor off the supercharger.

d The carburettor was stripped down and cleaned.

e I reattached the carburettor to the supercharger.

f The cylinder heads were removed.

g We removed and cleaned the pistons.

h The cylinder heads were sandblasted.

i The pistons and cylinders were refitted.

j The cylinder heads were reattached.

k (Phil) reassembled the supercharger.

l The supercharger was remounted onto the engine.

m The mounting display frame was painted.

n The engine was transferred to its display position.

Listening

A 🔊 Listen for general information (CD1 Track 21)

- Set the task for individual work and pairwork checking.

Answers

1 The engine came from a film company.

2 He had to do a lot of work.

3 Yes (the engine is running perfectly now).

B 🔊 Listen for specific information (CD1 Track 21)

- Give students time to read through the list of problems and repairs.

- Get students to listen to you pronounce the list of problems. In pairs, students should practise saying the phrases; they will need to recognise them for the listening task.

- Set the task. Point out that the problems are mentioned in the order they appear in the list on page 93.

- Play the recording once. Pairs compare answers.

- Feed back orally, onto an OHP if possible.

Answers

The problems mentioned by the speaker are:

1 water leaking

2 oil leaking

4 carburettor flooding

7 valve springs damaged

9 cylinder head leaking again

10 cylinder blocks out of alignment

C ⊙ Listen for detailed information (CD1 Track 21)

- Again allow students time to read through the list and to listen to you say the phrases.

- Set the task.

- Play the recording. Pairs compare answers. Feed back onto the OHP if possible.

Answers

Problems	Solutions
water leaking	new cylinder head gasket fitted
oil leaking	brass tubes made, old ones replaces
carburettor flooding	carb. stripped, inspected, reassembled
valve springs damaged	valve and all springs replaced
cylinder head leaking again	engine stripped down completely
front cylinder blocks out of alignment	front blocks machined to fit

Extension

- Review vocabulary from the lesson connected with disassembling and reassembling an engine, e.g., *strip down, refit*.

- Drill the pronunciation.

- Students must use the lists in Listening Exercise 2 as the basis for a question and answer role-play in which one of them takes the part of the man who bought the Merlin engine from the film company, e.g.,

   A: *What did you do when the cylinder head was leaking?*

   B: *Well, that was easy to repair. I just fitted a new gasket.*

   A: *And what about the carburettor problems?*

- If you have time, play the tapescript of CD1 Track 21 as students follow the tapescript in their books. Work on a very short section and ask students to identify sentence stress.

- Now have the students do the Workbook exercises for Unit 5, Lesson 6.

Workbook answers

Language: multi-word verbs

A Find the multi-word verbs

Answers depend on the students.

B Mark the sentences right (✓) or wrong (✗)

1 ✗	4 ✗
2 ✗	5 ✗
3 ✗	

C Make correct example sentences of your own.

Answers depend on the students.

Writing: completing check sheets

A Complete check sheet

Answers depend on the students.

Objectives

To practise reading and interpreting labelled drawings.

To practise describing airflow through a gas turbine engine.

Language

Gas turbine parts, e.g., *compressor*, *exhaust*, etc.

Phrases for describing airflow through a gas turbine, e.g., *fuel is added to the compressed air*.

Passive structures with *by* + agent.

Dependent prepositions, e.g., *expel from*, *exert pressure on*.

Vocabulary and speaking

A Focus on spelling

- Use a few examples from previous lessons on the board to demonstrate the task of supplying missing letters. Note that most of the missing letters are vowels, as these often cause difficulty.

- Set the task for pairwork. Point out that students know some words from previous lessons, and that others are new. Encourage students to look in the Glossary to find the new words – two of them can be found on the diagram in the 2nd stage of the exercise.

- Feed back onto the board. Drill the pronunciation of the words, including the correct stress.

Answers

1	propeller	**5**	shaft
2	gearbox	**6**	turbine
3	exhaust	**7**	compressor
4	combustion chamber	**8**	intake

B Label a diagram

- Set the task for individual work, using dictionaries as necessary, and pairwork checking. Monitor and assist.

- Feed back orally.

Answers

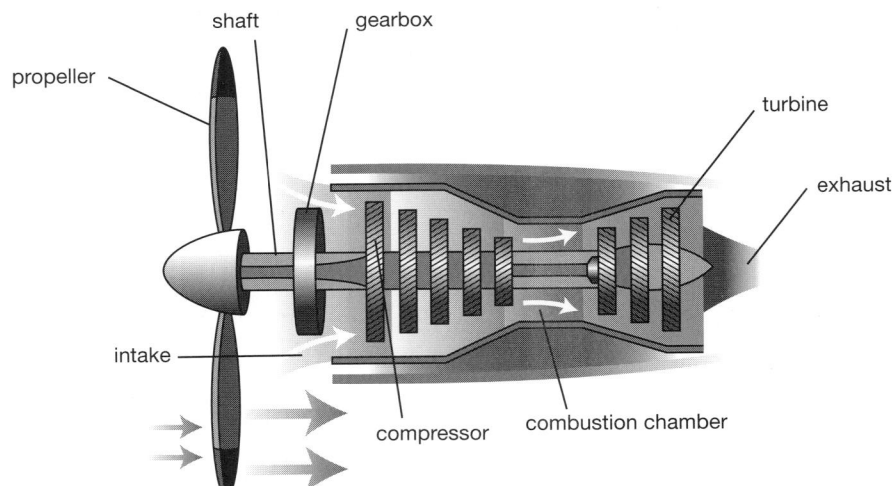

C Discuss a cutaway diagram using key vocabulary

- Give students time to read through the list of phrases. Clarify any vocabulary problems.

- Set the task for pairwork. The expressions are to help scaffold the task if students are unable to say very much about the diagram. They do not have to use all of them or put them in full sentences. Do not confirm or correct ideas yet – students will read to check.

Suggested answer

See the text on Course Book pages 94–95.

Reading

A Read the text to check ideas

- Set the task for individual work and pairwork checking.

- Feed back orally, if possible using a drawing or OHP of the engine diagram.

B Underline expressions in the text

air goes into
is drawn into

the compressor acts on the air
[air is] compressed by the compressor

the air is mixed with the fuel
fuel is added to the compressed air

the hot gases drive the turbine
the hot combustion gases provide power to the turbine

push the turbine blades
exert[ing] pressure on the turbine blades

the propeller uses the power
the propeller is driven

the exhaust gases go out *the exhaust gases are expelled*

Language

A Complete phrases with a preposition

- Elicit and put on the board examples of prepositions, including *in, to, on* and *from*.

- Set the task for pairwork. Point out that the answers are in Reading Exercise B; alternatively, you may prefer students to cover Reading B and work from memory, and then check their answers.

Answers

draw something *into*

add something *to*

provide power *to* something

exert pressure *on* something

expel something *from* somewhere

B Focus on use of *by* in the text

- Present passive forms are widely used in technical English, and are often accompanied by the preposition *by* and the agent.

- Set the first task for individual work and pairwork checking: ask students to circle all the incidences of *by*.

Answers

Paragraph 1

Sentence 2:

This engine is a type of gas turbine which has a propeller very similar to the ones that are used <u>by</u> piston engines, but which is driven <u>by</u> the combustion of gas in a single combustion chamber instead of several cylinders.

Paragraph 2

Sentence 2:

Air is drawn into the intake and compressed <u>by</u> the compressor.

Sentence 3:

Fuel is then added to the compressed air in the combustion chamber and is ignited <u>by</u> a spark.

Sentence 4:

The hot combustion gases expand through the turbine, to provide power to the turbine <u>by</u> exerting pressure on the blades, causing the central shaft to rotate.

Sentence 5:

Some of this rotary power drives the compressor, and the propeller is driven by the remaining power via a reduction gearbox.

• Put the six sentences above on the board. Establish that the first sentence is passive. Pairs read and discuss the rest.

Answers

All the verb structures are passives except *provide power to the turbine by exerting pressure*. This use of *by* is also common in explaining how systems work.

Speaking

A Discuss the Pratt & Whitney PT6 engine

• Refer students to the diagram on Course Book page 95. Ask them to find and point to the different parts of the engine as you say them.

• Set questions *1–3* for pairwork. Feed back orally.

Answers

1 The air intake is at the back of the engine.

2 The air travels from right to left, through the compressor and the combustion chamber which are on the right of the reduction gearbox and propeller.

3 The propeller and gearbox are on the left, next to the exhaust rather than the air intake.

The Pratt & Whitney PT6 engine

PT6 engines are noted for reliability and ease of maintenance. They are unusual because the air intake is at the back of the engine, which means it is much easier to access the turbine and combustion chamber for maintenance, overhaul and repair.

The air enters the engine through the air intake; it is then compressed by a multi-stage compressor and fed to the combustion chamber, where it is mixed with fuel and ignited. The hot gas expands through two turbine stages; the first drives the compressor and the accessories; the second, mechanically independent from the first, drives the propeller shaft by means of a reduction gearbox. Finally, the hot gas is discharged through the exhaust ducts.

B Ask and answer questions about the PT6 engine

• Students work in pairs asking each other questions about the engine, e.g.,
  What does the X do?
  What happens after X?
  Where is the X?

• Alternatively, get students to ask you their questions. Answer them using the text above.

• If you have time, use an OHP or the drawing of the engines in the lesson to elicit the airflow/operation of gas turbine engines.

• Encourage students to use the expressions students underlined in Reading 2.

Answers

Answers depend on the students.

• Now have the students do the Workbook exercises for Unit 5, Lesson 7.

Workbook answers

Reading and vocabulary: engine parts

A Complete the gaps with vocabulary from the box

1 the air intake

2 compressor

3 the combustion chamber

4 fuel

5 turbine

6 the propeller shaft

7 the exhaust ducts

Pronunciation: longer sounds

A Different vowel sounds

1	sh<u>a</u>ft	**5**	w<u>eigh</u>t
2	eng<u>i</u>ne	**6**	fl<u>oo</u>d
3	bl<u>o</u>ck	**7**	specif<u>i</u>c
4	b<u>o</u>re		

Lesson 8 — Turbo stats

Objectives

To practise intensive reading: a description of how an engine works.

To recognise and produce numbers and simple calculations, including decimals and fractions.

To practise listening for very specific information.

Language

Numbers, e.g., *3.142* and *35,000*.

Expressions for airflow through a turbojet, e.g., *rate of rotation; temperature of combustion*.

Speaking and reading

A Discuss true/false statements

- Set the task for pairwork. Stress that students should apply their general knowledge; they are not expected to have learnt these things in the course so far.

- Feed back orally, but do not confirm or correct students' ideas yet.

Answers

No answers at this point.

B Read the text to check answers

- Set the task for individual work and pairwork checking. Point out to students that although the text is quite dense, they are reading only to answer the three questions in Exercise A.

Answers

1 False (Lines 2–3)

2 True (Line 4)

3 True (Lines 7 and 11): The text says that most turbines are turbofans. In a turbofan, the *fan* acts as a propeller.

C Read the text again to establish the position of the fan

- Set for pairwork. Students should read through the text carefully and locate the position of the fan on the diagram.

- Check comprehension by eliciting how a turbofan is different to a turboprop (it is quieter and more efficient).

Answers

The fan is to the left of the compressor, at the front of the engine.

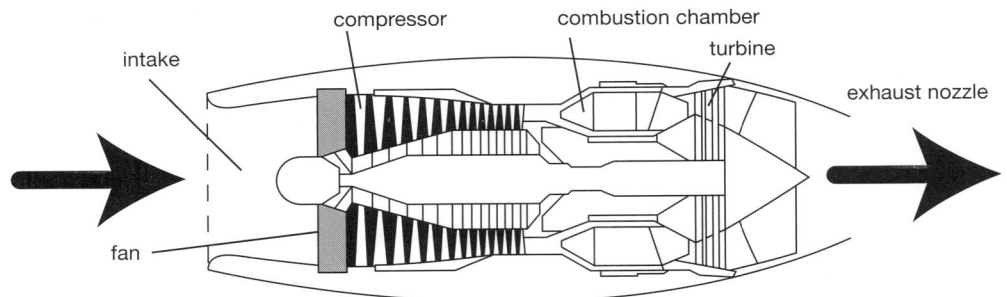

Listening and speaking

A Practise using numbers and mathematical terms

- Write some numbers on the board and elicit, correct and drill the pronunciation as necessary. Include fractions, decimals, millions, etc., as in the exercise. If students are weak in this area, extend the activity by dictating some numbers for students to write.

- Set the exercise for pairwork. Monitor.

- Feed back orally, e.g.,

  four thousand five hundred
  (US Eng also *forty-five hundred*)

  eleven million

  forty

  fourteen

  one hundred and fifteen point eight

  one and three-quarters

  one (a) hundred and sixty-seven
  (US Eng *one (a) hundred sixty-seven*)

  three point one four two

  four hundred and ninety thousand

  nine-sixteenths

- Review the mathematical symbols +, -, x, ÷ and their pronunciation in English.

- Point out that fractions can be expressed with *over*, e.g., *one over seven*, and multiplication can be expressed using *by*, e.g., *distance equals speed by time*.

Suggested answers

4,500	forty-five hundred
11,000,000	eleven million
40	forty
14	fourteen
115.8	one hundred and fifteen point eight
1 ¾	one and three-quarters
167	one hundred and sixty-seven
3.142	three point one four two
490,000	four hundred and ninety thousand
9/16	nine-sixteenths

B Practise numbers for the listening task

- Refer students to the left-hand column of the table and to the Skills Box. Clarify as necessary.

- Set the task for pairwork. Get pairs to read aloud the numbers in the left-hand column of the table.

- Feed back orally.

C 🔊 **Circle the number you hear (CD1 Track 22)**

- Set the listening task: students must circle the correct number in each section according to the speaker.

- Play the recording. Pairs compare answers. Feed back orally.

Answers

See below.

D 🔊 **Match numbers and categories (CD1 Track 22)**

- Ask students to read the list of phrases in the right-hand column of the table. Go over any vocabulary that students are unsure about, e.g., *core*, *bypass*.

- Play the recording. Pairs compare answers. Feed back orally.

Answers

35,000 ft	cruising altitude of Boeing 747-400
12,000 lb	thrust from each engine when cruising
700 lb/sec	air drawn into engine
80%	air bypassing engine core
150 psi	pressure of compressed air
850 °F	temperature of compressed air
1¾ lb/sec	fuel injected into compressed air
2,000°F	temperature of combustion
3,300 rpm	rate of rotation of fan
9,500 rpm	rate of rotation of compressor
1,400 ft/sec	exit velocity of exhaust gas
1,000°F	temperature of exhaust gas

E **Ask and answer questions about the information in the table**

- Draw attention to the example in the Course Book. Elicit other possible questions about the numbers and quantities in the table, e.g., *How much (air bypasses the engine core)? How fast (does the fan rotate)? What is the temperature of (the exhaust gas)?*

- Set the task for pairwork. Monitor and check pronunciation of the numbers.

- If you have time, do further work on vocabulary, focusing on the table in Listening and speaking 3, OR say the phrases from the table in Listening and speaking 3 as students read them. Make some mistakes; students must tell you where your mistakes are.

Answers

Answers depend on the students.

- Now have the students do the Workbook exercises for Unit 5, Lesson 8.

Workbook answers

Language: fractions and large numbers

A **Match the fractions with the words and the pie chart**

1 a quarter = ¼ = pie chart 2
2 a fifth = ⅕ = pie chart 4
3 a third = ⅓ = pie chart 3
4 half = ½ = pie chart 1
5 three-quarters = ¾ = pie chart 5
6 five-sixths = ⅚ = pie chart 6

B **Practise saying the fractions**

Answers depend on the students.

C **Cross out the sentence which is wrong**

1 There are hundreds and hundreds of different components on the market.

~~There are hundred and hundred of different components on the market.~~

2 ~~They will cost about nine hundreds euros each.~~

They will cost about nine hundred euros each.

3 Modern birds have existed for thousands of years.

~~Modern birds have existed for thousands years.~~

4 Five hundred times four is two thousand.

~~Five hundred times four is two thousands.~~

5 The company has spent a million dollars on this design.

~~The company has spent million dollars on this design.~~

Language: making comparisons

A Underline the expressions in the text

like	(line 8)
the difference is that	(lines 9–10)
it differs from X in that	(line 13)
in the same way	(line 12)
but	(lines 12 and 15)
similar to	(line 15)
the same	(line 15)

B Write sentences using set expressions

Answers depend on the students.

Lesson 9   Fuel

Objectives

To extend skills in scanning text to check data.

To discuss the properties of different kinds of fuel and changes in design since the 1940s.

Language

Vocabulary for fuel properties, e.g., *volatility*, *flash point*.

Present perfect verb structure (including passive) to discuss changes in design and specifications).

Vocabulary and speaking

A Match vocabulary for properties of fuel to definitions

- As an introduction, revise the noun-adjective pairs for materials from Unit 1, Lesson 3, e.g., *malleability*, *ductility*. Elicit the fact that words ending in –*ity* are nouns.

- Put the words on the left on the board and elicit ideas as to what they might mean. Students may be more familiar with other forms of the words, e.g., *lubricate* (verb) rather than *lubricity* (noun), *volatile* (adj.) rather than *volatility*.

- Set the task and use new pairs to compare answers.

- Do not feed back yet, as students will find definitions of some of the properties in the reading text.

Answers

No answers at this point.

Reading

A Read to check answers

- Lead into the text by eliciting what information it gives us. Students should skim the text to find out [information about the properties of jet fuel].

- Students should read again more carefully to check their answers to the Vocabulary exercise. Set a time limit and do not allow the use of dictionaries.

Answers

1	h	**5**	c
2	b	**6**	a
3	g	**7**	d
4	e	**8**	f

Speaking

A Discuss questions about jet fuel properties with a partner

- You may wish to do some work on vocabulary now, although the students' understanding of the items in the exercise on page 98, and their application of this to their own knowledge, should be enough for them to answer the questions.

- Set the task. Feed back orally. Encourage the use of the vocabulary from page 98.

Answers

1 Kerosene has optimal characteristics; petrol is too volatile, although it has the advantage of being light. Diesel is less volatile than petrol but too heavy for aircraft use. Kerosene lies between the two.

2 Unsafe fuels may: have high volatility (they may explode easily); have a low freezing point (they would become solid at high altitudes); have poor lubricity; be conductive (electrical current may create heat and so danger of fire); be corrosive (corroded tanks

and/or lines would result in leaks); have a high sulphur content. Uneconomical fuels may: have a low net heat of combustion (produce relatively little energy per unit burned); be corrosive (require frequent replacement/maintenance of fuel tanks/lines); have a low flash point (a relatively small amount of energy is required to ignite it).

Language

A Form questions about things that have changed over time

- Work through the Language Box with the group. Use the example sentences to highlight the fact that this aspect is used when looking at non-specific past time and changes over time.

- Highlight the form of the present perfect in questions. Draw attention to the way that the auxiliary *have* is separated from the main verb by the subject in many *wh*-questions (just as in questions with *do* and *did*), i.e., *Why has X changed?* Elicit and contrast the passive form, i.e., *What changes have been made?*

- Set the task for individual work and pair-checking.

Answers

1 Why have fuel specifications developed since the 1940s?

2 How has fuel become safer over the last 50 years?

3 What materials have been invented over the last century?

4 How have aircraft shapes changed since the 1950s?

5 Over the last five decades, what major changes (modifications) have been made to engines?

B Discuss answers to the questions

- Set the task. Stress that students must use their knowledge to suggest answers to the questions – the answers are not contained in the reading text.

- Feed back orally.

Suggested answers

1 New engines have been invented. New demands for speed and performance have been put on designs, e.g., airliners, high-altitude military aircraft, high-performance jet fighters.

2 Developments include: increased standardisation, monitoring and checking of fuels. Quality of fuel has improved: fewer harmful chemicals are used (most vehicles now use lead-free fuel). CO_2 emissions have been reduced. There is greater use of additives to ensure a lower freezing point, higher lubricity, etc.

3 Strong but light materials such as titanium began to be used in the 1960s. Composites are the most significant new aircraft materials of this century. They are formed of resins, such as epoxies and polyamides, mixed with reinforcements, such as glass, boron and carbon fibres.

4 More triangular-shaped aircraft with delta wings were designed in the 1960s, e.g., Concorde. Modern aircraft have started to be designed with smoother, more continuous contours to minimise drag. Modern design ideas include C wing shapes and shape-changing wings.

5 Engines have become more fuel-efficient and can perform more effectively. Car engines have become lighter, smaller and able to use different fuels. Exhaust emissions have been greatly reduced and engine temperatures are now controlled more effectively.

- If you have time, ask students to alternately cover the definitions and vocabulary items in the table on page 98 and try to remember them, OR exploit the text further: set comprehension questions; read out the text with errors for students to spot, practise scanning for individual items.

- Now have the students do the Workbook exercises for Unit 5, Lesson 9.

Workbook answers

Language: word relationships

A Write sentences

Answers depend on the students.

Language: parts of speech

A Mark the words N or V

1	ignite	V	6	maintenance	N
2	contain	V	7	specify	V
3	combustion	N	8	corrode	V
4	freeze	V	9	ease	N
5	tolerance	N	10	develop	V

B Write the verb form for the nouns and the noun form for the verbs

1	N = ignition	6	V = maintain
2	N = containment	7	N = specification
3	V = combust	8	N = corrosion
4	N = freeze	9	V = ease
5	V = tolerate	10	N = development

C Write sentences using all the words

Answers depend on the students.

Objectives

To practise further skills of listening for key information; note-taking; interpreting technical drawings and reading from notes.

To discuss possible problems associated with aircraft fuel supply.

Language

Pump types: *delivery jet*; *rotating vane*.

Vocabulary for describing pump operation: *flow*; *outlet*.

Speaking and listening

A Discuss potential problems with fuel

- Elicit a few ideas about what the most serious problems might be.

- Refer students to the list on page 100. Ask them to read through the list to see if any of their ideas are there. Feed back orally.

- Make a note on the board of the problems that students feel are the most serious.

Answers

- All of *1–7* might certainly represent problems, with the exception of 6, but are minimised by technology:

  1 Fuel certainly is heavy, but lighter fuels, e.g., kerosene, minimise this problem.

  2 Fire and explosion are a serious danger when carrying flammable substances in flight. Secure fuel systems reduce the risks.

  3 Maintenance is important to avoid corroded or damaged fuel tanks (particularly where kerosene is used), which could lead to fire if fuel were ignited by, e.g., electrical sparks.

  4 Frozen fuel would be disastrous. Specially designed fuels are used according to the service ceiling of different aircraft.

  5 Uneven weight of fuel in different areas of the aircraft would cause imbalance. Pumps are used to distribute fuel throughout the tanks in flight. This is dealt with in the lesson.

  6 Fumes from fuel, rather than the smell itself, may be harmful to ground crew.

  7 Another potential danger; instruments tell the pilot how much fuel he has.

B Listen to see if ideas are mentioned (CD1 Track 23)

- Play the recording and encourage students to note down the problems mentioned.

- Feed back in pairs, then orally as a class.

Answers

The instructor talks about points *2* and *5*.

C Activate schemata: identify key features of pumps

- Set the task for pairwork. Encourage students to read the labels carefully.

- Feed back orally. Clarify any unknown vocabulary in the labels, e.g., *vane*, *flange*.

Answers

1 2 (note the electric wires)
2 3
3 5 (note *filter* in the labelling)
4 1 (the flap valve)
5 4 (the vanes rotate)

D 🔊 **Listen to descriptions of pumps and match drawings to titles (CD1 Track 24)**

- Refer students to the list *a–e* after the drawings on page 101. Read them aloud so that students become familiar with them – they will need to recognise them in the text.

- Set the task. Point out that the text is fairly long. This is to simulate lecture-style listening.

- Play the second part of the recording. Pairs compare answers. Feed back orally.

Answers

1 (b) delivery jet pump

2 (e) electric booster pump

3 (d) transfer jet pump

4 (c) high-pressure gear pump

5 (a) main (rotating vane) pump

E 🔊 **Listen and complete notes (CD1 Track 24)**

- Give students time to read through the table.

- Set the task. Point out as usual that students should aim to use note forms, including symbols and abbreviations of words and grammatical structures.

- Play recording 2 again.

- Pairs compare answers. Monitor and help with note forms.

- Feed back onto the board.

Suggested answer

pump	notes
main (rotating vane) pump	heart rotating vanes push fuel, power from engine valve relieves press, 2 jobs: supplies h.p.g.; drives t.j. and d.j.
transfer jet pump	tube, 'ventura' principle – sucks fuel, transfer fuel wing tanks to collector tanks
high-pressure gear pump	supplies combustion chamber, has filter to clean fuel, power from engine: as engine turns, pump rotates + supplies more fuel
delivery jet pump	ventura tube, supplies fuel from collector tank
electric booster pump	28V runs from battery until engine starts, PC9: 1 in each collector tank, fixed, sealed, in liquid

F **Use notes to describe the pumps to a partner**

- This exercise gives students further practice in reading from their own notes, with the added benefit that their notes will improve as they discover the deficiencies in them.

- First ask pairs to practise while looking at their notes and at the pictures.

- Gradually get students to work without the pictures – partners give each other the names of the pumps they describe.

Answers

Answers depend on the students.

- If you have time, build a description of one or two of the pumps from memory on the board, OR get students to prepare a short talk on an aspect of the work in Unit 5.

- Now have the students do the Workbook exercises for Unit 5, Lesson 10.

Workbook answers

Writing: comparison

A **Write a paragraph**

Answers depend on the students.

Language: direction

A Underline the sentences in the tapescript

See tapescript to CD1 Track 24.

B Complete the gaps using the expressions

1	supplies; to	**5**	sucked up
2	go out of	**6**	from; to
3	push; through	**7**	push; through
4	in and out of		

Language: present perfect sentences

A Expand the sentences

1 Since 1990, the Eurofighter project has developed quickly.

2 Semi-monocoque airframes have been common since the 1930s.

3 Over the last two years, they have designed several new aircraft.

4 For 50 years, airliners have used kerosene fuel.

5 Smaller, slower aircraft have continued to use turboprop engines.

6 Jet engines have kept the same basic design for the last 50 years.

B Write *by* where necessary

1	by	**4**	unnecessary
2	unnecessary	**5**	by
3	by	**6**	unnecessary

Reading: fuel systems

A Underline the key points

Answers depend on the students.

Unit test

A test for this unit is available at:

http://www.garneteducation.com/reps/documents/1248/SDT-u5-test.pdf

Contact your local Garnet Education representative for information about how to access these resources.

Objectives

To practise reading specifications data.

To discuss options and changes to design characteristics.

To work further on writing note forms.

Language

Reduced passive note forms.

Noun/verb forms, e.g., *develop* – *development*.

Speaking

A Discuss considerations in car design/manufacture

- Elicit a few ideas from the group about both *1* and *2*.

- Put the list: *materials*, *size*, etc., on the board and check students understand all the items.

- Set the task for pairwork. Feed back orally.

Suggested answers

1 small size, simple systems, using new technology which speeds up production

2 low fuel consumption, simple construction, cheap materials, no 'extras' showroom models, no new technology that might mean costly development or retooling of factories.

Reading

A Read to check ideas

- Set the task for individual work and pairwork discussion.

Students

Answers depend on the students.

B Read for detail

- Give students time to read through *1–9* and clarify any vocabulary problems.

- Do *1* with the group as an example.

- Set the task for individual work. Pairs compare answers. Feed back orally.

- There may be more than one answer in each case.

Answers

Exercise B	Exercise A
1	1, 6, 3 (in design brief)
2	3, 5
3	4, 5
4	4, 5
5	4, 5
6	3, 7
7	7
8	2
9	7

Vocabulary

A Word-building

- Review the difference between a noun and a verb. On the board, use examples from previous lessons to practise transforming one to the other, and highlight noun endings such as *~ment*, *~ion*, *~ity*.

- Do the first item as a class.

- Set the task for pairwork. Students can use dictionaries, but ensure you monitor to help and assist with correct dictionary use.

Suggested answers:

contain – containment; container

measure – measurement

increase – increase

design – design

develop – development

produce – production; productivity

connect – connection

fit – fit

support – support

assemble – assembly

B Complete the sentences

- Again, do the first sentence with the group, then set the task for individual work. Pairs compare their answers.

- Feed back orally.

Answers

1 Issigonis' _design_ was innovative in its use of existing technology.

2 The first Mini went into full _production_ in 1959.

3 In front-wheel drive, there is no _connection_ between the engine and the rear wheels.

4 The wheels are _supported_ by independent rubber suspension.

5 The _development_ of the car had to take a maximum of three years.

6 The cabin needed to _contain_ seating for four people.

7 Issigonis had to _increase_ the dimensions of the original design.

8 The passenger cabin had to _fit_ into 1.8 m of the overall length.

9 _Assembly_ of the prototype was complete by late 1957.

10 3 m x 1.2 m x 1.2 m were the outside _measurements_.

C Make sentences using other parts of speech

- Elicit one or two possibilities onto the board, e.g., _design (v) – Rolls Royce design aircraft and car engines._

- Stress the fact that students should not limit themselves to car design, but should consider aircraft or industrial products such as computer hardware.

- Encourage students to look back through Units 1–5 for ideas.

- Set the task for pairwork. Monitor closely and assist. Feed back onto the board.

Answers

Answers depend on the students.

Writing

A Note-taking

- This exercise again practises reduced forms appropriate to notes.

- On the board, draw the following table and write in the first item in each column.

design brief problem	design solution
overall length reduced	engine turned sideways

- Elicit the fact that, as usual, the note form of passives omits the auxiliary _be_.

- Set the task for individual work and pairwork checking.

- Feed back onto the board.

- Now have the students do the Workbook exercises for Unit 6, Lesson 1.

Workbook answers

Language: sentence structure

A Put the parts of the sentence into the correct order

1 This computer program can understand you if you speak to it.

2 Some of the large assemblies are built in Germany.

3 Materials development will allow more efficient designs in the future.

4 The component is fitted at right angles to the main assembly.

5 Does it use mains power or does it need batteries?

6 It is designed so that it can be easily repaired or replaced.

7 The prototype was faster than this one but much more complicated.

8 When did the nanotechnology project begin?

Language: relative position

A Match the meanings of these words and expressions for position

1	c	5	a
2	e	6	g
3	f	7	b
4	h	8	d

B Use the expressions to complete the sentences

1	against	5	set into
2	next to	6	mounted on
3	just below	7	Just before the end
4	at right angles	8	between

Spelling and pronunciation

A Complete these words using *a, e, i, o, u*

1 capacity

2 research

3 development

4 speed

5 technology

6 weight

7 innovation

8 height

9 kilometre

10 engine

11 vertical

12 alloy

B Check your spelling against the word lists

Answers depend on the students.

C Put the words into groups with syllables

One syllable

speed

weight

height

Two syllables

research

engine

alloy

Three syllables

vertical

Four syllables

capacity

development

technology

innovation

kilometre

Objectives

To review terminology for different types of technical drawings.

To practise using prepositions of position and expressions with *of*.

Language

Expressions with *of*, e.g., *the use of*, *an amount of*.

Prepositions *in*, *on*, *through*, *without*, *from*, *at*.

Reading

A Read descriptions of different types of drawings

- Elicit the names of kinds of drawings that students have already seen in Units 1–6.
- Set the task for individual work and pair-checking.
- Feed back orally. Discuss advantages and uses of particular drawing types.

Answers

The picture shown is an orthographic projection.

B Complete the descriptions

- Give students time to read through *1–7*, and clarify any vocabulary problems.
- Set the task for individual work and pairwork checking. Feed back orally.

Answers

1 Installation drawings
2 Perspective or pictorial drawings
3 Sketches
4 Orthographic projections
5 Assembly and exploded drawings
6 Cutaway drawings
7 Sectional drawings

C Find different types of drawing on other pages

- Direct students to the pages and elicit what type of drawing each page shows.
- See if students can come up with some more examples of different types of drawings on other pages in the book. They can do this in pairs or small groups and feed back to the class.

Answers

page 68	installation drawings
page 82	assembly/exploded diagram
page 100	cutaway drawings
page 101	cutaway and sectional drawings
page 102	sketch and pictorial image

Language: prepositions

A Complete the sentences

- Ask students to cover the descriptions in the reading section.
- Review the meaning of the prepositions if you think it is necessary. Then set the task for pairwork. Monitor and assist. Allow students to use their dictionaries.
- When they have finished, students should check in the descriptions in the Reading section.

Answers

1 the location of the parts <u>on</u> the completed aircraft
2 as if it was cut <u>through</u> the middle

3 the point <u>at</u> which it has been cut

4 they are separated <u>from</u> each other

5 assemblies <u>on</u> the completed aircraft

6 materials used <u>in</u> the component

7 made <u>without</u> the use of instruments

8 they can be seen <u>in</u> relation to each other

9 an overview of the assembly <u>from</u> different angles

10 it is clearly indicated <u>on</u> the drawing

11 as they look <u>in</u> 3D

B Scan the text for expressions with *of*

- Put the graphic on the board. Elicit one or two more and add them to the graphic.

- Set the task for individual work and checking in pairs.

Answers

the use of

the location of

the outside of

part of

a view of

the faces of

the details of

an impression of

a bit of

an amount of

dimensions of

an overview of

the reference number, part number, quantity, description of

(different) types of

C Vocabulary extension

- Again, work through the example with the group. Elicit ideas for *a description of*.

- Set the task for individual work and pairwork checking. Encourage the use of dictionaries.

- Feed back onto the board.

Suggested answers

the use of:	technology, force
the location of:	a place, an item in storage
the outside of:	a building
part of:	an assembly, a process, a team
a view of:	an assembly
the faces of:	a cube
the details of:	a design, an assembly, a job, a contract
an impression of:	costs
a bit of:	data
an amount of:	time, money, pressure, energy, data
dimensions of:	a component, a part, an assembly
an overview of:	a project, a job
a description of:	a machine, a process, a place, a person
(different) types of:	materials, processes, design, skills

D Sentence writing

- Elicit an example sentence from a strong student for one of the combinations on the board, e.g., *The dimensions of a manufactured component must always be exactly the same.*

- Set the task for individual work. Point out that students do not need to write sentences for all the items. Set a target of five or six. Monitor closely and assist.

- Students compare sentences. Encourage individuals to memorise their sentences.

- If you have time, build on the board descriptions of each kind of drawing from students' memory.

- The following may be of interest to students for research purposes:

  http://en.wikipedia.org/wiki/Engineering_ drawing

  with links to *cross section – orthographic projection*.

Answers

Answers depend on the students.

- Now have the students do the Workbook exercises for Unit 6, Lesson 2.

Workbook answers

Language: word-building

A Complete the table

Noun	Adjective	Opposite adjective
smoothness	smooth	rough
heaviness	heavy	light
transparency	transparent	opaque
width	wide	narrow
strength	strong	weak
corrosion	corrosive	fortifying
malleability	malleable	brittle

B Choose the correct form of the word

1 ~~corrosive~~/corrosion
2 transparency/~~transparent~~
3 strength/~~strong~~
4 ~~malleability~~/malleable
5 ~~width~~/wide; weight/~~heavy~~
6 smooth/~~smoothness~~

Writing: definitions

A Complete the definitions with a compound noun

1 surface area
2 orthographic projection
3 mechanical properties
4 O ring
5 take-off weight

B Find a compound noun

Answers depend on the students.

C Write short definitions

Answers depend on the students.

Lesson 3 Hand tools: Review

Objectives

To review language for the use and care of hand tools.

To practise comparing and contrasting different types of tools.

Language

Vocabulary for features and uses of multi-tools: *wire strippers, pliers, grip, cut*.

Comparisons: *more/-er + adj. + than/as … as*.

Vocabulary

A Activate vocabulary

- Elicit the names of tools onto the board. Refer students to Unit 2, Lessons 1–2, for further ideas.

- Elicit what each one does, e.g., *cut, grip*.

- Set the task for pairwork. Students should use their dictionaries as necessary. Monitor and assist.

- Feed back onto the board. Go over new vocabulary.

Suggested answers

grip:	pliers, vice, spanner
loosen/tighten:	screwdriver, spanner
cut:	saw, wire cutters, scissors, knife
beat:	hammer
make a hole:	drill, pin, punch, reamer (makes holes larger)
measure:	calipers, tape measure, ruler
file:	file
open:	bottle (corkscrew) or can opener
strip:	wire strippers
turn:	lathe
mark out:	blade, sharpie

B Matching

- Check that students understand the differences between the four illustrations. Point out that the Phillips and the pozidrive bit are quite similar, but that the Phillips bit has slightly rounded corners so that it will 'slip out' if you attempt to tighten the screw too much.

- **Note:** If you want to go into more depth about the different features and uses of the various screwdriver types, look at the section on screwdrivers at:

  http://en.wikipedia.org/wiki/Screw

- Set the task for pairwork. Students should use dictionaries.

- Feed back orally.

Answers

1	slotted bit	3	pozidrive bit
2	Phillips bit	4	hex bit

Reading

A Read and identify the attachments

- Clarify the meaning of *attachment* in this case: a tool that is part of a multi-tool.

- The students will know many of the words in the multi-tool descriptions and be able to work out others from previously encountered vocabulary. Point this out to them.

- Set the task for individual work and pairwork checking.

Answers

Answers depend on the students.

B Scan to compare specifications of two multi-tools

- Elicit from or remind students of techniques for scanning: looking for initial letters, numbers, capitals, units, etc. Set this as a timed activity to encourage students to scan rather than read all of the words.

- Direct students to the conversion tables at the back of the book, or put *1 oz = 28.35 g* and *1 in = 25.4 mm* on the board.

Answers

1 the Leatherman (4 in = 102 mm approx.)
2 the Leatherman (6 oz = 170 g approx.)
3 the Leatherman
4 the Victorinox
5 they both have wire strippers, pliers, screwdrivers, bottle opener

Speaking

A Discuss, describe and compare the multi-tools

- Elicit a few ideas for question *1*.

- Set the task for groups of three.

- Point out that there are no correct answers here, but that students should speculate on the relative usefulness in different contexts of the two multi-tools.

1 The Leatherman is a strong steel tool that folds into a pair of pliers; the Victorinox folds away completely into a rectangular unit.

2 People doing unspecialised – probably civilian-life – maintenance or repairs. The Victorinox includes more lightweight and mundane 'emergency' attachments, e.g., the tweezers and toothpick.

3 Yes, they include sharp knife blades and scissors that must be handled with care and put away correctly.

4 By keeping them clean, oiled; blades sharpened. They should be kept dry. Attachments should not be used for jobs they are not designed for, e.g., opening tins with screwdriver blades, or used for jobs which require more heavy-duty, or specialised, tools.

Writing

A Writing instructions

- Elicit from the group how the wire stripper is used and build a short description on the board, e.g., *First, decide how much of the plastic sleeve on the wire you want to remove. Then grip the plastic with the wire strippers and pull it to the end of the wire.*

- Students work individually to write descriptions of their own for some of the other tools. They can use dictionaries, but ensure you monitor and assist.

- When they have finished, ask students to read and suggest amendments to each other's writing.

Answers

Answers depend on the students.

B Compare and contrast the multi-tools

- Review comparative structures with the group.

- Elicit one or two comparisons onto the board, e.g., *The Victorinox has more attachments, but it doesn't look as strong as the Leatherman.*

- Students work individually to continue the comparisons, using the ideas from the reading and speaking sections – size, weight, types, number and variety of attachments, uses, users, etc.

Answers

Answers depend on the students.

- Now have the students do the Workbook exercises for Unit 6, Lesson 3.

Workbook answers

Language: verbs

A Words that can be used as verbs

Pupils should circle:

bend	file
reverse	adjust
calibrate	align
grip	transmit
corrode	work

B The second and third forms of the verbs

reverse	reversed	reversed
calibrate	calibrated	calibrated
corrode	corroded	corroded
file	filed	filed
adjust	adjusted	adjusted
align	aligned	aligned
transmit	transmitted	transmitted
work	worked	worked

Writing: question forms

A Expand to form correct sentences

1 What does it consist of?

2 How does it operate?

3 Is it a modern or old invention?

4 What possible problems will it cause?

B Write short notes about a tool

Answers depend on the students.

C Connect your answers to make a paragraph

Answers depend on the students.

Spelling and pronunciation

A/B Correct the spelling

1	tighten	6	knob
2	resistance	7	thread
3	measure	8	conductor
4	wrench	9	fastening
5	rotation	10	gearbox

C Put the words into groups according to the numbers of syllables

One syllable

wrench

knob

thread

Two syllables

tighten

measure

gearbox

Three syllables

resistance

rotation

conductor

fastening

Lesson 4   Tools and techniques: Review

Objectives

To review concepts in the safe use of tools.

To practise listening for main ideas and for details.

To practise dealing with and filling in a form.

Language

Conditionals, e.g., *If you dropped it, it might hurt your foot.*

Speaking

A Discuss potential accidents that misuse of tools can cause

- Review the meaning of the words in the box.

- Highlight on the board the grammar of the two example sentences. Review it if necessary.

- Set the task for pairwork. Feed back orally or onto the board for further language analysis.

Answers

Answers depend on the students.

Listening and reading

A 🔘 **Listen for main ideas and circle the correct answers (CD2 Track 1)**

- Give students time to read through *1–4* and clarify any vocabulary problems.

- Set the task and play Part 1 of the recording once.

- Students compare answers in pairs. Feed back orally.

Answers

1	c	**3**	a
2	a	**4**	b

B 🔘 **Scan an accident report form (CD2 Track 1)**

- Ask students to look at the form quickly to find out what kind of form it is. Set a time limit. Feed back orally.

Answer

The form is an Accident/Incident Report Form.

- Ask students to cover the form for the moment. Give them time to read through *a–h*, and clarify vocabulary problems.

- Stress that students are to look only at the titles of the boxes in the form. Set a time limit. Students compare ideas in pairs. Feed back orally to the whole class.

- Students should read through the form more slowly before listening. There is no need to work through all the unfamiliar vocabulary, as this is a listening task, and the students should already know the meaning and pronunciation of all of the items necessary to complete it.

- Set for individual work and pairwork checking. Play the recording once.

- Play the recording again if necessary. Again pairs compare answers.

- Feed back orally.

Answers

Form for the reporting of accidents, incidents and occupational ill health
Please complete all sections and tick boxes as appropriate.

Details of injured person:

Name   _Francis Robert Day_

Occupation _fitter_   Age _28_

Male ☑   Female ☐

Employee ☑   Contractor ☐   Pupil ☐   Client/Public ☐

Date accident happened _26th November, 2006_ Time _11.15_

Address and tel. no. of premises where accident happened

Date notified _Same_   To whom _supervisor_

Names and addresses of any witnesses 1 _John's, Mr.D, Mr. Green's_
2

Type of injury/ill health (give dates):

None	/ /	
Death	/ /	
Over 3-day absence	/ /	
Fracture	26 / 11 /	
Dislocation	/ /	
Burn	/ /	
Scald	/ /	
Cut/scratch	/ /	
Puncture wound	/ /	
Bruise/swelling	/ /	
Loss of limb	/ /	
Loss of sight	/ /	

Partial loss of sight	/ /	
Concussion	/ /	
Shock	/ /	
Poisoning/Gassing	/ /	
Internal injury	/ /	
Hearing impairment	/ /	
Disease	/ /	
Irritation	/ /	
Strain/sprain	/ /	
Other _____		
First-aid treatment	26 / 11 /	

Agent of injury:

None	☐
Handing/lifting	☐
Hit by moving object	☐
Hazardous substance	☐
Electricity	☐
Pressure system	☐
Machinery (powered)	☐
Heat or cold	☐
Machinery (hand-held)	☑
Animal/insect	☐
Slip, trip or fall	☐
Human	☐
Other _____	☐

Site of injury:

Head ☐   Chest ☐   Abdomen ☐   Back ☐   Internal ☐

	Left	Right
Eye	☐	☐
Ear	☐	☐
Face	☐	☐
Neck	☐	☐
Foot	☐	☐
Upper arm	☐	☐

	Left	Right
Elbow	☐	☐
Lower arm	☐	☐
Wrist	☐	☐
Hand	☐	☐
Ankle	☐	☐
Toes	☐	☐

	Left	Right
Finger	☑	☐
Upper leg	☐	☐
Knee	☐	☐
Lower leg	☐	☐
Shoulder	☐	☐
Hip	☐	☐

Name and address of doctor

Phone/Fax _____   E-mail _____

Cause of incident:

None	☐
Unsafe environment	☐
Unsafe machinery	☐
Unsafe stacking	☐
Unsafe system of work	☐
Misuse of equipment	☐
Manual handling	☐
Horseplay	☐

Brief description of incident: (attach additional sheets if required)

Operative injured by chuck key left in hand drill

Recommendations/observations on remedial measures to prevent recurrence: (indicate dates, etc.)

Date incapacitation commenced _____   Date of return (if known) _____   Total absence _____

Any property/asset damage   Yes ☑   No ☐   Estimated costs _____

Signature of person completing form _____   Print name _____

Workplace address and occupation of person completing form _____

I hereby declare that the information contained in this report is true and no material information within my knowledge in regard thereto has been withheld.

Investigating officer/Supervisor signature _____   Date _____

Designation _____   Tel _____

Original	To be retained locally
Duplicate	Send to Health and Safety Team

Extension: Speaking and writing

A Writing guidelines

- Refer students back to the Speaking section on page 108. Elicit ideas as to how accidents can be avoided when using the tools mentioned.

- On the board, build sentences using the expressions *Before –ing, make sure that …; Ensure that … If you …; Never/Always …*

- Students write their own guidelines for some of the tools. Monitor and assist as usual. When they are ready, students should exchange and comment on each other's work.

B Role play

- In pairs, students should note down briefly the circumstances of an imaginary workshop accident.

- Elicit onto the board questions for each section of the form, e.g., *Where did it happen? What is his full name? How can we avoid this problem in the future?*

- In new pairs, students ask and answer the questions and fill in an Accident Report Form. They can reuse the one on page 109 or draw up a simplified version of their own.

- If you have time, exploit the Tapescript for further listening practice and pronunciation work.

- Now have the students do the Workbook exercises for Unit 6, Lesson 4.

Workbook answers

Reading and writing: fuel maintenance instructions

A Read the fuel system text and underline examples

Pupils should underline:

ensure	carry (do not)
display	refuelling/defuelling
exclude	check

B Decide whether the sentences are true or false

1	T	4	F
2	T	5	T
3	F	6	F

C Choose a situation and write similar instructions

Answers depend on the students.

Writing: abbreviations

A Write in full

1 Take note: minimum temperature 18°C

2 pressure should not vary more than plus or minus 1 pound per square inch

3 ratio not less than 0.5:1

4 6 feet multiplied by 186 millimetres

5 dimensions: (height) 110 centimetres multiplied by (width) 124 centimetres

B Write these sentences using abbreviations

1 e.g., carbon + oxygen = carbon dioxide

2 g-force > 4

3 max. speed approx. 300 kph

4 $22 \div 7$ = approx. 3.14

5 supp = 26 V alt. current

Language: movement

A Complete the explanations

1	against	6	through
2	into	7	from/to
3	over	8	into
4	to	9	up and down
5	off	10	in

Objectives

To review language for parts of an aircraft and flight manoeuvres.

To raise awareness of effective note-taking.

To develop reading skills.

Language

Vocabulary for aircraft parts, e.g., *fuselage*, wing and manoeuvres, e.g., *roll*, *bank*.

Vocabulary

A Activate vocabulary

- Draw two circles on the board; one with *parts of an aircraft structure* and one with *flight manoeuvres* written inside. Elicit one or two vocabulary items for each. Students add as much vocabulary as they can in their notebooks.

Model answer

Parts of an aircraft structure: wings, tail, foreplane, power plant, landing gear, fuselage, skin, struts, flaps, fins, propeller, rudder, skin, longerons, ailerons, frames, bulkhead, spars, stringers, ribs

Flight manoeuvres: bank, roll, yaw, take off, land, climb, turn, dive, hover, fall, rotate, spin, loop, cross

B Discuss the role of aircraft parts in manoeuvres

- Set the task for pairwork. Students should concentrate on the roles of the wings, flaps, rudder and ailerons. Refer students back to Unit 4 if they are unsure of the roles of these parts.

- Feed back orally.

Answers

Answers depend on the students.

Reading and writing

A Scan the text to find ideas and words from the vocabulary section

- Set a time limit so that students read quickly.

- Feed back, eliciting the fact that the text describes projects for aircraft with wings that can move and change shape.

Answers

Answers depend on the students.

B Read the text more carefully and decide which project is shown in the illustration

- Set the task for individual work and pairwork checking. As usual, point out to students that the exercise is to scan the text.

- Compare answers in pairs.

Answer

The illustration shows the 'batwing'.

C Read for detail

- Set the task for individual work and pairwork checking. Stress that it is not important for students to understand all of the words at the moment.

Answers

1 The possibilities of shape-changing theory (T):

- Tiny heaters could perhaps be attached to the wing. When a shape change is complete, these would heat the SMP and so reseal the joints around the new wing shape.

- If a pilot could change the wing shape and size quickly enough, shortening one wing slightly and raising or lowering the other, it would be possible to turn, bank and dive, much like birds do.

- Possible smart materials, such as SMPs, and new types of actuators, would allow extension.

- A batwing would save weight and reduce drag, since there would be no need for ailerons, rudders or tail fins.

- Theoretically, if the aircraft's whole wing was turned into a reshapeable control surface the pilot could simply twist the whole wing to climb, roll, etc.

2 The new design (D):
- The new wing will bend and fold without open joints.

- The plan is that 'Shape Memory Polymer' coverings will be used to seal joints. As the wing changes shape, the polymer will stretch with the movement.

- When this design is in operation, wing sections will move in and out of the fuselage.

- The aim is that the wing will be able to change its surface area by more than 150%.

- A series of 'sliding skins' will disconnect, extend to create new wing shapes, and then reconnect.

- SMPs will be used to cover joints.

3 Facts about existing technology (F):
- We know that gaps in a wing joint cause drag, and that this rips a wing apart at high speeds.

- This type of shape-changing has been tried before: the US, Europe and the USSR all developed aircraft in the 1960s and 1970s that could swing their wings forward or back for take-off, diving or cruising.

- 'Shape Memory Polymers' have existed since the 1930s. These stretch, and then resolidify when heated.

D Note-taking

- Use the example provided on page 111 to elicit and build up notes for Project 1 on the board. Stress to students that there is no correct way to reduce the information. The aim is to be as concise as possible.

- Set for individual work and pairwork checking.

- Feed back onto the board and discuss with students the different possibilities they offer.

Suggested answers

Project 1

Facts

gaps – drag: v. dangerous @ high speed

Theory

tiny heaters heat SMP to reseal joint after shape change

if pilot could change height + shape of wing quickly $\longrightarrow$ turn/bank, etc., like birds

Design

new wing – will bend and fold without open joints

Shape Memory Polymers (SMP) cover wing: stretch when change shape

Project 2

Facts

not new – 60s/70s (US, Euro, USSR)

Theory

new mats, e.g., SMPs + actuators $\longrightarrow$ extension

Design

wings will move in + out of fuselage

wing will change 150% surface area

Project 3

Facts

SMPs since 30s: stretch + resolidify w heat

Theory

batwing = less drag/weight: no control surfaces

if whole wing reshapeable ⟶ twist whole wing to climb, etc.

Design

sliding skins will disconnect, move, reconnect

SMPs will cover joints

Speaking

A Give a description of an aeronautic project, using notes as a prompt

- This exercise tests the effectiveness of students' notes.

- Use the notes for Project 1 on the board to elicit a description of it, including the associated facts and the theory.

- In pairs, students practise using their notes to describe one or more of the Projects 1 and 2 without referring back to the text.

- After they have spoken, encourage them to amend their notes, using the text, to improve their usefulness.

- If you have time, do further work on vocabulary from the reading text.

- Students may be interested in researching the technologies in this lesson at:

  http://www.nextgenaero.com/

  http://www.afrlhorizons.com/

- Now have the students do the Workbook exercises for Unit 6, Lesson 5.

Workbook answers

Language: prefixes

B Make new words by matching the prefixes with the words

un-safe

im-practical

pre-flight

in-adequate

re-paint; re-use

de-activate

dis-use

anti-clockwise

C Complete the exercise with words from Exercise 2

1	impractical	5	anticlockwise
2	repaint	6	reuse
3	deactivate	7	Pre-flight
4	unsafe	8	inadequate

Language: *might, will, could, would*

A Complete the blanks with *will*, *would*, *might* or *could*

1	would	5	could
2	will	6	might
3	might	7	might
4	will	8	could

Writing: describing machines

A Think of a part of a machine you know well and write sentences

Answers depend on the students.

Objectives

To review types of mechanical stress.

To practise drawing simple sketches.

To practise reading for main ideas.

Language

Types of mechanical stress, e.g., *shear*, *tension*, etc.

Vocabulary

A Make phrases connected with force

- Exploit the illustration. Elicit ideas for what it shows and write on the board *parts can fail under excessive loads*.

- Ask students to find the two halves of the phrase in the lists.

- Set the task for individual work and pairwork checking. Feed back orally. Clarify vocabulary problems.

Answers

pull something – apart

a squeezing or – crushing force

wings are subjected – to compression forces

the force is – resisted by structural strength

skin panels are – riveted together

one force acts against – another

one piece of material slides – over another

the propeller – pulls the plane forward

parts can fail – under excessive loads

B Sketch the meanings

- Do one or two more sketches on the board with the group. Point out the arrow, which shows where the stress occurs.

- Set the task for individual work and pairwork checking. Students should choose only one or two phrases to sketch, as the task will be over-long otherwise. Also, stress that the task here is to draw sketches, so students need not worry about the level of their drawing skills.

- When they are ready, students can work in pairs and use their sketches to elicit from each other the phrases in Exercise 1. You may wish to get some students to draw their sketches on the board to elicit the phrases from the group.

Answers

Answers depend on the students.

Reading

A Review five types of mechanical stress

- Elicit the name for a *squeezing or crushing force* (compression).

- Set the task for pairwork. Students can check their ideas against Unit 3, Lesson 2.

- Feed back orally.

Answers

Answers depend on the students.

B Reading for main ideas

- Ask students to read the text at the bottom of page 112 only. Elicit the type of stress described (tension).

- Set the rest of the reading task for individual work and pairwork checking.

- Feed back orally.

Answers

1	Tension	4	Shear
2	Compression	5	Bending
3	Torsion		

Speaking and writing

A Discuss how stresses operate in day-to-day situations

- Go through the list of situations, clarifying any problem vocabulary, e.g., *hinge*.

- In pairs, students discuss the forces and stresses which are operating in the day-to-day situations in the list. Point out that they should try to use the vocabulary from this lesson, e.g., *squeezing water out of a cloth*; *torsion is operating*; *the two ends of the cloth are turned in opposite direction*s.

- If you have time, review the phrases in Exercise A on page 112 OR do further vocabulary work on phrases from the reading texts.

- Now have the students do the Workbook exercises for Unit 6, Lesson 6.

Workbook answers

Language: prepositions

A Match the directions

up – down

forwards – backwards

from side to side – up and down

horizontal – vertical

anticlockwise – clockwise

over – under

Writing: instructions

A Complete the sentences

Answers depend on the students.

Language: spelling

A Complete these words using *a, e, i, o, u*

1 compression   5 material

2 bending   6 fuselage

3 solution   7 experiment

4 torsion   8 structural

B Put each word into the correct column according to its stress pattern

Oo	Ooo	oOo	oOoo
bending	fuselage	compression	experiment
torsion	structural	material	
		solution	

Lesson 7 — Controls: Review

Objectives

To review descriptions of control systems.

To give a brief talk which contextualises the language in the lesson.

Language

Nouns and verbs for describing control systems, e.g., *hydraulic fluid*, *cylinder*, *force* (v), *transmit* (v).

Reading

A Read a description of a brake system and label a diagram

- On the board, review the basic flight controls: *rudder, elevators, ailerons, pedals*.

- Tell students they are going to read about a braking system.

- Ask students to read through the text once. Set the general questions: *What control does the pilot use? (a pedal); how is the force transmitted to the brakes? (by mechanical linkages and hydraulic cylinders).*

- Refer students to the vocabulary box. They should check the meanings in pairs using dictionaries where necessary.

- Tell students to reread the text individually and label the drawing. Pairs check their answers.

- Feed back onto the board, or an OHP if possible.

Answers

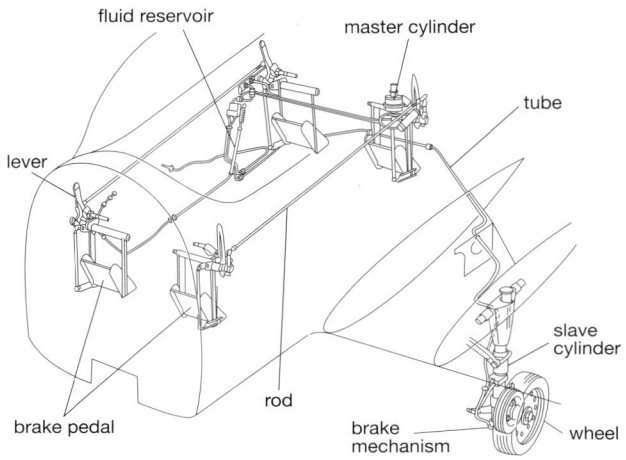

fluid reservoir
master cylinder
tube
lever
slave cylinder
brake pedal
rod
brake mechanism
wheel

Vocabulary

A Match verbs with similar meanings

- Ask students to cover Column 2 of the table, and to underline in the text the verbs in Column 1. In pairs, they check the meaning.

- Individually, they then match the verbs in Columns 1 and 2. Pairs check answers. Feed back onto the board.

Answers

operate – make sth. work	force – push
connect – join	fit – install
transmit – send	function – work

B Check understanding of the verbs in context

- Do the first sentence with the class.

- Set the task for individual work and pairwork checking.

Answers

1/2

a   wrong: *repairs*

b   correct

c   wrong: *transferred/taken/sent*

d   wrong: *contains*

e   correct

f   correct

g   correct

h   wrong: *filled*

i   correct

j   wrong: *pushes/moves*

Writing

A Make sentences using verbs

- Tell students to cover the explanation on page 114. Elicit a sentence about the braking system using the first verb in the list e.g., *Rods and tubes join the brake pedals to the cylinders.*

- Encourage students to write similar sentences using the other verbs. (Do this orally if you prefer.)

Answers

Answers depend on the students.

B Write a sequenced explanation of how another system works

- Students look again through the explanation of the braking system in the reading section.

- Ask them to choose another control system – *flaps* or *rudder*. They should look back at Unit 4 to review how it works.

- Working in pairs, students then write a numbered or bullet-pointed explanation of the operation, including the pilot's actions, linkages, hydraulics and the effect the control surface has on the aircraft's flight.

- Monitor closely and assist.

Answers

Answers depend on the students.

- Now have the students do the Workbook exercises for Unit 6, Lesson 7.

Workbook answers

Language: categories

A Put the words into the correct category

linkages	control surfaces	fasteners
cable	rudder	bolt
rod	flap	screw
tube	aileron	rivet

B Add one more word to each category

Answers depend on the students.

Reading: brake fluid

A Read the text and put the missing words in the gaps

1 hydraulic
2 overheat
3 pedal
4 temperatures
5 boiling point
6 cylinder
7 reservoir
8 seals

B Write four sentences about working with brake fluid

Suggested answers

You should keep it clean and free of air bubbles.

The brake fluid level should be checked regularly.

If the level drops noticeably over a short period of time, the brake system should be examined.

Containers of brake fluid shouldn't be left open.

Lesson 8   Hydraulics: Review

Objectives

To understand and complete a Job Sheet.

To practise reading to choose the correct word for a context.

Language

Reduced passive note forms using capital letters, e.g., *SPARK PLUGS CLEANED AND REFITTED*.

Words similar in shape or meaning, e.g., *damage/break; check/control*.

Speaking and vocabulary

A Check pronunciation of vocabulary connected with hydraulics

- Write on the board one or two single- and multi-syllable words which students are very familiar with. Review the notion of words having a stressed syllable.

- Set the task for pairwork. Point out that recognising these words will be important for the listening section. Feed back onto the board.

Answers

check	/tʃek/
control	/kənˈtrəʊ/
replace	/rɪˈpleɪs/
repair	/rɪˈpeə/
damage	/ˈdæmɪdʒ/
break	/breɪk/
inspect	/ɪnˈspekt/
open	/ˈəʊpen/
leak	/liːk/
flood	/flʌd/
pressure	/ˈpreʃə(eə)/
heat	/hiːt/
empty	/ˈemptiː/
dismantle	/dɪsˈmæntl/
refill	/ˈriːfɪl/
refit	/ˈriːfɪt/

B Check meaning of vocabulary

- Work through the first pair or two pairs of words as a class. Set the task for pairwork.

- Feed back orally, focusing on how the words in each pair collocate with other words, e.g., you can *empty* a *container* such as a bin, but *dismantle* a *machine* or *gadget*, such as a computer.

Answers

Answers depend on the students.

Listening and writing

A 💿 Listen for main idea (CD2 Track 2)

- Use the diagram to review vocabulary connected with brakes, e.g., *disc*, *cylinder*, *fluid*, *pad*.

- Give students time to read through the three problems. Clarify vocabulary difficulties. Elicit the meaning of R/H and L/H (*right-hand* and *left-hand*).

- In pairs, students identify the parts of the diagram in the list of problems.

- Play the recording once. Pairs check answers. Feed back orally.

Answers

1 3

2 1

3 2

B Transfer the information

- Set the task for individual work. Monitor to ensure accurate copying and use of capital letters.

- As feedback, show or write the *fault* column of the Job Sheet on the board or OHP.

Answers

fault
1   BRAKE MASTER CYLINDER LEAKING
2   LEAK IN FLUID LINE L/H FRONT BRAKE
3   R/H FRONT DISC DAMAGED + OUT OF ALIGNMENT

C 💿 Listen for more detail (CD2 Track 2)

- As before, allow reading time and clarify vocabulary.

- Replay the recording. Pairs check their answers. Feed back by adding the *cause* column of the table onto the board.

fault	cause
1 BRAKE MASTER CYLINDER LEAKING	BRAKE FLUID ESCAPING. AIR GETTING IN
2 LEAK IN FLUID LINE L/H FRONT BRAKE	DAMAGED TUBE
3 R /H FRONT DISC DAMAGED + OUT OF ALIGNMENT	EXCESSIVE HEAT DUE TO OVERWORK – L/H BRAKE AT LOW PRESSURE

D 📀 **Listen and take notes (CD2 Track 2)**

- Write *SPARK PLUGS CLEANED AND REFITTED* on the board. Elicit that this is a reduced note form of *The spark plugs have been cleaned and refitted*, and that the verb *be* has, as usual, been omitted.

- Point out that the *corrective action* column will require students to use the list of verbs at the top of page 117. Check that students understand these.

- Play the recording once more. Set the task for individual work and pairwork checking. Monitor the pairwork and assist with the formation of the reduced passive note form. Remind students that they should use the verbs at the top of the page in their answers.

Answers

JOB SHEET: 20,000-MILE SERVICE		
fault	**cause**	**corrective action**
1 BRAKE MASTER CYLINDER LEAKING	BRAKE FLUID ESCAPING; AIR GETTING IN	• CYLINDER REPLACED
2 LEAK IN FLUID LINE L/H FRONT BRAKE	DAMAGED TUBE	• FLUID EMPTIED • TUBE REPLACED • SYSTEM REFILLED WITH FLUID
3 R /H FRONT DISC DAMAGED + OUT OF ALIGNMENT	EXCESSIVE HEAT DUE TO OVERWORK – L/H BRAKE AT LOW PRESSURE	• DISC (SURFACE) MACHINED • DISC STRAIGHTENED • OTHER BRAKES CHECKED

Vocabulary

A Read to choose correct meaning in context

- Point out that the underlined items in the tapescript are the same as the ones students discussed in speaking and vocabulary.

- Set the task for individual work and pairwork checking.

- Feed back orally.

Answers

leak	damaged
replace	inspected
empty	pressure
refilled	checked

- If you have time, students can listen again to mark the intonation and stress on a part of the Tapescript on page 117. They can then role-play a short section OR do further work on passive forms and their reduced note forms.

- Now have the students do the Workbook exercises for Unit 6, Lesson 8.

Workbook answers

Language: verbs and nouns that go together

A Add two possible verbs to complete each phrase

Answers depend on the students.

Reading and writing: interpreting notes

A Expand the notes

There was a disc out of alignment. We checked and straightened the disc.

The brake light came on because the brake pads were worn and cracked. We replaced the pads.

Objectives

To review language and concepts connected with engines.

To practise transferring information into a flow chart.

Language

Vocabulary for oil filter systems, e.g., *filter*, *housing*.

Reading and speaking

A Activate vocabulary

- As an introduction, briefly review the main components of an internal combustion engine.

- Set the questions for small groups. Encourage students to draw on their knowledge of cars and motorcycles, and to make informed guesses.

- Feed back orally.

Answers

1 Vehicle oil filters can become blocked with particles of dirt in the oil. If this happens, the engine loses its lubrication system, causing overheating and eventual malfunction.

2 Aircraft need effective oil filters because of the danger that a malfunctioning engine would represent during flight.

B Read two texts and give each one the correct title

- Tell students to ignore the diagrams for the moment. Check understanding of the term *bypass*.

- Set the task for individual work and pairwork feedback. Check understanding by asking when it is necessary to use the *bypass operation* (when the filter is blocked).

Answers

The first text (on page 118) is Normal Operation; the second (also on page 118) is *Bypass Operation*.

C Study and identify diagrams

- Refer students to the two diagrams of the oil filter. Ask them to study the labels. Elicit the meaning of the black arrows in the diagrams (they indicate the direction of the oil flow). Point out that they should look carefully at the direction of the oil flow.

- See whether students can identify which diagram goes with which text.

Answers

Diagram 1: *Normal Operation*

Diagram 2: *Bypass Operation*

Writing

A Transfer information from the text to a chart

- Give students time to read the eight sentences. They reread the texts to put the sentences into the right places in the chart, and check their answers in pairs. Monitor and assist. Feed back onto the board.

- On the board, write the full sentence: *The oil from the oil pump enters the filter assembly*, and the note form of the same sentence as it appears in the chart. Elicit how the note form is made, i.e., by deleting the repetition of *oil*, the obvious fact that it is the *filter* assembly, and the article *the*.

- Do the same with the sentence *If the filter is blocked …* In the note form, the *if* condition is replaced by *filter blocked*: and the solution; long nouns are also shortened; articles and non-essential repetition are again deleted; *through* and *and* are shortened to *via* and +.

- Set the task for individual work and pairwork checking. Monitor and assist. Feed back onto the board.

Answers

The order of the sentences is:

1 The oil from the oil pump enters the filter assembly.

2 The oil enters the filter element housing.

3 The oil passes through the filter element and into the central core.

4 The clean oil then flows out of the core into the engine.

5 Any solids which were in the oil stay on the outside of the filter element.

6 If the filter is blocked, the oil flows into the filter through a spring-loaded bypass valve and a bypass filter.

7 The clean oil passes into the core.

8 The oil flows into the engine in the normal way.

Suggested note forms:

1 oil from pump enters assembly

2 oil enters filter element housing

3 oil into central core via element

4 clean oil flows out of core into engine

5 solids stay on outside of element

6 filter blocked: oil flows in via bypass valve + filter

7 clean oil into core

8 oil flows into engine in normal way

- If you have time, read out the text making deliberate mistakes which students must spot, OR using the completed chart, students work on building up as full a description as they can of the *Normal* and *Bypass* operation of the filter system.

- Now have the students do the Workbook exercises for Unit 6, Lesson 9.

Workbook answers

Language: the oil filter assembly

A **Find words in the wordsearch**

spring	core
element	filter
tank	valve
housing	bypass

B **Label the diagram**

See Course Book page 119.

Language: the passive voice

A **Choose the correct verb form (passive or active voice) for the sentences**

1 consists/~~is consisted~~

2 ~~locates~~/is located

3 flows/~~is flowed~~

4 ~~deposits~~/is deposited

5 ~~closes~~/is closed

6 becomes/~~is become~~

7 filters/~~is filtered~~

8 ~~change~~/be changed

Language: *pass* and *flow* + preposition

A Complete the sentences

The verbs *pass* or *flow* are acceptable in all cases. The preposition which follows the verb in each case is given here:

1 through
2 out of
3 into
4 through
5 into

Engine servicing: Review

Objectives

To review concepts and verbs connected with aircraft maintenance.

To practise reading technical advertising information for main ideas and details.

To practise giving explanations about servicing and maintenance using the passive voice.

Language

Present passive forms, e.g., *all components are removed*.

Vocabulary and speaking

A Review verbs for describing maintenance

- Ensure that students understand all the items *1–8* and the items in the box. Drill the pronunciation.

- Elicit ideas for the first item in the box. Ask students to justify and explain their ideas.

- Set the task for pairwork. Encourage discussion and justification of ideas. Feed back orally.

Suggested answers

a damaged selector: replace, check

an engine: dismantle, inspect, strip, reassemble

a reported fault: check

a reworked component: replace, check, inspect, fit

a new compressor: inspect, check, fit

a faulty pump: clean, replace, dismantle, inspect, check, strip, reassemble

a filter: all

a finished job: inspect

a dirty spark plug: clean, replace, dismantle, strip, reassemble

a leaking gasket: replace, inspect, check

a hydraulic tube: clean, replace, inspect, check, fit

Reading

A Read for gist and match headings to paragraphs

- Elicit or get students to check in their dictionaries the difference between *maintenance*, *service* and *overhaul* (*maintenance* (nU) refers generally to routine inspection and repair work; *service* (nC) to obligatory periodic maintenance, and *overhaul* (nC) (v) to a comprehensive operation involving complete stripping of assemblies).

- Refer students to the short text on page 120. Hold a general discussion about what service, maintenance and overhaul might involve specifically, how much time each might take, and what type of environment and technology might be appropriate to each.

- Check students understand the headings in the list.

- Set the task for individual work and pairwork checking.

Answers

1 Engine stripping
2 Inspection and parts ordering
3 New parts quality control
4 Final assembly
5 Post-assembly test and final inspection

B Read for detail and complete the gaps with verbs

- Work through the first few gaps with the whole group.

- Set the task for individual work and pairwork checking. Monitor and assist.

Answers

1 stripped, inspected, removed
2 checked, replaced
3 inspected, dismantled, fitted
4 rebuild, assembled
5 fitted, tested, sent

Speaking and grammar

A Identify grammar in context

- Elicit the location of the first item. Elicit that all the verbs except for one (rebuild) are used in the passive voice.

- Set the task for individual work and pairwork checking.

- Feed back orally, or onto an OHP if possible.

Answers

replaced	line 10
subjected	line 26
cleaned	line 2
ordered	line 10
dismantled	line 16
sent	line 28
fitted	lines 17 and 25
stripped	line 2
removed	line 3
assembled	line 21
tested	line 25

B Find further examples of the passive voice

- Again, set for individual work and pairwork checking. Feed back orally and/or onto the OHP.

Answers

Paragraph 1
are received
are discarded
(are) labelled

Paragraph 2
is checked

Paragraph 3
are unpacked
(are) inspected
are disassembled
(are) inspected

Paragraph 5
is immobilised
(is) packed

C Identifying noun-verb collocation

- Work through the verbs in Paragraph 1 with the group.

- Set the task for individual work and pairwork checking. Monitor and assist.

- Feed back onto the board. Use an OHP if possible.

Answers

All noun + *is/are* verb correspondences in the text:

Paragraph 1

are received; (are) stripped down: engines

(are) cleaned: the parts

are removed; (are) sent: all components

are discarded; (are) labelled: parts requiring mandatory replacement

Paragraph 2

is checked: the engine

are ordered: all parts to be replaced

Paragraph 3

are unpacked; (are) inspected: all new parts received

are disassembled: these parts

(are) inspected: factory supplied cylinder kits

Paragraph 4

is assembled: each engine

Paragraph 5

is fitted; (is) tested: each reassembled engine is subjected; is immobilised;

(is) packed; (is) sent: the engine

D Describe the engine servicing procedure using the passive voice

- Using the list of passive verb structures on the board, elicit and build up a description of the activities involved in Engine Stripping. Stress to students that it is not important that they remember the text verbatim – correct and appropriate use of the noun + verb phrases is the target here.

- Set the task for pairwork. Feed back orally.

Answers

Answers depend on the students.

- Now have the students do the Workbook exercises for Unit 6, Lesson 10.

Workbook answers

Language: verbs connected with engine maintenance

A The odd word out

1 block – the others all mean *movement* of *liquid*

2 dismantle – the others have a negative meaning

3 split – the others all mean *look at*

4 install – the others all mean *to solve a problem*

5 remove – the others all mean *put back together*

6 overhaul – the others all mean *end*

7 strip – the others all mean *put together*

B Words that contain the sound /ɪ/

Suggested answers

destroy	refit
inspect	rebuilt
split	finish
fix	link
install	strip

C Practise saying the words

Answers depend on the students.

Language: sentence structure

A Match the sentence halves

1 d		**4** a	
2 e		**5** b	
3 c			

Writing: instructions

A Write sentences giving advice

Answers depend on the students.

Unit test

A test for this unit is available at:

http://www.garneteducation.com/reps/
documents/1250/SDT-u6-test.pdf

Contact your local Garnet Education
representative for information about how to
access these resources.

Objectives

To raise awareness of language used in posters and signs giving warnings, instructions and information.

To practise reading signs and posters for information.

Language

Vocabulary connected with rules and warnings: *mandatory*, *hazard*.

Imperatives and modals of obligation.

Vocabulary and speaking

A Discuss the functions of signs

- Brainstorm places where students have noticed signs in their school or workplace.

- Elicit the functions of the different signs, e.g., prohibition, rules, warnings, information (about procedures and events). See if students can remember the wording of any signs. Point out or elicit that modal verbs and imperatives are often used.

Answers

Answers depend on the students.

B Match words with near synonyms

- Set the task for individual work and pairwork checking. Monitor to check for any problems.

Answers

No answers at this point.

C Discuss slight differences in meaning

- Set for pairwork and feed back as a whole class. Focus on differences in style and register; *precaution* is more formal than *protection*, etc.

Answers

prevent – stop

prevent (v) [+ someone + from doing something] - more formal; *prevention* happens before someone starts the action; someone can be *stopped* even if they have begun the action

warning – caution

a *caution* (n, v) may be less strong than a *warning* (n)

hazard – danger

danger (n): used in more general situations; *hazard* (n): often used to refer to chemical and environmental dangers

mandatory – regulation

mandatory (adj) used particularly in formal registers; *regulation* (adj, n) – more general use

precaution – protection

protection (n, unc.) is defence against danger and is often physical; *precaution* (n, c) is an advance action to defend against potential danger

provide – give

provide is more formal usage

remove – take off

take off is less formal

assess – evaluate

both mean to calculate the value of something; *assess*: often used when talking about income/tests/risks; *evaluate* is often used in reference to ideas/maths

prohibit – forbid

prohibit is a more formal term

Reading

A Look quickly at the posters and identify their purpose

- Students should be able to do this without reading much of the text. The use of colour, symbols and illustrations should help them. Set a time limit.

Answers

1 Poster <u>1</u> gives information about the correct way to do something.

2 Poster <u>4</u> explains the responsibilities of the company and the workers.

3 Poster <u>2</u> gives simple instructions in case of an accident.

4 Poster <u>3</u> describes different types of safety warnings.

B Scan the posters for specific information

- Set the task for individual work and pairwork checking.

- Feed back orally. Pinpoint the language in each poster that conveys the information.

Answers

1 a Employers must train people in how to use safety equipment: T

  b PPE needs to be specialised for some tasks: T

  c Employers are responsible for putting away safety equipment: F

2 a An ambulance should be called in every case of a burn or scald: F

  b You should not try to take off any injured person's clothes: T

  c You should take off the person's watch: F

3 a A sign that tells you where fire equipment is kept is blue: F

  b A yellow sign means some kind of danger: F

  c First aid equipment will be found near a green sign: T

4 a There are five stages to lifting correctly: F

  b If you can, you should use a machine to lift things: T

  c You should look for dangers in your way before you lift: F

C Focus on words that go together (noun-verb collocations)

- Before students read, write the verbs on the board and elicit nouns that commonly go with them, e.g., *provide – training, equipment; consult – a doctor, a manager, a manual*

- Set for pairwork. Ask students to underline the verb-noun combinations they find.

- Feed back by eliciting examples. Draw attention to the use of modal verbs, e.g., *The employer must provide suitable PPE free of charge*, and imperatives, e.g., *assess the situation*.

Answers

provide:	information, training, PPE
consult:	employees
report:	loss, defects, damage
maintain:	PPE, signs
prohibit:	behaviour
assess:	a situation

Extension: write sentences giving warnings and instructions

- Drill the sentences from the previous exercise for correct pronunciation.

- Ask students to write similar sentences giving warnings, instructions or rules about another topic, such as looking after tools. They should use the verbs from the lesson.

- Now have the students do the Workbook exercises for Unit 7, Lesson 1.

Workbook answers

Language: parts of speech

A Label the words v (verb), a (adj), or n (noun)

1	verb	**5**	verb	**8**	noun
2	verb	**6**	noun	**9**	verb
3	noun	**7**	noun	**10**	verb
4	noun				

B Complete the table

noun	verb	adjective
adjustment	adjust	adjustable
protection	protect	protective
assessment	assess	assessable
identification	identify	identifiable
maintenance	maintain	maintainable
provision	provide	providable
prohibition	prohibit	prohibitable
requirement	require	requirable
warning	warn	warning
injury	injure	injurable

Writing: nouns, verbs and adjectives

A Rewrite the sentences

Answers depend on the students.

B Write a short phrase using the adjectives

Answers depend on the students.

Lesson 2   Safety at work

Objectives

To practise listening for key (stressed) words in a conversation and work out meaning from context.

To practise using language for warnings.

Language

Vocabulary for workshop safety: *flammable*, *caution*.

Pronunciation

Sentence stress.

Speaking

A Review information on safety colour-coding

- Set for pairwork and whole-class feedback.

Answers

Answers depend on the students.

B Look at the signs

1 mandatory (must do): blue

2 prohibition (do not do): red

3 safety measures and first aid: green

4 warning (dangerous): yellow

C Discuss the meaning of different signs

- Students discuss their ideas in pairs or small groups.

- Feed back, eliciting as many places as possible where each sign might be seen.

Suggested answers

Use ear protectors when operating this machine

Mandatory: This would be placed on noisy machinery, e.g., a ground power unit or even next to roadworks.

Do not clean or oil moving machinery

Prohibition: This sign might be used in a workshop which needs to be kept clean all the time, e.g., a sheet metal shop for aluminium bending and forming.

Emergency shower

Safety measures and first aid: This would be placed in an area where chemicals are used, e.g., a battery shop, in case there was an accident involving acid or other dangerous chemicals.

Caution slippery surface

Warning: This could be found on any public walkway that is being cleaned, or where there is a spillage.

Flammable

Warning: This might be found in, or near a garage, or fuel tank.

Caution: very hot water

Warning: Could be found above a hot tap.

Emergency stop

Safety measures and first aid: This might be found where there is fast-moving machinery that would need to be stopped very quickly, e.g., a machine shop.

Guards must be used

Mandatory: This could be found in a machine shop where there are cutters, lathes, etc., flying metal.

Do not switch on

Prohibition: This could be found in a machine shop or an area where there is high-voltage equipment. It is likely that the equipment is being repaired.

Caution: wet paint

Warning: This poster would be placed anywhere where painting has been carried out.

Listening

A Gist listening: identify the situation (CD2 Track 3)

- Play the recording once and monitor as students number the correct signs.

- If they find the task difficult, encourage pair-checking and/or play the recording a second time before checking the answers.

- Elicit where students think each conversation is taking place and why. Do not confirm answers at this stage.

Answers:

1 Caution: very hot water

2 Do not clean or oil moving machinery

3 Emergency shower

4 Flammable

5 Do not switch on

6 Caution wet paint

B Intensive listening to pick out words and expressions in a stream of speech (CD2 Track 3)

- Pause the recording after each key phrase or between each conversation to allow students time to write phrases and/or sentences. Alternatively, with a very small group, the students could choose when to stop the recording by saying *stop* whenever necessary or selecting one student to operate the player.

- Be prepared to play the recording two or three times, allowing time for pair-checking after each hearing. Do not insist on everyone writing full sentences.

- Feed back by writing their ideas on the board and then correcting.

- Draw attention to the use of informal language such as *we don't want …* and *they could do with …*

- You may wish to give out the Tapescript at this point and look at the use of pronoun references and/or other colloquial language. See if students have revised their ideas about where each conversation took place.

Answers:

1 You won't make that mistake again in a hurry.

2 We don't want another accident like last week.

3 The sump plug fell out while I was underneath.

4 Keep in a cool place, well away from sources of ignition.

5 There's no protection on it at all.

6 They could do with a few more windows open.

Pronunciation and speaking

A Focus on sentence stress on 'message' words

- Ask students to look at the skills box and check understanding of what sort of words constitute 'message' words.

- Model the conversations, exaggerating the stress slightly. Students should then work in pairs to practise the conversations *1* and *2* with the correct stress.

Answers

Answers depend on the students.

B Identify stressed words in a conversation and practise it

- In pairs, students look at the Tapescript of the Listening exercise. They should choose a conversation and underline the words that were stressed.

- **Note:** If you are short of time, students could mark the stresses on the sentences they wrote in Listening B instead.

- Pairs practise reading the conversation or sentences with appropriate stress.

Answers

Answers depend on the students.

Writing

A Write short dialogues giving warnings

- Elicit which signs were not mentioned in the Listening exercise.

- Draw attention to the prompts for the dialogue and elicit possible ways to complete the conversations.

- Students work alone or in pairs to write their ideas. Stress that their dialogue should illustrate only *one* sign.

Model answer

A: What are you <u>doing</u>?

B: I'm going to use the <u>lathe</u>.

A: Be <u>careful</u>: make sure you put the <u>guard</u> into position before you use it.

B: Yes, <u>don't</u> <u>worry</u>.

A: We don't want an <u>accident</u> like we had <u>last</u> month.

B Practise the dialogue

- Monitor while students mark stressed words and practise the dialogue. Model correct stress for any pairs who are not sure which words should be stressed.

Answers

Answers depend on the students.

C Identify the sign referred to in the dialogue

- Students should perform their dialogues to another pair, who guess which sign it refers to. Alternatively, with a small group, ask pairs to perform their conversation in front of the class.

Answers

Answers depend on the students.

- Now have the students do the Workbook exercises for Unit 7, Lesson 2.

Workbook answers

Language: prohibition

A Meaning of expressions

1	c	**4**	a
2	a	**5**	b
3	b	**6**	c

B Choose four signs and write short sentences

Answers depend on the students.

Language: collocations with *make* and *do*

A Write *make* or *do* in front of each phrase

1	make	**7**	make
2	do	**8**	do
3	do	**9**	make
4	make	**10**	do
5	do	**11**	make
6	do	**12**	make

B Make sentences of your own

Answers depend on the students.

C Memorise combinations with *make* and *do*

Answers depend on the students.

Pronunciation: long and short vowel sounds

A Put the words into the correct column of the table

short vowels	long vowels
fl<u>a</u>mmable	c<u>au</u>tion
w<u>a</u>sh	mach<u>i</u>ne
ign<u>i</u>tion	em<u>er</u>gency
prot<u>ec</u>tion	<u>g</u>uard
w<u>oo</u>d	rem<u>o</u>ve

B Practise saying the words

Answers depend on the students.

C Circle the words with the same vowel sound

Pupils should circle:

1 dry

grind

mind

supply

sign

right

2 away

ventilate

replace

change

paint

mistake

face

Objectives

To practise reading and listening to information in order to complete a table.

To practise speaking: giving information about fire and extinguisher types.

Language

Vocabulary for materials, elements and chemicals, e.g., CO_2, *magnesium*, *foam*.

Use of prepositions *for* and *by*, e.g., *Can be used for Class A and B fires.*

Speaking and reading

A Discuss types of fire and ways of extinguishing them

- Elicit a few ideas. Then set for pair or group discussion.
- Feed back, checking understanding of and writing up useful vocabulary on the board.

Suggested answers

1 by cigarettes, electrical faults and sparks, overheating of gases or flammable liquids

2 smother it, put water on it (via a hose or bucket), use a fire extinguisher

3 water, foam, chemicals (CO_2), powder

B Focus on vocabulary: materials and fire types

- Set the matching activity for individual work. Encourage students to use a dictionary to check unknown vocabulary.
- Conduct brief feedback, checking terms such as: *flammable, residue, titanium, propane, thinner.*

Answers

Class A:	2
Class B:	3
Class C:	4
Class D:	1

C Jigsaw reading and information swap

1

- Check students understand what should go into the table. Ask students to close their books and draw the headings for the table on the board. Read the paragraph headed *Common types of fire extinguisher* and elicit what should go under each heading.

2

- Organise students into groups of three, and clarify which paragraph each one should read (foam, dry powder or CO_2).

3

- Students should take turns to give information to other two group members about their paragraph (in their own words as far as possible). As they swap information, the group attempt to complete the table.
- Elicit ideas and draw or show an OHP of the completed table on the board.

Answers

colour	type	solids	flammable liquids	flammable gases	electrical equipment	cooking oils and fats	notes
red	water	✓	✗	✗	✗	✗	cools burning material, reduces O_2
cream	foam	✓	✓	✓	✗	✓	forms film on surface of fire; not recommended for home use
blue	powder	✓	✓	✓	✓	✓ (up to 1,000 volts)	knocks down flames, deposits layer of powder
black	CO_2	✗	✓	✓	✓	✓	prevents O_2 coming into contact with the material

D Exchange information on ways of extinguishing different types of fire

- Lead into the speaking exercise by reading the example given and suggesting one or two other possible fire situations, e.g., a car engine.

- Students should work in pairs to think of more situations and suitable extinguishers.

- Conduct brief feedback. Elicit whether students have ever used a fire extinguisher.

Answers

Answers depend on the students.

Vocabulary

A Focus on use of *for* and *by*

- Model two sentences; one with *for* and one with *by*, e.g., *Who is this manual for? (end-user or purpose)* and *Who is this manual by? (creator or agent)* Clarify the difference in meaning.

- Ask students to look through the text to a) underline uses of *for* and *by* and b) identify words used with *for* and *by* to make fixed expressions, i.e., *suitable/unsuitable for*, *used for*, *(not) recommended for* [+ noun phrase or present participle], *work(s) by* [+ present participle]

- Set the exercise for individual work and whole-class feedback.

Answers

1 for (lines 2, 4, 5, 7, 11 and 12)

   by (lines 2, 5, 9 and 13)

2 suitable (for); work (by); can be used (for); recommended (for); work (by); recommended (for); work (by); ideal (for); unsuitable (for); work (by)

B Complete the sentences

1	by	3	for
2	for	4	for; for

- Now have the students do the Workbook exercises for Unit 7, Lesson 3.

Workbook answers

Language: opposites

A Form the opposite of the words using *in~*, *un~* or *ir~*, and write a sentence

1	unavailable	6	irregular
2	unsuitable	7	irresponsible
3	irrelevant	8	unsafe
4	inflexible	9	uncomfortable
5	inaccurate		

Note: Sentences depend on the students.

Language and writing: expressions with *of*

A Write sentences with expressions from the Course Book

Answers depend on the students.

Pronunciation: long vowel sounds

A Put the words into groups according to the underlined vowels sound

group 1 (safe)	group 2 (no)	group 3 (now)
available	code	down
flames	O$_2$	powder
Class A	foam	thousand
paint	notes	
	sodium	

Lesson 4 — Safety procedures

Objectives

To review concepts in safety procedures.

To practise listening for details and taking notes.

To practise using language for obligation.

Language

Semi-modal verbs: *have to, need to* and *be allowed to.*

Vocabulary for safety equipment, e.g., *torch, mask, life jacket.*

Speaking

A Make a list of procedures and equipment

- Set the task for individual work. Go round and monitor to check for any problems.

Answers

No answers at this point.

B Compare your list with a partner

- Ask the students to compare their ideas with a partner.

- Conduct brief feedback. Make a list of the different safety equipment the students suggest. Ask them how each piece of equipment is used and what it is for.

Answers

Answers depend on the students.

Listening

A **Listen for key words and ideas (CD2 Track 4)**

- Ask students to read through the list of safety equipment again.

- Play the recording while students tick the items that are mentioned.

- Students compare answers in pairs. Feedback orally.

Answers

signs, fire extinguishers, oxygen bottles, oxygen masks, first aid kits, passenger information cards, torches, sick bags, life jackets, megaphone, seats, seat belts, safety harnesses

B Listen and complete notes (CD2 Track 4)

- Give students time to read through the list of items and the notes on the inspection procedures.

- Set the task and play the recording again.

- Students compare answers in pairs. Give time to write in extra notes.

- Feedback orally. Project the answers onto an OHP if possible.

Answers

signs	1 make sure they are there 2 check every one can see/read
fire extinguishers	1 right number 2 in right places 3 up-to-date service tags
oxygen masks/bottles	1 drop down correctly 2 in case of smoke in cabin
torches	1 ensure there are enough 2 batteries must be fully charged
sick bags	1 in case of turbulence 2 keep cabin clean
life jackets	1 ensure one is under each seat 2 in good condition
megaphone	1 must be at least one 2 crew to give instructions in an evacuation
seats	adjust up and down
seat belts	1 must work properly 2 not twisted or frayed

Grammar

A Find examples of language of obligation

- Ask students to read the examples in the Language Box.

- Clarify the difference between *need to* and *has to* (necessity vs. strong advice).

- Then ask them to find further examples in the Tapescript.

- Feedback orally.

Answers

have to

I <u>have to</u> check that there's one [life jacket] under each seat.

Oh, yes, of course <u>there has to be</u> at least one megaphone.

… if the <u>plane has to be</u> evacuated …

… there's another very important thing <u>I have to</u> check.

need to

I <u>need to</u> make sure that there are the right number of fire extinguishers.

… the torches – I <u>need to</u> make sure there are enough of them.

… the crew <u>need to</u> give instructions to the passengers.

be allowed to

Aircraft – or airlines – <u>are not allowed</u> to fly if they don't have all the right safety equipment.

… the passengers <u>are allowed to</u> board.

It's the one thing <u>I'm not allowed to</u> tell you.

B Manipulate language of obligation

- Point out that more than one answer may sometimes be possible.

- Set the task for individual work and pair-checking. Go round and monitor to check for any problems.

- Feedback orally.

Answers

1 has to

2 are not allowed to

3 need to/have to

4 not be allowed to

5 is not allowed to

6 need to

7 have to/need to

8 have to

Speaking

A Practise using language for obligation

- Set for pairwork. When they are practising in pairs, go round and monitor for language quality.

- Conduct brief feedback and correct any errors in the use of *have to*, *need to* and *be allowed to*.

- If there is time, ask students to give similar descriptions about the duties involved in another job that they are familiar with, using *have to/need to* and *be allowed* to.

- Now have the students do the Workbook exercises for Unit 7, Lesson 4.

Workbook answers

Language: countable and uncountable nouns

A Put the nouns into the correct column of the table

countable nouns	uncountable nouns
seat belt	equipment
hazard	damage
injury	first aid
fire extinguisher	oxygen
emergency	information
warning	advice
sign	protection
torch	training
life jacket	safety
passenger	

B Choose four nouns from each list and write sentences

Answers depend on the students.

Language: *in case*

A Underline the structures

in case there is smoke

in case all the lights go out

in case of an emergency

B Complete the sentences with *in case* or *in case of*

1 In case of

2 in case

3 in case

4 IN CASE OF

5 in case

Objectives

To review language to describe types of damage.

To practise reading for gist.

To write short damage reports using vocabulary from the text.

Language

Vocabulary associated with damage, e.g., *corrosion*, *wear*, *contamination*.

Introduction

- Lead in to the lesson by discussing belts in general and what happens to them when they wear out. Introduce vocabulary such as *fading*, *cracking*, *chafing*, *tearing*.

Reading

A Skim a text and choose a title

- Set the task for individual work and pairwork checking. Point out to the students that they should look at the text quickly.

- Feed back orally.

Answers

Maintenance of restraints is the most suitable title.

B Match types of damage with definitions

- Elicit or point out that the words in the left column are nouns (although *cut*, *tear* and *wear* can also be verbs) and draw attention to their endings.

- Set the task for individual work and pairwork checking. Point out that the students should use the text as well as their dictionaries to help them work out the meaning of unknown words.

- Feed back orally. Check understanding of new words such as *distortion*, *contamination*, *deterioration*. Go over the pronunciation.

Answers

1 g (damage with a sharp tool, e.g., a knife)
2 l (damage caused by pulling, e.g., paper, in two directions)
3 j (damage caused by rubbing against other objects)
4 d (the threads of the edge of the material are loose)
5 c (marks caused by other substances)
6 h (loss of colour over time)
7 e (bending, twisting, change of shape)
8 b (becoming longer)
9 k (long, narrow breaks or splits)
10 f (rust or oxidation)
11 i (general damage caused by wear over a long time)
12 a (general loss of shine, damage to surface)

Speaking and writing

A Discuss damage shown in the pictures

- Students should use the pictures as prompts and use the vocabulary from the previous exercise.

- Discuss the first picture as a group and the other pictures for pairwork. Monitor and deal with any problems.

Answers

Picture 1: shows fraying

Picture 2: shows distortion of end fittings

Picture 3: shows staining

Picture 4: shows breaks in stitching

B Identify belt that is described in the report

- Clarify any problem vocabulary, e.g., *fixings*, *discard*.

- Ask students to read the damage report carefully. Ask questions to check comprehension, e.g., *Are there any problems with the fixings? (No.) Are they going to repair the seat belt or throw it away? (Throw it away.)*

Answers

The damage report describes the belt in Picture *a*.

C Write damage reports to describe problem

- Set the task for individual or pairwork. If you prefer, ask different pairs to complete reports about different pictures. Tell students they can refer to the text for useful vocabulary.

- While they are writing, go round and monitor for language quality.

- Conduct brief feedback. Elicit students' ideas and write up a model on the board.

Answers

Answers depend on the students.

- Now have the students do the Workbook exercises for Unit 7, Lesson 5.

Workbook answers

Spelling and pronunciation

A Correct the words that are wrong

1	procedure	7	maintenance
2	harness	8	examine
3	correct	9	inspection
4	correct	10	correct
5	chafing	11	individual
6	correct	12	correct

B Mark the stress

1	pro<u>ce</u>dure	7	<u>main</u>tenance
2	<u>har</u>ness	8	ex<u>a</u>mine
3	con<u>ta</u>mination	9	in<u>spec</u>tion
4	<u>sa</u>ddle washer	10	in<u>tact</u>
5	<u>chaf</u>ing	11	indi<u>vi</u>dual
6)	re<u>straint</u>	12	dis<u>tor</u>tion

Language: ~*en*

A Make verbs from the adjectives

sharp – sharpen

long – lengthen

wide – widen

loose – loosen

short – shorten

strong – strengthen

weak – weaken

B Check the spelling in your dictionary

Answers depend on the students.

C Complete the sentences using the verbs

1	loosen	5	weaken
2	shorten	6	widen
3	strengthen	7	lengthen
4	sharpen		

Language: nouns and adjectives

A Make adjectives from the nouns

contamination – contaminated

fading – faded

distortion – distorted

elongation – elongated

chafing – chafed

fraying – frayed

corrosion – corroded

wear – worn

deterioration – deteriorated

cut – cut

cracking – cracked

tear – torn

break – broken

B Use the adjectives above to describe damage

Answers depend on the students.

Lesson 6   Fuel + heat + O_2 = fire

Objectives

To review language for fire safety.

To practise reading for detail and compare fire extinguishers.

To complete a table comparing types of fire extinguishers.

Language

Vocabulary associated with gases, e.g., *suffocation, environmentally friendly.*

Speaking

A Activate background knowledge: discuss a diagram showing how fires operate

- Before students look at the diagram, elicit the elements necessary for a fire: *an energy source (heat), fuel vapour and oxygen.*

- Set the task for pairwork. Explain that the diagram is known as a combustion triangle.

- For the second part of the discussion, encourage students to think of specific fires, e.g., *a burning candle, a bonfire, a house or forest fire, electrical fires*, etc. They should discuss how they are started and extinguished.

- Feed back orally, with reference to specific types of fires and extinguishers.

Answers

The fuel vapours and oxygen mix ⟶ the heat energy ignites the mixture ⟶ a reaction occurs and creates fire ⟶ sufficient energy must be produced by the flame to transfer to the supply of fuel and oxygen to maintain the fire.

B Discuss the three main elements in the diagram

Answers

If any of the elements is removed, the fire is extinguished.

If fuel is used up, the fire will go out, e.g., in a forest fire, trees can be cut down and removed from the path of the fire to stop it from spreading.

If heat is removed, a fire cannot start or continue, e.g., some types of fire can be doused with water, creating steam that dissipates the heat.

If oxygen is removed, the fire will quickly use up the available oxygen and go out, e.g., a fire can be smothered mechanically or with chemicals to remove the oxygen.

Reading and writing

A Read for gist: identify which gas halon is compared with

- Set the task for individual work and pairwork checking.

- Feed back orally. Elicit reasons why halon is used on aircraft.

- Highlight some key vocabulary in the text, e.g., *react with, disperse, displace, suffocate.*

Answer

Carbon dioxide (CO_2)

B Read and differentiate characteristics of halon and CO_2

- Ask students to read through the list of characteristics before they look at the text again.

- Ensure they understand the task and what to write in the gaps.

- When they have finished, ask them to compare their answers with a partner.

- Feed back orally.

Answers

1	neither	6	CO_2
2	CO_2	7	H
3	H	8	H
4	CO_2	9	CO_2
5	H	10	both

C Read for detail and complete the table

- This exercise requires students to read the text in some detail, so give them plenty of time to complete the table.

- Monitor their work and give help as required. Encourage pair-checking.

- Feed back orally onto the board or OHP.

Answers

	CO_2	halon
class of fire	B, C	A, B, C
how it works	displaces oxygen	chemical reaction with oxygen
open spaces	unsuitable	unsuitable
closed spaces	?	highly effective
storage pressure	high	low
weight of extinguisher	heavy	light < 1.2 CO_2
size of extinguisher	large	small < ½ CO_2
damage to humans	danger of suffocation	lower risk of suffocation
amount of gas	enough to replace all oxygen	much less than CO_2
environment	harmful	harmful (to ozone layer)
cost	?	more expensive than others

Pronunciation

A Find words that fit the stress pattern

- Direct attention to the first highlighted word in the text: *extinguishing*. Encourage students to say it with the correct stress. Ask them how many syllables it contains (4) and which one takes the main stress (the second).

- Elicit the stress patterns shown in the table and make sure students understand where *extinguishing* should go (middle column).

- Set the activity for individual work and pair-checking.

- Feed back, eliciting the words from each column and checking that students pronounce them correctly. Project the completed table onto an OHP or write it up on the board.

Answers

Ooo	oOoo	ooOo
atmosphere	extinguishing	interrupting
dangerous	capacity	suffocation
military	unsuitable	radiation

- Now have the students do the Workbook exercises for Unit 7, Lesson 6.

Workbook answers

Language: phrases with *which*

A Complete the sentences in your own words

Answers depend on the students.

Language: comparison/contrast

A Write sentences comparing O₂ and halon extinguishers

Answers depend on the students.

Writing: note-taking

A Find symbols for the words or expressions

1 therefore = ∴
2 approximately = ≈
3 less than = <
4 plus or minus = ±

Lesson 7   Safety first

Objectives

To practise reading an advertisement for main ideas.

To review vocabulary for fuel systems.

To practise summarising and writing bullet points.

Language

Vocabulary associated with fire danger and fire prevention on aircraft, e.g., *volatile*, *combustible*.

Reading and vocabulary

A Vocabulary prediction exercise

- Tell students the missing words in each phrase are high-frequency, known words.
- Ensure they understand that each dash represents one letter.

- Set the task for individual work and pairwork checking.
- Students should check answers and spelling of the words by skimming the text to find them.

Answers

1	oxy(gen)	6	crew
2	air	7	fire
3	maintenance	8	mixture
4	flow	9	tanks
5	fuel		

B Discuss unfamiliar vocabulary

- Unfamiliar words will probably be *ullage*, *vulnerability* and *sloshing*. They may be able to guess the meanings of all three from their contexts in the text.
- Clarify as necessary. Emphasise that these words are low-frequency and do not need to be learned.

Answers

No answers at this point.

C Read and match headings to paragraphs

- Check students understand the meaning of *adaptability* and *risk*.

- Set the task for individual work and pairwork checking.

Answers

1 Risk	**4** Features
2 Solution	**5** Specifications
3 Adaptability	

D Read and match to correct summary of main ideas

- Set the task for individual work and pairwork checking.

- Feed back orally.

- Ask students why they have made their choices and what is wrong with the other two summaries.

Answer

2 best summarises the information in the advertisement;

1 and 3 contain inaccuracies and irrelevant information.

Writing

A Summarise the information

- The aim is for students to write a brief summary of the first four sections, so discourage them from simply copying out sections.

- Refer them to the example and remind them that they should use note form. Elicit features that can be used, e.g., *abbreviations, symbols and omission of articles*, etc.

- Set the task for individual work and pairwork checking.

- While they are writing, go round and monitor for language quality.

- Feed back examples of notes from students and show the model answer on the board or on an OHP.

Model answer

1 Ullage vapours volatile esp. in helicopters as fuel in fuel tank decreases.

2 OBIGGS inerts system against threats – provides N enriched gas? eliminates fuel/air mix.

3 Designed to meet reqs. of specific aircraft.

4 Excellent operating and maintenance features.

5 2.2 lbs/min @ 40 psi; 28 V DC (25 W); 26 x 22 x 33 cm; 13.6 kg.

- Now have the students do the Workbook exercises for Unit 7, Lesson 7.

Workbook answers

Vocabulary and writing: measures

A Match the abbreviations to the words

mm = millimetres

lbs/min. = pounds per minute

psi = pounds per square inch

V = volts

W = watts

" = inches

cm = centimetres

lbs = pounds

kg = kilos

B Practise pronouncing them

Answers depend on the students.

C The meaning of abbreviations

Answers depend on the students.

Language: prepositions

A Complete the sentences with the correct preposition

1 up to and including	**5** into
2 to	**6** against
3 by	**7** for
4 in	**8** to

Writing

A Write comments in note form in the table

Answers depend on the students.

Lesson 8   Hidden problems

Objectives

To practise listening for detail.

To complete a summary using key vocabulary.

To ask and answer questions about a technical problem.

Language

Vocabulary associated with oxygen equipment, e.g., *hose*, *flexible*, *tough*.

Vocabulary

A Match words with near synonyms

- Look at the first set of words together with the class: *hose/tube/line*. Emphasise that they are similar but not identical in meaning. Elicit that *hose* and *line* are often used when we talk about transporting liquids and gases in systems. They are used particularly when talking about science and engineering. *Tube* is a more high-frequency word that is used to describe any hollow cylinder.

- Set the task for individual work and pairwork checking.

Answers

1 tube, line	**5** hard, rigid
2 inner, interior	**6** soft, pliable
3 outer, exterior	**7** stop, close off
4 crimped, folded	

B Discuss slight differences in meaning

- Feed back, focusing on words that have more specialised meanings, but without going into too much depth.

Answers

hose – tube, line

hoses are often soft and flexible; *lines* often have small diameters, *tubes* can be almost any material, both soft and hard

internal – interior, inner

interior and inner can only precede a noun and are more general, e.g., *interior design*; *inner layer*; *internal* often applies to science:

the human body, e.g., *internal organs*, and to the insides of machines, parts and so on

external – exterior, outer

as for *interior/internal/ inner*

bent – crimped, folded

crimped is more specific than *bent* or *folded*; *bent* can be used to describe many different things and degrees of curve; *fold* is used to describe any material with a sharp bend in it; but *crimped* refers specifically to a fold in a hose or tube

tough – hard, rigid

tough often refers to how strong a material is; *hard* refers to its solidity; *rigid* refers to how much it resists bending

flexible – pliable, soft

flexible is more abstract, can also be applied to people and broadly means *adaptable*; *pliable* means *able to bend or change shape easily without breaking* (opposite of *rigid*) *soft* is the opposite of *hard* and describes materials such as cloth and foam rubber

block – close off, stop

these are very similar in meaning: *block* can be caused by accident, while people generally *stop* and *close* something off deliberately, e.g., an oxygen supply

Listening

A Listen for gist and answer questions (CD2 Track 5)

- Use the picture to briefly review vocabulary, e.g., *high altitude, oxygen equipment, mask, tube*, etc.

- Ask students if they know how the pilot is provided with oxygen at high altitudes. Refer them to the picture on page 137. Clarify what the acronyms LOX and OBOGS stand for (*liquid oxygen* and *on board oxygen generating system*) and explain that the OBOGS system replaced the old LOX system. Explain that with OBOGS,

oxygen reaches the pilot through a *line* or *hose* that comes from a *concentrator*, i.e., machine that concentrates O_2 from the air.

- Give students time to read the three questions before you play the recording.

- Play the recording once.

- Pairs check answers. Feed back orally.

Answers

1 OBOGS

2 A maintenance problem: the mechanics were not handling the concentrator correctly.

3 A general problem: 9 out of 12 aircraft had the problem.

B Listen for specific information: identify correct sentences (CD2 Track 5)

- As before, allow reading time and clarify vocabulary.

- Replay the recording. Pairs check their answers.

- Feed back onto the board. Check comprehension with concept-check questions, e.g., *How were the LOX hoses different?*

Answers

No answers at this point.

C Check your answers

Answers

1, 4, 5, 8, 10

D Complete a summary paragraph with vocabulary from the box

- Students should be familiar with all the vocabulary in the box, except, possibly, *sleeve*. Set the task for individual work and pair-checking.

- Go round and monitor; give help if required.

- Feed back orally.

Suggested answers

The concentrator on the OBOGS supply was tested and had no defects. However, when the <u>hose</u> on one of the aircraft was inspected, the metal braided <u>sleeve</u> was broken, and the true condition of the <u>internal hose</u> could be easily seen. It was found to be crimped, which meant that the flow of oxygen was partly <u>blocked</u>. Several of the other aircraft were also found to have <u>bent</u> or <u>crimped</u> oxygen lines. The reason was that the mechanics who maintained the system thought that the OBOGS line was <u>flexible</u>, like the old LOX lines. In fact, OBOGS lines are <u>rigid</u>. To remove the concentrator, they bent the inner plastic <u>lines</u> forcefully out of the way. This action crimped the hose, although the <u>damage</u> was not seen because the <u>external</u> sleeve remained <u>intact</u>.

Speaking and writing

A Ask and answer questions about the problem on the recording

- Refer students to the diagram. Elicit a few example questions about the problem, e.g., *What happened to the pilots at high altitudes? What did the inspector do?*

- Ask students to write at least three questions about the problem individually.

- They should ask and answer the questions with a partner.

- Feed back orally.

Answers

Answers depend on the students.

- For further information on OBOGS, refer students to:

  http://www.airliquidadvancedtechnologies. com/en/our-offer/aeronautics/on-board-oxygen-generating-systems.html

- Now have the students do the Workbook exercises for Unit 7, Lesson 8.

Workbook answers

Pronunciation: syllables

A Group the words according to the number of syllables they have

One syllable	Three syllables
flow	forcefully
sleeve	concentrate
crimp	modify
hose	easily
	external
Two syllables	protective
damage	consulted
problem	oxygen
inspect	carefully
inner	maintenance
replace	nitrogen
system	generate
defect	separate

B Mark the stressed syllable

Note: Words with only one syllable do not have a stressed syllable.

Two syllables	modify
damage	**ea**sily
problem	ex**ter**nal
in**spect**	pro**tec**tive
inner	con**sul**ted
re**place**	**ox**ygen
system	**care**fully
defect	**main**tenance
	nitrogen
Three syllables	**gen**erate
forcefully	**sep**arate
concentrate	

Language: past tenses

A Complete the sentences with the past tense verb/passive voice

1 The old system did not give a warning when there was low oxygen in the line.

2 Pilots experienced a lot of problems with the old LOX system.

3 All the other aircraft have been inspected.

4 The manuals were consulted and the equipment was tested.

5 The inner hose could be seen easily.

6 He discovered that the maintenance technicians' technique was not correct.

7 The oxygen system changed before he joined the trip.

8 After this, technicians became much more careful.

Writing: incident report

A Write a brief report in note form

Answers depend on the students.

Lesson 9   Split-second action

Objectives

To practise listening for detailed information and identify the sequence of events in a text.

To practise explaining the purpose of different stages/events in a sequence.

Language

Expressions of purpose, e.g., *x enables y to do … x provides y with …*

Vocabulary

A Introduce the topic

- Ask students to work in pairs and discuss what they can see in the illustrations and how they are different from each other.

- Feed back orally and write key vocabulary on the board, e.g., *ejection seat*.

Answers

Answers depend on the students.

B Label the diagrams

- Set the task for individual work and pairwork checking.

- Feed back orally.

- Focus on the word stress in pa̲rachute, re̲straints and se̲parate, as well as the pronunciation of *drogue*.

Answers

Answers depend on the students.

Reading

A Read to identify the sequence of events

- Ask students to read through the notes.

- Clarify any vocabulary problems, e.g., *limb restraint cords*.

- Ask students to predict the correct order of events and compare with a partner.

- Elicit but do not confirm ideas yet.

- Students should read the text and check if their predictions were correct.

- Feed back orally, clarifying any vocabulary difficulties.

Answers

7, 2, 3, 1, 6, 5, 4

B Scan the text for numbers and quantities

- Set the task for individual work and pairwork checking.

- Feed back orally.

- Ask students to read out each item in its full form, e.g., *zero point five seconds*, and correct any pronunciation problems.

Answers

160 m/h	speed (160 miles per hour); speed that the ejection seat is thrust from the plane
30 G	gravity (1 G is equal to the force of gravity); the force that an average man has to withstand
0 m/h	speed (zero miles per hour); speed of seat before ejection
0.5 sec.	time (zero point five seconds); time delay before the drogue parachute fires
65 kg	weight (65 kilos/kilograms); weight of an average man
0.25 sec.	time (zero point two five seconds); time it takes for the ejection seat to accelerate to 160 m/h
2,000 kg	weight (two thousand kilograms); the weight that a 65-kg man flying through the air in an ejection seat equals

Vocabulary and speaking

A Focus on expressions of purpose

- Students have to scan the text to locate the verbs for expressing purpose.

- Set the task for individual work and pairwork checking.

- Feed back on the board.

Answers

provide (lines 5–6)
... the ejection gun **provides** the initial thrust for seat ejection ...

provide (line 11)
A delay mechanism **provides** a time delay ...

sustain (line 13)
The thrust of the ejection is **sustained** by a rocket motor ...

ensure (line 12)
... this **ensures** that the pilot and navigator do not collide during ejection ...

enable (line 14)
... to **enable** the parachute to deploy ...

B Ask and answer questions about the ejection system

- Ask students to prepare two or three questions individually about the different elements in the ejection seat process.

- Then they work in pairs and ask and answer the questions.

- Feed back orally.

Answers

Answers depend on the students.

- Refer students to more information about ejection seats on the Internet. One commercial site is:

  http://www.martin-baker.com/

- Now have the students do the Workbook exercises for Unit 7, Lesson 9.

Workbook answers

Language: prepositions

A Put the word in the correct place

1 The seat firing handle is connected directly **to** a breech unit.

2 The ejection gun unit is in turn attached **to** the aircraft structure.

3 The ejection seat is mounted **on** the guide rails and ejection gun unit.

4 The thrust is provided **by** means of three cartridges.

5 2,000 kg weight is equivalent **to** 30 g.

6 It is only made possible **by** flying with special safety clothing.

7 Ejection might sometimes be initiated **at** zero speed at ground level.

8 Crews have to release themselves **from** the parachute harness.

Language: noun forms

A Complete the table

verb	noun
deploy	deploy
collide	collision
propel	propeller
restrain	restraint
activate	activation
reinforce	reinforcement
stabilise	stabilisation

B Fill the gaps with nouns from the table

1 restraints

2 collision

3 activation

C Use the words to write sentences of your own

Answers depend on the students.

Lesson 10   Plan B

Objectives

To practise speaking: discuss ideas about back-up systems.

To practise working out the meaning of vocabulary connected with back-up systems.

Language

Vocabulary related to back-up systems, e.g., *accumulator, oil head*.

Language for obligation: *should, need to, have to*.

Vocabulary and speaking

A Discuss and compare systems

- Write the phrase *power failure* on the board.
- Check students understand all the expressions in the box.
- Ask them to work with a partner and discuss which of the systems are important and why.
- Feed back orally, asking the students to give reasons for choosing certain items and reasons for not choosing others.
- **Note:** There are no single correct answers, but the most important system in the list is the *back-up system*. This will enable some of the other systems, e.g., landing gear, to work during an emergency power failure.

Answers

Answers depend on the students.

B Prioritising systems

- Ask students to choose the two most important items in the list and then to agree with their partner on which two are the most important.

- Feed back orally and see if the whole class agrees. Listen to the reasons for their choice.

Answers

Answers depend on the students.

C Find synonyms for words in the text

- Ask students to skim the text to confirm that back-up systems are important for survival.

- Set the vocabulary task for individual work and pairwork checking.

- Feed back orally.

Answers

back-up – secondary

take over – act as replacements

goes down – fails

normally – usually

essential functions – basic jobs

enables – makes it possible for

emergency – unplanned

D Check to see if your ideas were correct

- Ask students to work with a partner to discuss whether their ideas in Exercise 1 were mentioned.

- Feed back orally.

Answers

Answers depend on the students.

Reading

A Scan the text and find words and phrases in the box

- Tell students that they do not need to know the meaning of the words in the box. They should find them first and then try to work out meaning from context and by guessing the meaning of new phrases from the words that they already know.

- Set the task for individual work and dictionary/Glossary checking.

- Feed back orally, checking understanding of *accumulator* and *transducer*.

- Check students can pronounce these words and phrases correctly.

Answers

accumulator	line 2
oil head	line 2
charging value	line 4
hydraulic lines to the landing gear and flaps	lines 8–9
pipeline from the system pressure line	lines 10–11

B Label the diagram

- Set the task for individual work and pairwork checking.

- Feed back on the board.

Answers

1 pipeline from the system pressure line

2 oil head

3 hydraulic lines to the landing gear and flaps

4 accumulator

5 charging valve

Speaking

A Discuss possibilities for back-up systems

- Students discuss the different possibilities in pairs.
- Feed back orally.
- Ask students to give reasons for their choice.

Suggested answers

air brakes – a manually operated system

radio – battery-operated system

high-altitude oxygen – back-up emergency system

rudder and ailerons – a second identical system

- Now have the students do the Workbook exercises for Unit 7, Lesson 10.

Workbook answers

Language: two-part verbs

A Nouns that go with the two-part verbs

1 find out

2 put out

3 take off

4 switch on

5 carry out

6 take over

7 fall out

8 come into contact with

9 open out

B Write sentences using the noun/verb combinations

Answers depend on the students.

C The difference between the pairs of verbs

Answers depend on the students.

D Mark the stress

1 pro<u>vi</u>de/prot<u>ect</u>

2 <u>trans</u>port/trans<u>mit</u>

3 de<u>ter</u>mine/dis<u>cov</u>er

4 at<u>tach</u>/con<u>nect</u>

5 rep<u>ort</u>/as<u>sess</u>

Language

A Match the sentence halves

Foam extinguishers work – by forming a layer on the material.

The crew activate the ejection system – by pulling a handle.

The equipment inerts the system – by providing a flow of oxygen.

The OBIGGS system works – by concentrating oxygen from the air.

Halon works – by reacting with the oxygen in the vicinity of the fire.

B Complete the sentences using *by*

Answers depend on the students.

Reading: different text types and styles

A Where each text comes from

1 encyclopedia

2 aircraft manual

3 accident report

B Read the text and answer

1	1	4	3
2	3	5	2
3	2	6	1

C Discuss differences in style or meaning

Answers depend on the students.

Unit test

A test for this unit is available at:

http://www.garneteducation.com/reps/
documents/1251/SDT-u7-test.pdf

Contact your local Garnet Education
representative for information about how to
access these resources.

UNIT 8 Lesson 1 — Looking for damage

Objectives

To build vocabulary connected with tyres.

To practise listening for details and completing a form.

To focus on pronoun references in a text.

Language

Use of *some* and *any*, *something/anything*.

Pronoun: *it*.

Speaking

A Discuss features of tyres

- Write the word *tyre* on the board and elicit associated vocabulary from students, e.g., *pressure*, *inflate*, *deflate*.
- Set discussion activity for pairwork.
- Feed back as a whole class. Elicit vocabulary such as *bump*, *skid*, *blow out*, *burst*.

Suggested answers

1 If the tyre is too hard, it could burst or blow out on impact with the ground.

2 If the tyre is too soft, it results in a bumpy ride.

3 It is important to check for tyre damage to avoid steering problems and skidding (caused by worn tyres). Tyre checks also help to avoid burst tyres caused by cracking, bulging or something puncturing the tyre.

Vocabulary

A Categorise words connected with tyres

- Write the first word: *tread*, on the board and elicit which column it should go in (parts of a tyre). Use the diagrams to elicit where it is on the tyre.
- Set the task for individual work and pairwork checking.
- Go round and monitor to check for any problem words.
- Feed back as a whole class. Use the diagram to clarify tyre parts.

Answers

parts of a tyre	problems with tyres
tread	misalignment
bead	burst
sidewalls	damage
layers	vibration
	cracking
	wear
	puncture

B Mark the stress

- Set for pairwork and feed back as a whole class.
- Model the pronunciation of the vocabulary items and ask students to repeat.

Answers

misa<u>lign</u>ment

<u>side</u>walls

<u>da</u>mage

vib<u>ra</u>tion

<u>cra</u>cking

<u>la</u>yers

Listening

A 🔘 Listen and complete the job sheet (CD2 Track 6)

- Ask students to read the job sheet carefully and help with any unfamiliar vocabulary.

- Play the recording once and monitor as students complete the job sheet.

- If they find the task difficult, encourage pair-checking and/or play the recording a second time before checking the answers.

- Feed back as a whole class.

Answers

Tread depth measured – Yes

Tyre deflated – No

Foreign body removed – No

Wheel replaced – No

Wheel flipped – No

Beads checked – Yes

Sidewalls inspected – Yes

B 🔘 Listen and complete the remarks section (CD2 Track 6)

- Encourage students to write notes, not full sentences, and ask them to focus on possible further work that may need to be carried out on the tyre.

- Pause the recording from time to time to allow students time to write phrases and/or sentences. Alternatively, with a very small group, the students could choose when to stop the recording by saying *stop* whenever necessary or selecting one student to operate the player.

- Be prepared to play the recording two or three times, allowing time for pair-checking after each hearing. Feed back by writing their ideas on the board and then correcting.

- Focus on key terms such as *bulge*, *tread gauge* and *groove*.

Possible answers

- treads are more worn on one side so the tyre may need flipping

- nose wheel gear may need checking for misalignment

- workshop will have to inspect the tyre as there is a bulge

C 🔘 Pronoun reference: use of *it* in the text (CD2 Track 6)

- Set the task for individual work and pairwork checking.

- Play the recording again so students can check their answers.

- You may wish to give out the tapescript at this point and look at the use of pronoun references.

- Feed back with the whole class.

Answers

1	the tread gauge	**4**	the tyre
2	the tyre	**5**	the tyre
3	the wheel		

Language

A Complete the sentences

- Ask students to read the information in the Language Box.

- Check they understand this particular difference between the use of *some* and *any*.

- Set the task for individual work and pairwork checking.

- Give out the tapescript (if you haven't already done so) so that students can check their answers.

Answers

1	anything	**3**	something
2	anything; any	**4**	some

- Now have the students do the Workbook exercises for Unit 8, Lesson 1.

Workbook answers

Language: quantity

A Find the expressions in the text

plenty of
> Speaker A's fourth turn (There's **plenty of** wear left on these treads.)

no other
> Speaker A's fifth turn (If there's **no other** problem, we can just flip it.)

a lot of
> Speaker A's second turn (That can sometimes do **a lot of** damage when the plane comes down quite hard.)

a bit more … than
> Speaker B's fourth turn (Oh yes, the treads look **a bit more** worn on the right-hand side **than** on the left.)

B Use the expressions to complete the sentences

1 a lot of
2 a bit more … than
3 no other
4 a bit more … than
5 plenty of
6 no other
7 a lot of
8 a lot of

Spelling: AmE/BrE

A Complete the tables

1

AmE spelling	BrE spelling
traveled	travelled
tire	tyre
vapor	vapour
meter	metre
aluminum	aluminium
gage	gauge
maneuver	manoeuvre

2

AmE word	BrE word
flashlight	torch
wrench	spanner

Language: *some* and *any*

A Complete the sentences

1 somewhere
2 Something
3 anybody
4 somebody
5 anywhere
6 something
7 anything
8 something

Assemblies and systems

Objectives

To practise reading and evaluating explanations of a system.

To discuss and interpret diagrams of a landing gear assembly.

Language

Vocabulary related to landing gear assembly, e.g., *strut*, *slip ring*, *barrel*.

Expressions containing *of*, e.g., *the demands of*.

Vocabulary

A Discuss ideas about the function of a landing gear assembly

- Check students understand the four vocabulary items.

- Set for brief pairwork discussion.

- Elicit ideas, but do not explain the function of the landing gear if students do not have any idea how it works.

Answers

Answers depend on the students.

B Discuss diagrams showing operation of the landing gear assembly

- Elicit what type of drawings these are and what they show.

- Encourage students to read the labels carefully and check the meaning of new words in their dictionaries or in the Glossary.

- They should speculate about how the assembly operates, but again, do not clarify too much. They will read a short explanation in Exercise C.

Answers

Answers depend on the students.

C Match sentences halves to create an explanation of the landing gear assembly

- Set the task for individual work and pairwork checking.

- Feed back with the whole class. Show the sentences on an OHP or write them on the board, one by one. Check comprehension of key words, such as *cylinder barrel*, *actuator*, *reciprocate*.

Answers

1	d	**4**	b
2	e	**5**	a
3	c		

Reading

A Evaluate two texts to decide which one is clearer

- Check students understand the task. Elicit differences which are immediately apparent to students, e.g., the layout: use of bullet points and underlining in Text 2.

- Set the task for individual work and pairwork checking.

- Feed back with the whole class. There is no set answer: some students may find the first text easier to follow, because it contains less detail and fewer technical words and phrases; others may prefer Text 2, because of the clearer layout.

Answers

Text 1 is written as continuous prose, and it could be argued that it focuses more on the operation of the complete assembly.

Text 2 uses bullet points. Each point describes a feature or individual component of the assembly.

B Focus on differences in the vocabulary used in each text

- Do an example together with the class. Then set the task for individual work and pairwork checking.

- Feed back with the whole class.

Answers

drawing - illustration

impact - shock

under pressure - compressed

annular space - the space around the piston

barrel - cylinder

loads - demands

uneven - non-level

reciprocation - operation

return - recoil

strut - leg

Vocabulary

A Focus on how expressions with *of* are used in the text

- Set the task for individual work and pairwork checking.

- Feed back with the whole class. Establish that the phrases are all noun phrases. Point out that these are common combinations.

Answers

consisting of

a cylinder of

the demands of

the operation of

a combination of

the compression and extension phases of

the return of

B Complete sentences with expressions using *of*

- Check understanding of *in case of*.
- Set the task for individual work.
- Feed back with the whole class.

Answers

1 in case of

2 a combination of

3 instead of

4 the demands of

- Now have the students do the Workbook exercises for Unit 8, Lesson 2.

Workbook answers

Language: prefixes and suffixes

A The meaning of 'non' words

1 a substance that isn't metallic

2 an inspection that isn't destructive

3 a valve that doesn't have a return

4 something that isn't technical

5 something that isn't essential

B Write sentences

Answers depend on the students.

C Make another word from the words

1 assemble – assembly

2 press – pressure

3 combine – combination

4 replace – replacement

5 vibrate – vibration

D Underline the correct forms

1	displacement	5	taxiing
2	restriction	6	extension
3	container	7	damping
4	compression	8	disassemble

E Choose words and use them in sentences

Answers depend on the students.

Language: prepositions

A Delete the underlined words

Pupils should delete:

1	in	5	into
2	between	6	to
3	up/on	7	into
4	at		

Writing: explaining problems and solutions

A Write answers to the questions

Answers depend on the students.

Objectives

To practise intensive listening for very specific information.

To focus on the language of measurement and calculation (Charles' Law).

Language

Vocabulary for different units of measurement, e.g., *cubic in.*, *Pa*.

Vocabulary – key verbs to indicate increase and decrease.

Vocabulary

A   Match verbs and definitions

- Introduce the topic by asking students to look at the picture of the balloon. Elicit what the balloon can be filled with and how gases differ from liquids.

- Tell them to read the sentence at the beginning of the unit.

- Set the task for individual work and pairwork checking.

- Feed back with the whole class.

Answers

1   d

2   b   (**Note:** We can also say *go up*)

3   e

4   c

5   a

B   Further practice of vocabulary: verb/noun collocation

1

- Set the first task for individual work and pairwork checking.

- Feed back with the whole class. Elicit ideas for other things that can expand and contract.

Answers

increase and decrease
temperature, volume, pressure, cost, height

expand and contract
a balloon, a tyre, halon, wood

2

- Students should be familiar with most of the measurements from previous units. Elicit the answers to Exercise *b* from the whole class.

Answers

psi	pounds per square inch
Pa	pascals
K	degrees Kelvin
C	degrees Celsius
F	degrees Fahrenheit
cm³	cubic centimetres
cubic in.	cubic inches
m³	cubic metres

3

- Ask students to work with a partner and say each of the units aloud to focus on the pronunciation.

- Check the pronunciation with the whole class, making sure the words are stressed correctly, e.g., Celsius, Fahrenheit, Kelvin.

Answers

See answers in Exercise B2.

Speaking and listening

A Speaking: discuss the picture

- Set for pairwork and feed back ideas to the whole class. Do not confirm yet, as students will hear more details in the listening.

Answers

Answers depend on the students.

B Listen and identify specific information about parts of a balloon (CD2 Track 7)

- Ask students to read the complete list of points for each section.

- Tell them that items are not necessarily mentioned in the order in which they appear on the page.

- Play the recording once and check progress.

- If necessary, play the recording again.

- Feed back with the whole class.

Answers

the burner	it burns propane
the envelope	it is filled with hot air; it measures about 10 x 20 m; it is blue
the flap	it is at the top of the balloon
the basket	it can carry four passengers; it is tied to the ground

C Read about Charles' Law and guess the missing information

- See whether students know or remember the information about Charles' Law from the recording.

- Elicit suggestions for the first gap and then set for pairwork discussion.

- Feed back by eliciting ideas, but do not confirm yet.

Answers

No answers at this point.

D Listen and complete the missing information (CD2 Track 7)

- Play the recording so students can check their answers.

- **Note:** The paragraphs are a summary of the explanation on the recording, so students do not need to write the exact words (particularly for the final three gaps).

- Feed back with the whole class.

Answers

1 volume; proportional; temperature

2 hot; cold; dense; rises

3 cools (down); contracts; sinks; open the flap; escapes; is speeded up

Reading and speaking

A Read and discuss the formula for the Combined Gas Law

- Explain the Combined Gas Law to the students as follows: The Combined Gas Law is a combination of Charles' Law and Boyle's Law. Boyle's Law states that if you decrease the volume of a specific quantity of gas, the pressure increases. Thus, the pressure of a gas is inversely proportional to its volume.

- Direct attention to the formula. Elicit or explain that it enables you to work out the temperature of a gas in relation to its pressure and volume.

- Set the exercise for pairwork discussion and feed back to the class.

Answers

1 P = pressure, V = volume and T = temperature

2 The initial pressure times the initial volume divided by the initial temperature equals the subsequent pressure times the subsequent volume divided by the subsequent temperature.

- Now have the students do the Workbook exercises for Unit 8, Lesson 3.

Workbook answers

Language: opposites

A Write the opposite

1 contract – expand
2 cool – heat
3 rise – lower
4 push – pull
5 increase – decrease
6 outside – inside
7 upwards – downwards

B Write sentences

Answers depend on the students.

Reading: numbers and abbreviations

A Check the meaning of the abbreviations

–56°C	fifty-six degrees Centigrade/Celsius
CO_2	carbon dioxide
7.5 psi	pounds per square inch
21.3.99	twenty-first of March, 1999
N	nitrogen
43,000 ft	forty-three thousand feet
2,000 kg	two thousand kilos (kilograms)
55 m	fifty-five metres
12,000 m	twelve thousand metres
21 hrs	twenty-one hours

B Complete the blanks

12,000 m
21 hrs
21.3.99
55 m
2,000 kg
43,000 ft
–56°C
7.5 psi
CO_2
N

Compressed gas

Objectives

To review language related to pump and compressor systems.

To practise reading detailed instructions from a manual.

To focus on sequencing language.

Language

Vocabulary – technical verb + noun collocations, e.g., *prime a pump*.

Adverbials of time, e.g., *before, after, once, while*.

Vocabulary and speaking

A Complete the phrases

- Elicit sentences containing the word *arm*. Establish that it can be used as a verb as well as a noun. Elicit nouns that it can be used with, e.g., *arm a soldier, arm an alarm system*.

- Draw attention to the other words in the box and elicit that they are all verbs used for instructions in manuals and technical literature.

- Set the task for individual completion and pair-checking.

- Feed back with the whole class. Check understanding of any unfamiliar verbs, e.g., *prime, cock* and *purge*.

Answers

1	return	7	fill
2	use	8	correct
3	check	9	cock
4	consult	10	set
5	leave	11	arm
6	purge	12	prime

Reading

A Discuss uses for compressed gas

- Exploit the picture. See if students can explain what a compressor does (from Unit 6). Get them to check with the Glossary.

- Elicit uses of compressed gas.

Possible answer

Compressed gas is used in refrigeration, heating and air conditioning systems as well as engines and hydraulic systems.

B Reading for details

- Give students plenty of time to read the extract from the maintenance manual. Clarify that they do not need to understand all the instructions or vocabulary, e.g., *shutdown tattle tales*. They should look for parts that are shown on the diagram.

- Feedback with the whole class.

Answers

The oil pump, vent and bypass valves are shown on the diagram.

C Read for specific information

- Ask students to read the questions carefully and check that they understand.

- When they have finished, ask them to compare their answers with a partner.

- Feed back with the whole class.

- Highlight the use of *by + ~ing* to answer the question *How?*

- Deal with any vocabulary problems as necessary.

Answers

1 by adding oil; by pushing the lever in the direction of gas flow; by opening the discharge block valve while slowly closing the bypass valve

2 to ensure all the air has been removed from the system; to be sure they are armed; to be sure it is correct

3 it shows the oil requirements; it is the required water temperature; it is the time required for the start run

Language and speaking

A Find examples of time expressions

- Ask students to read the Language Box.

- Use the examples given to clarify the difference between *after* and *once* (the latter means *as soon as*). Highlight the use of present perfect after *after* and present tense after *once*.

- Then ask them to find examples in the text (there are six in total).

- Feedback orally.

Answers

after	'After Starting'; After the engine has warmed up …
once	Once the unit is running …; Once the bypass valve is closed …
while	… begin loading the unit by opening the discharge block valve while slowly closing the bypass valve.
finally	Finally, close the vent valve …

B Practise asking and answering questions using time expressions

- Draw attention to the example in the Course Book and ask another question to the whole class, e.g., *When do you return the timer to the zero position? (Once the unit is running.)*

- Listen to some examples with the whole class and correct any errors.

Answers

Answers depend on the students.

- Now have the students do the Workbook exercises for Unit 8, Lesson 4.

Workbook answers

Language: sequence

A Write a sentence to explain the sequence of operations

Answers depend on the students.

Language: directions

A Complete the sentences using expressions of direction

1 through

2 to and from

3 in the direction of

4 from

5 to

6 into

7 upwards / downwards

8 anticlockwise

Writing: instructions

A Write brief notes using imperatives

Answers depend on the students.

Objectives

To review verbs that describe a change of physical state.

To practise intensive listening: following an explanation of how a refrigeration system works.

To raise awareness of and practise pausing at the end of sense units.

Language

Vocabulary associated with refrigeration, e.g., *condenser*, *refrigerant*, *vapourise*.

Vocabulary and speaking

A Check understanding of verbs connected with a change in physical state verbs

- **Note:** It might be an idea to point out that there is very little difference between the words *gas* and *vapour*. *Gas* is commonly used to describe a substance that appears in its gaseous state under standard conditions of pressure and temperature, and *vapour* to describe the gaseous state of a substance that normally appears as a liquid.

- Use one or two examples from the list of verbs in the box to illustrate a change of physical state, e.g., *evaporate*: liquid ⟶ gas, e.g., when heat is absorbed; *liquefy*: solid ⟶ liquid, e.g., when heated.

- Set the task for pairwork and feed back as a whole class.

Answers

Answers depend on the students.

B Pronunciation: identify stressed syllables

- Do one example with the whole class, e.g., *eva̲porate*.

- Ask students to work with a partner, say the words aloud and mark the stressed syllables.

- Check the answers with the whole class.

- Drill the pronunciation of the more difficult words, e.g., *evaporate*, *vapourise*, *liquefy*.

Answers

e̲vaporate	ab̲sorb
li̲quefy	cool d̲own
so̲lidify	heat ṵp
va̲pourise	exp̲and
con̲dense	con̲tract
give off h̲eat	

Listening and speaking

A Prediction: decide whether statements are true or false

- Focus attention on the diagram. Ask questions, e.g., *What are the blue and red lines? What is the difference between the blue and the red sections? What do the arrows signify? What are V1 and V2?*

- Set the task for pairwork. Encourage students to guess answers they are unsure of.

- Feed back as a whole class, but do not correct their predictions at this stage as they will hear the correct answers on the recording.

Answers

No answers at this point.

B 🔊 Listen to check ideas (CD2 Track 8)

- Ask students to read through the list of statements again.

- Play the recording through once.

- Check their progress and play the recording again if necessary.

Answers

No answers at this point.

C Discuss answers

- When they have discussed their answers with a partner, feed back with the whole class.

- If any of the answers are still unclear, play the recording again and stop at the part of the recording in question.

Answers

1	T	5	F	8	T
2	T	6	F	9	F
3	T	7	T	10	T
4	T				

Pronunciation

A Focus on sense units

- Ask students to read through the examples in the Skills Box.

- Point out that beginnings and ends of sense units are often marked by words like *and* or *but* and by commas and full stops in written English.

B Read and mark pauses in a text

- Ask students to read the extract and mark where they think each sense unit ends.

- Listen to their suggestions, but don't correct them at this stage.

Answers

No answers at this point.

C ☉Listen and identify pauses (CD2 Track 9)

- Play the recording or read the section aloud so that they can check and amend their answers.

Suggested answer

Well OK/basically a heat pump uses the fact that when a liquid evaporates /it needs to absorb heat from its surroundings/and when a vapour condenses/it gives off heat/The best example is an ordinary domestic refrigerator/ The inside is cold/but the grill at the back is warm/The heat is removed from inside the refrigerator/and then given off into the room/ which is why you can't cool down a room by leaving the refrigerator door open/Because when you leave the door open/the refrigerator works hard to try to cool the room/and at the same time it's giving off the same heat back into the room/heating the place up.

D Practise reading the text aloud with appropriate pauses

- Ask them to work with a partner and take it in turns to read the text aloud, pausing appropriately.

- Listen to one or two examples with the whole class.

Answers

Answers depend on the students.

Speaking

A Practise asking and answering questions about how the system works

- When students have practised asking and answering questions about the diagram in pairs, listen to a few examples with the whole class and correct any errors.

Answers

Answers depend on the students.

- Now have the students do the Workbook exercises for Unit 8, Lesson 5.

Workbook answers

Language and writing: *too* and *enough*

A Think about the meaning of the words

1 enough; too
2 enough; too
3 too; enough

B Tick the sentences that are correct

1 correct
2 correct
3 incorrect
4 correct
5 incorrect
6 incorrect
7 incorrect
8 correct

C Underline the useful expressions with *too* and *enough*

more than enough time

it's not too late

far too much

not nearly enough space

D Write sentences of your own

Answers depend on the students.

Language: prepositions

A Complete the sentences

1 into
2 from
3 down
4 together
5 up; from; to
6 inwards
7 off
8 from/to

Reading: clarification of terminology

A Decide whether the statements below are true or false

1 False
2 True
3 True
4 True
5 False
6 False
7 False
8 True

On cloud nine

Objectives

To focus on language and concepts connected with changes in temperature and humidity, e.g., *relative humidity, water vapour.*

To practise detailed reading to understand definitions.

Language

Word-building: verbs and related noun forms, e.g., *contain, content.*

Vocabulary related to measurement of humidity, e.g., *hygrometer.*

Introduction

- Discuss weather conditions in different parts of your country. Introduce key vocabulary such as *humidity*, *evaporation*, *moisture* and *condensation*.

Vocabulary and speaking

A Revision of vocabulary and spelling

- See if students can remember some of the verbs associated with change in physical state from the previous lesson. Set the task for individual work and pairwork checking.

- Feed back with the whole class.

Answers

evaporate	expand
condense	contract
absorb	temperature

B Complete the table

- Set the task for individual work and pairwork checking.

- Feed back with the whole class using an OHP or the board. Point out unusual spelling patterns, e.g., *absorb* but *absorption*.

Answers

verb	noun
condense	*condensation*
absorb	absorption
humidify	*humidity*
weigh	weight
evaporate	*evaporation*
contain	content
measure	*measurement*
read	reading
moisten	*moisture*

C Check comprehension of key vocabulary

- Set for pairwork discussion and monitor to check for any problem words.

- Go over any problem words with students. They will need to understand them in the reading tasks that follow.

- Focus on the pronunciation of some of these words, especially *absorption* and *moisture*.

Answers

Answers depend on the students.

Reading

A Read for gist and match titles to paragraphs

- Ask students to look carefully at the five paragraph headings first.

- Point out/elicit that there are five headings but only four paragraphs, and tell students to skim the text and match the headings to each paragraph.

- Check the answers with a partner.

- Feed back with the whole class.

Answers

Paragraph 1 – Absolute humidity

Paragraph 2 – Relative humidity

Paragraph 3 – Measuring humidity

Paragraph 4 – Atmospheric humidity

- **Note:** The heading *Controlling humidity* is not used.

B Read for detail: choose correct definitions for terms in the text

- Set the task for individual work and pairwork checking.

- Feed back with the whole class.

- If you have time, focus on some of the vocabulary in the text. Words that could be highlighted include *moisture, vapour* and *saturate*.

- Note: *Moisture* generally refers to the presence of water, often in trace amounts; *vapour* is the gaseous form of a substance that is generally liquid, e.g., water. When vapour is saturated with moisture, it becomes liquid.

- Apart from the meaning, focus on the stress of these words and others in the text, particularly the common stress pattern in *hygrometer* and *thermometer*.

Answers

1 a device for measuring relative humidity using a comparison

2 the water content of a volume of air at saturation point

3 the combination of water vapour and air in the atmosphere

4 the mass of water in a volume of unsaturated air compared with its absolute humidity

- If you have time, ask a few questions about the diagrams, e.g., *What is the absolute humidity of air at 10 degrees C according to Figure 1?* (9.4 g/m³.)

- Now have the students do the Workbook exercises for Unit 8, Lesson 6.

Workbook answers

Language: temperature and humidity

A Complete the paragraph with vocabulary from the lesson

absorb	saturation
atmosphere	absolute
vapour	humidity
condenses	temperature

Writing: expanding sentences

A Expand the sentences

Suggested answers

1 The mass of vapour that a volume of air contains is dependent on the temperature of the air.

2 The difference between the two thermometer readings is equal to (or *equals*) the relative humidity.

3 There is a maximum amount of moisture the air can hold.

4 Condensation found on car windows is due to the atmospheric humidity being high.

5 It is not practicable to weigh air, so a device called a hygrometer is used.

6 The higher the temperature, the more moisture the air can hold.

7 Water vapour and air intermix in the atmosphere.

8 A specific volume of air that contains 2.5 g per m³ of water vapour at 10 degrees C is fairly dry.

Language: parts of speech

A Rewrite the sentences using the words given

1 Desert regions have no humidity.

2 Saturation is the point at which the air cannot absorb any more vapour.

3 As long as it is hot, humidity will not fall.

4 There is a difference between absolute humidity and relative humidity.

5 The temperature falls at the point of evaporation.

Objectives

To review vocabulary connected with heating and cooling systems.

To practise reading and transferring information to a schematic.

To extend language for describing reason and purpose.

Language

Vocabulary connected with air conditioning systems, e.g., *thermostat fan, reheater*.

Expressions for describing reason and purpose, e.g., *so that, in order to*.

Introduction

- Elicit whether students have air conditioning in their homes and whether they have had any problems with it. Highlight the stress pattern – *air conditioning*. Point out that it is often known as *AC*.

Vocabulary and speaking

A Match words and definitions

- Set the task for individual work and pairwork checking.

- Feed back orally, checking the meaning of the vocabulary by eliciting other equipment or systems that have the same parts, e.g., *filter*: an oil pump also has a filter.

Answers

1 f	**5** g
2 a	**6** b
3 d	**7** e
4 c	

B Look at air conditioning

- Activate schemata: discuss what students already know about heating/cooling systems

- Ask students to work with a partner and discuss these questions. If there is no AC system, they could look at a fan heater or look at the diagram on page 155 (Figure 2).

- Feed back with the whole class.

Answers

Answers depend on the students.

Reading

A Interpret the schematic

- As students look at the schematic, ask them to read the labels and look at the direction of the air flow. Establish that it moves from left to right and is mixed and then cooled in the AC system.

- Elicit how students think the numbered boxes should be labelled, e.g., *fan, filter, reheater*, but do not confirm yet.

Answers

No answers at this point.

B Read the text and finish labelling the schematic

- Give enough time for students to read carefully and label the boxes before checking answers in pairs. Monitor and note any problems.

- Feed back, building up the schematic on the board, eliciting what each component of the system does.

- Recap by asking students to give a description of the whole process, encouraging them to use sequencers, e.g., *first*, *next*, *then*, etc.

Answers

1	filter	**3**	reheater
2	thermostat	**4**	fan

Writing and grammar

A Focus on expressions for describing reason and purpose

- Write *air conditioning* and *a comfortable inside temperature* on the board. Ask students to look at the first sentence of the text on pages 154 and 155 and establish that the phrase *is a method of providing* is used to describe the purpose of air conditioning.

- Set the task for individual work and pairwork checking.

- Feed back with the whole class, eliciting the whole sentence in each case and writing it on the board.

Answers

which	(lines 9 and 12)
this causes	(line 12)
so that	(line 10)
is a method of	(line 3)
in order to	(line 13)
to	(line 10)
because	(line 18)

B Word grammar

- Choose one of the expressions and draw attention to where it comes in the sentence and what follows it, e.g., *this causes* forms the subject and verb at the beginning of a clause or sentence, and is followed by a noun or noun phrase.

- Explain that students should analyse the other phrases in the same way. Set the task for individual work and pairwork checking.

- Feed back with the whole class.

- Focus on the different forms that follow each of these expressions.

Answers

is a method of
follows the subject (noun or noun phrase) and is followed by an *~ing* form

which
starts a new (relative) clause and is followed by a verb

so that and **because**
are linkers followed by new clauses

to and **in order to**
come at the end of a clause and are followed by the bare infinitive of the verb

this causes
forms the subject and verb at the beginning of a clause or sentence and is followed by a noun or noun phrase

C Write sentences using expressions from Exercise A

- Ask students to write full sentences to answer these questions.

- Tell them to try and use some of the expressions for describing reason and purpose highlighted in Exercise A of this section.

- Set the task for individual work and pairwork checking.

- Feed back with the whole class.

Answers

1 The moisture in the air condenses because it passes across a cooling coil.

2 A hygrostat is necessary in order to measure the humidity.

3 The correct temperature is ensured through the use of a thermostat *or* a thermostat is used in order to ensure the correct temperature.

D Write questions and give answers

- Ask students to look at Figure 2 again and try to work out what it shows. They should work individually and prepare three or four questions about the AC system.

- Then ask them to work with a partner and ask and answer their questions.

- Listen to some examples with the whole class and correct any errors.

- Now have the students do the Workbook exercises for Unit 8, Lesson 7.

Workbook answers

Writing: AC units

A Correct the information

The basic function of an AC unit is to provide a comfortable inside atmosphere. The unit consists of a reheater, a cooler, a hygrostat, a thermostat, a mixer, a filter and a fan which circulates the currents of air. The hygrostat measures and controls the humidity of the air by comparing measurements against the required values, and the air can be cooled or heated further if necessary. The air can be reheated. Moisture is removed from the conditioned air, so a pipe or outlet is necessary.

Language: word order

A Reorder the words in these sentences

1 The function of the compressor is to keep the coils cool.

2 A filter is used so that the freshened air contains no dirt.

3 The fan sends the air into the room or air-conditioned area.

4 The hygrometer and the thermostat measure humidity and temperature respectively.

5 The measured values are compared with the required values.

6 An AC provides a comfortable atmosphere by cooling and drying and recirculating air.

7 The air is cooled to condense excess moisture.

8 The reheater and cooler are activated as required.

Language: compound nouns

A Match the words to make multi-word (compound) nouns

1	d	**5**	c
2	g	**6**	h
3	b	**7**	e
4	a	**8**	f

B Write five sentences about the importance of nouns

Answers depend on the students.

Objectives

To practise detailed reading to find advantages of a system.

To focus on language of comparison.

Language

Vocabulary connected with air flow, e.g., *circulation*, *ventilate*.

Comparative expressions, e.g., *as much as*, *more than*, *less than*.

Vocabulary and pronunciation

A Check the meaning of a list of words

- Encourage students to use dictionaries to carry out this task. They should check the words they do not know and make notes, e.g., of synonyms to help explain each one to their partner.

- If possible, they should find examples of the words used in context.

Answers

No answers at this point.

B Explain the words to a partner

- Stress that students should try to paraphrase or explain the words in English rather than translate. When they have finished working with a partner, feed back with the whole class.

- Ask for examples of *contaminants* and where *filters* are used.

Answers

replenish	to refill or to bring something back to its previous level by replacing what has been used
percentage	a part of a total in relation to 100
dilute	to make a solution weaker by adding water or another liquid
contaminant	a substance that makes something polluted, dirty or even poisonous
ventilate	to allow air into a space or a room
circulation	the continuous flow or movement of a liquid or gas within a system or area
replace	to remove something and put something else in its place
filter	a device for separating and removing solid particles from liquids and gases
recirculate	to allow something to move around a system or area again
interval	a short period of time

C Divide words into groups according to syllable number

- Clarify that there will be three groups with three or four words in each group. Set the task for individual work and pairwork checking.

- Feed back with the whole class.

Answers

2 syllables   dilute, replace, filter

3 syllables   replenish, percentage, ventilate, interval

4 syllables   contaminant, circulation, recirculate

D Pronunciation and stress

- Ask students to work with a partner.
- Tell them to mark the stressed syllable in each word and practise saying the words to each other.
- Feed back with the whole class.

Answers

re<u>ple</u>nish	circu<u>la</u>tion
per<u>cen</u>tage	re<u>place</u>
di<u>lute</u>	<u>fil</u>ter
con<u>ta</u>minant	re<u>cir</u>culate
<u>ven</u>tilate	<u>in</u>terval

Reading

A Read for gist

- Ask students to read the question carefully.
- Tell them to find the answer as quickly as possible by looking at the text. Set a time limit.
- Feed back orally. Elicit what features of the text made the task easier, e.g., the layout and diagram, the headings, reading the first (topic) sentence of each paragraph.

Answer

2   aircraft

B Read to find advantages of cabin AC systems

- Give students plenty of time to do this exercise.
- Ask them to note down their answers.

Answers

No answers at this point.

C Compare answers with a partner

- Compare answers with a partner.
- Feed back onto the board.

- If you have time, focus on some of the other vocabulary in the text, e.g., *randomly, seldom, cruising altitude* and *fungi*.

Suggested answers

It has a higher percentage of outside air.

The system has a high air-change rate.

It has high-efficiency filters.

The outside air is cleaner than at ground level.

Air circulation is continuous.

Levels of pollutants such as fungi and bacteria are lower than in normal indoor environments.

Levels of carbon dioxide are lower.

Grammar and writing

A Find comparative adjectives in the text

- Elicit examples of comparative adjectives. Set the task for individual work and pairwork checking.
- Feed back with the whole class.

Answers

lower	used to compare percentage of outside air, air-change rate and levels of pollutants
cleaner	used to compare quality of outside air
faster	used to compare the rate of consumption of oxygen

B Complete sentences comparing two items

- Do the first sentence together with the class. Elicit several different ideas.
- Set the task for individual work. Encourage students to be as imaginative as possible. The answers do not have to be related to the topic of the lesson. The aim is to practise comparative forms.

Answers

No answers at this point.

C Compare answers with a partner

- Get students to compare their ideas.
- Feed back with the whole class and choose the most interesting sentences.
- Accept anything that is grammatically correct and makes sense.

Suggested answers

1 A good night's sleep is equal to a weekend break from work.
2 An air ticket to New York costs as much as or more than a week's holiday at the seaside.
3 On average, girls study much more than boys.
4 Synthetic fibres are as strong as or even stronger than natural fibres.
5 These days, DVD players cost even less than video recorders did ten years ago.

Speaking

A Practise comparing different AC systems

- Direct attention to the graph and elicit what it represents.
- Set the task for individual work and pairwork comparison.
- Feed back with the whole class. Where the statements are false, elicit revised versions of the statements.

Answers

1 True
2 False (Aircraft change air at least twice as often as hospitals.)
3 True
4 False (Some aircraft change air three times as often as hospitals.)

Extension

- Students could be asked to look at other bar charts to compare items.

  http://www.nationmaster.com/index.php

  is a useful site that includes bar charts showing a variety of international statistics.

- Now have the students do the Workbook exercises for Unit 6, Lesson 8.

Workbook answers

Language: word-building

A Mark the words v, n, adj or adv

1	n	5	n
2	v	6	adj
3	n	7	adj
4	adv	8	adj

B Add two more examples

Answers depend on the students.

Language: nouns as adjectives

A Fill in the blanks

1 gas law
2 cylinder and piston assembly
3 piston reciprocation
4 runway surface irregularities
5 ground crew
6 aircraft maintenance manual
7 hot air balloon
8 airborne contamination level

Language and writing: expressions with *rate* and *level*

A Write sentences to describe the situations

Answers depend on the students.

Have a comfortable flight

Objectives

To practise listening for detail and taking notes.

To focus on the meaning and pronunciation of two-word combinations connected with cabin air systems.

Language

Compounds: noun + noun, e.g., *cabin floor*, and adj. + noun, e.g., *circular pattern*.

Pronunciation – long and short vowel sounds, e.g., *lower/hot*.

Vocabulary and pronunciation

A Discuss the meaning of key vocabulary

- Ask students to work in pairs and discuss the meaning of the compound nouns in the box and in the diagrams.

- Feed back with the whole class.

- **Note:**

  heat exchanger – a device for the efficient transfer of heat from one liquid to another

  lower lobe – a section of a passenger aircraft below the main cabin

  floor grilles – a device to allow ventilation into the cabin

  air ducting – piping used to move air around

Answers

Answers depend on the students.

Listening

A Identify the number of syllables in each compound

- Model the compound nouns from Exercise A in the Vocabulary and pronunciation section. Ask students to indicate the boundaries between the syllables with short vertical lines.

- Ask students to compare their answers with a partner.

- Feed back orally.

Answers

heat ex/chang/er	(4)
air duct/ing	(3)
ca/bin floor	(3)
fil/tered air	(3)
cir/cul/ar pat/tern	(5)
floor grilles	(2)
over/head out/let	(4)
out/flow valve	(3)
low/er lobe	(3)

B/C Focus on stress in compounds

- Set the task for individual work and pairwork checking.

- Feed back orally

- **Note:** Compound nouns consisting of noun + noun are generally stressed on the first element, e.g., *heat exchanger*, whereas expressions consisting of adjective + noun are often stressed on the second element, e.g., *lower lobe*. There are, however, exceptions, e.g., *cabin floor*, *circular*. Stress patterns are not always fixed. This is because new or distinguishing information is normally stressed, e.g., *Where in the cabin? It's on the cabin floor.*

Answers

com'pressor stages

'heat exchanger

'air ducting

cabin 'floor

special 'filters

'circular pattern

'floor grilles

overhead 'outlet

'outflow valve

lower 'lobe

D 🔊 **Listen to a talk and identify compounds (CD2 Track 10)**

- Tell students the words are not necessarily in the same order as in the book.

- Play the recording once and check that all students have heard the items in question. If necessary, play the recording a second time.

Answers

The words are heard in the following order:

heat exchangers	floor grilles
air ducting	lower lobe
cabin floor	outflow valve
filtered air	lower lobe
overhead outlets	cabin floor
circular pattern	

E **Interpret outline notes**

- Set the task for pairwork. Tell them that most of the missing words are the compounds from the Vocabulary and pronunciation section.

- Ask students to fill in as much information as possible.

- Check their progress, but do not correct their work at this stage. They will hear the correct version on the recording.

Answers

Answers depend on the students.

F 🔊 **Listen and complete the notes (CD2 Track 10)**

- Play the recording through once.

- Give students time to complete their notes.

- Play the recording again if the notes are incomplete.

Answers

No answers at this point.

G **Compare notes with a partner**

- Ask students to compare their notes with a partner.

Answers

No answers at this point.

H **Read the tapescript to check your answers**

- Refer students to the tapescript to check their answers. Clarify additional vocabulary, such as *microscopic particles*.

Answers

air comes from the compressor

flows via air ducting

further cooled by air conditioning units

mixed with filtered air from the cabin

air flows in a circular pattern

exits through floor grilles

½ air exhausted from the back of the plane /through outflow valve

mixed with outside air from the engine compressors

filters trap microscopic particles

Writing

A Identify abbreviations

- Elicit some of the abbreviations in the outline notes.

- Write them on the board and clearly establish what they mean, e.g., + = and; cab. = cabin; press. = pressure; w = with.

Answers

No answers at this point.

B Edit notes

- Ask students to work in pairs to review their notes.

- Feed back orally.

Answers

Answers depend on the students.

Pronunciation

A Identify long and short vowel sounds

- Go through the first line of the table together with the class. Say each word and elicit which is the 'odd one out' and why (it is the only short vowel).

- Set the task for individual work and pairwork checking.

- Feed back orally; encourage repetition of the words.

Answers

1	hot	4	engine
2	drawn	5	high
3	floor		

Speaking

A Practise describing a cabin air system

- Ask students to use their notes from the Writing stage.

- When students have finished working in pairs, review the cabin air system process with the whole class, eliciting each stage of the process from them, preferably without them looking at their notes.

Answers

Answers depend on the students.

- Now have the students do the Workbook exercises for Unit 8, Lesson 9.

Workbook answers

Writing: sentences with *keep* and *maintain*

A Write sentences

Answers depend on the students.

Language: verb forms

A Add the required verb endings

income	incoming
go	goes
leave	leaves
filter	filters
mix	mixes
come	comes
pass	passing
flow	flows

Language and spelling: AC systems

A Mark with a tick (correct) or a cross (wrong)

1	✓	5	✗	9	✓
2	✗	6	✗	10	✓
3	✓	7	✓	11	✗
4	✓	8	✗		

B Write sentences using the given verbs

Answers depend on the students.

Objectives

To practise describing and defining parts of an ECS (Environmental Control System) unit.

To raise awareness of compounds and long noun phrases.

Language

Long compound nouns, e.g., *firewall shut-off valve, air distribution nozzle*.

Defining relative clauses, e.g., *a gauge which shows oil temperature*.

Vocabulary I

A Focus on how compounds are formed

- Set the task for individual work and pairwork checking.
- Feed back orally.
- Highlight the ~*ed* endings that enable the words to function as adjectives.

Answers

1 pressurised nitrogen
2 a compressed fuel and air mixture
3 mixed air
4 a conditioned area
5 heated water
6 an expanded gas

Reading and speaking

A Discuss and speculate about an ECS unit

- Ask students to work with a partner and discuss the functions of the different parts of the ECS unit.
- Feed back orally. Correct any errors.

Answers

Answers depend on the students.

B Matching vocabulary with definitions

- Set the task for individual work and pairwork checking.
- Feed back with the whole class.

Answers

1 ram
2 temperature sensor
3 engine bleed air
4 outlets
5 water separator
6 heat exchanger unit
7 cooling air overboard
8 firewall shut-off valve
9 cooling turbine unit
10 flow control venturi

Vocabulary II

A Long compound nouns

- Ask students to read the information in the Language Box carefully.
- Focus on the 'reverse' idea, e.g., a *firewall shut-off valve* is a valve that shuts off a firewall; a *temperature control valve* is a valve which controls temperature.
- Set the task for individual work and pairwork checking.
- Check students can pronounce the words and phrases correctly.
- **Note:** In long compound nouns, the primary stress normally falls on the penultimate element, e.g., firewall *shut-off* valve, emergency packaged pressurised gas *delivery* line.

Answers

environmental control system

heat exchanger unit

overheat switch

temperature control valve

cooling turbine unit mixer

cooling turbine mixer

cooling turbine unit

water separator

ram air valve

temperature sensor

B Discuss the meaning of the compounds

- Set the task as pairwork.

- Feed back with the whole class, eliciting the compounds and checking that students say them with the correct stress.

Answers

1 a nozzle for distributing air

2 a grille for air to exit

3 a crew who maintain the computer systems

4 a system for controlling the environment

5 a line for delivering pressurised gas in an emergency

Writing and speaking

A Look at definitions and write compound nouns

- Draw attention to the structure of the first four definitions. Point out that they all use a relative clause.

- Set the task for individual work and pairwork checking.

- Feed back orally.

Answers

2 a landing gear hydraulic fuel reservoir

3 an oil temperature gauge

4 a tyre tread reinforcement ply

5 rear control surface linkages

B Describe items using relative clauses

- Model a definition of one of the items in the diagram, e.g., *It's a valve that/which controls temperature.* Students should identify the compound and its location on the diagram (the temperature control valve). Write the sentence on the board and highlight the relative clause.

- Set the task as pairwork. Students should take it in turns to give similar definitions of other items in the diagram. Their partner should locate and identify the compound.

Answers

Answers depend on the students.

- If you have time for further practice of relative clauses, students could write some definitions of their own.

- To learn more about aircraft environmental control systems, go to:

  http://en.wikipedia.org/wiki/Environmental _Control_System

- Now have the students do the Workbook exercises for Unit 8, Lesson 10.

Workbook answers

Writing: note-taking

A Reduce the following sentences by rewriting them in note form

1 Air under pressure drawn from engine.

2 Component removes water from air.

3 Atmospheric air forced into aircraft.

4 Device used block air-flow.

5 Two-stage assembly cools air.

B Complete the description of an ECS system

Answers depend on the students.

Language: ECS

A Find components in the diagram for the descriptions

1 bleed air valve
2 water separator
3 ram air valve
4 (firewall) shut-off valve
5 cooling turbine

B Tick the words you remember

Answers depend on the students.

C Write another word

Answers depend on the students.

Language: word order

A Rewrite sentences in the correct word order

1 The circulating air exits through grilles on either sides in the floor.

2 If you increase the volume of a gas, the pressure is reduced.

3 The mass of water vapour humidity is denoted in g/m3.

4 The actual readings are compared with the required values.

5 It needs to be tough enough to withstand a lot of stress.

6 There can be quite a bit of movement in the connectors.

7 A piston provides the spring to dampen shock loading.

8 The defogging nozzle prevents condensation in the cockpit.

Pronunciation: syllables

A Put the words into the correct column of the table

One syllable	Two syllables
dense	discard
mass	upwards
coil	ducting
worn	turbine
gauge	filters
	decrease

Three syllables	Four syllables
uneven	misalignment
orifice	humidity
thermostat	effectiveness
pneumatic	reciprocate
circulate	
exchanger	
compression	
inversely	
temperature	

B Practise saying the words aloud

Answers depend on the students.

Unit test

A test for this unit is available at:

http://www.garneteducation.com/reps/
documents/1252/SDT-u8-test.pdf

Contact your local Garnet Education representative for information about how to access these resources.

Objectives

To extend vocabulary for electrical cables and wiring.

To practise reading for gist and specific information.

To ask and answer questions about components of cables.

Language

Verb/noun collocations, e.g., *dissipate heat*.

Language to discuss different types of cables and situations, e.g., *What should you use?/What would be suitable if (there is a danger of mechanical damage)?*

Vocabulary 1

A Read the text and label the diagrams

- Check that students understand the differences between *wires*, *cables* and *busbars*, and elicit everyday uses for them.

- Elicit that the diagrams are cross-sections of cables.

- Encourage students to read the text and label the diagrams individually, but to check answers in pairs. They can either label one diagram or all three.

Answers

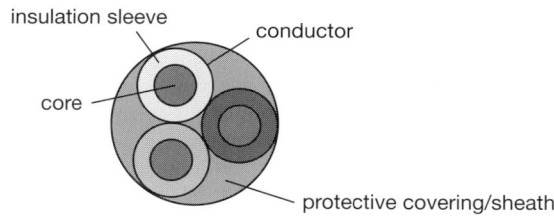

insulation sleeve

conductor

core

protective covering/sheath

B Match verbs and noun/noun phrases

- Point out that some of the verbs in the exercise can go with more than one noun phrase. Emphasise that students should look for the most likely combination.

- Follow the standard procedure for the matching activity: do the first one together as a class before setting for individual work and pair-checking.

- In feedback, go over the meanings of any unknown vocabulary, such as *dissipate*.

Answers

dissipate – heat

carry – a current

make use of – space

insulate – a conductor

protect something from – damage

consist of – several layers

separate – the conductor from its protective sheath

act as – a shock absorber

Reading

A Scan the text and circle the expressions

- Ask students to read quickly and circle the expressions from the previous exercise. Set a time limit.

- Feed back onto an OHP if possible.

Answers

core	(line 20)
conductor	(lines, 2, 3, 4, 5, 8, 11, 15 and 24)
insulation sleeve	(line 6)
protective covering or sheath	(line 17)

B Choose a title for each section of the text

- Students should read the four titles and then skim-read to find which one goes with each paragraph.

- Encourage brief pair-checking before they write each one in the correct place.

Answers

Section 1   Types of conductor

Section 2   Shape of conductor

Section 3   Protection

Section 4   Potential problems

C Careful reading and note-taking

- Go through the list of types/features of cables. Review vocabulary for properties of solids, such as *flexibility*, *conductivity*, etc., and go over any new vocabulary, e.g., *armour*.

- Encourage students to take brief notes in their notebooks for each feature. If necessary, do the first two as an example. Elicit answers and write them on the board in note form.

Answers

No answers at this point.

D Compare answers

- Students should discuss and compare their notes in pairs. Encourage them to ask and answer questions. Model a few examples with the class as follows before going into closed pairs.

  Q: *What should you use/What would be suitable (if you want flexibility)?*

  A: *Several smaller conductors.*

- Feed back suggestions to the class and write up brief summary notes on the board.

Suggested answers

lightness:
   aluminium (lighter than copper)

effective heat dissipation:
   flat cross-section (due to larger surface area)

flexibility:
   use of several smaller conductors (strands)

good conductivity:
   copper

space-saving:
   sector-shaped conductors

protection for the cable against the armour:
   bedding material

protection against corrosion:
   sheathing

a circular cross-section:
   use of filler

high current-carrying capacity:
   greater cross-section (larger diameter)

Extension: speaking

- Outline a few sample situations and encourage discussion on the best type of cable for each, for example:

  1 You need to install cables in a small space where ambient temperatures are high.

  2 You need to install cable around machines with moving parts and want it to be highly flexible.

  3 The cable does not need to move but must carry a high current and withstand possible impacts.

  4 Non-plastic insulation is required; the cable must be light.

- If you have time, recap by eliciting some of the verbs from the lesson, e.g., *dissipate*, *separate*, *protect*, *conduct*, *insulate*, *damage*. Elicit the noun (and/or adjective) form of each verb.

- Now have the students do the Workbook exercises for Unit 9, Lesson 1.

Workbook answers

Language: easily confused words

A Circle the correct word to complete each sentence

Pupils should circle:

1	conductive	**4**	dissipate
2	insulation	**5**	filler
3	sector-shaped	**6**	bedding

Writing: noun phrases and compounds

A Rewrite the sentences

Suggested answers

1 A protective covering prevents damage.

2 Emergency operating conditions put high demands on a cable.

3 The current-carrying capacity is different for different conductors.

4 Several types of material are used for insulation sleeves.

5 Breakdown voltage is affected by high temperatures.

Language: modal verbs *may* and *will*

A Expand the sentences

Suggested answers

1 Moving machinery may damage cabling.

2 A larger cross-section of cable will carry more current.

3 A single solid conductor will not bend as easily as several smaller conductors.

4 Liquids may corrode cabling.

5 In most cases, bedding will be used to separate armour insulation.

6 If insulation breaks down, it may start to conduct.

Lesson 2   Ohm's Law

Objectives

To practise intensive reading: follow explanations of voltage calculations and charts.

To practise working out voltage requirements using diagrams and graphs.

Language

Language for mathematical calculations, e.g., *I is V over R, so …*

Vocabulary for graphs, e.g., *column, curve, axis.*

Introduction

- You may wish students to find out about Ohm's Law, described in the Reading section, before the lesson, or research it yourself.

- Information about voltage, currents and resistance can be found on the Internet, e.g., at:

  http://jersey.uoregon.edu/vlab/Voltage/

  This includes an online experiment involving practical application of the calculations in this lesson (Ohm's Law).

http://www.noard.com/education1.htm helps to clarify the calculations in Reading and speaking 1c.

Vocabulary

A Match terms and definitions

- Set for individual work and pairwork checking. Students should be able to guess the answers from the clues in the text, even if they are not familiar with the terms.

- Feed back to the class or check with the Glossary.

Answers

voltage (volts):
  the difference in electrical potential between two points in an electrical or electronic circuit.

current (amperes):
  a steady movement of energy, e.g., the flow of an electrical charge.

resistance (ohms):
  the opposition of something to an electric current passing through it, so that the current changes into heat or another form of energy.

Reading and speaking I

A Understand the function of the text

- Set a time limit and encourage skim-reading.

- Elicit the answer (2) and why it is necessary to differentiate between them (in order to make calculations necessary to choose the appropriate conductor).

- If necessary, go through the examples to check comprehension. Try out the calculations in open class using some other numbers.

Answer

The answer is 2: how to differentiate between voltage, current, resistance and power.

B Practise language and calculations

- Set the task for pairwork. Ensure students understand that they need to make up the values themselves, and that they should not make their calculations too complex, e.g., by choosing very high numbers or fractions.

- Monitor and ensure that students understand the calculations and are pronouncing the numbers correctly.

- Feed back by going over any problems students have with expressing numbers.

Answers

Answers depend on the students.

Reading and speaking II

A Identify key terms for describing a graph

- The questions can be set for pairwork or discussed as a whole group.

- Feed back, using an OHT of the wire chart to point out the different sections of the graph if possible.

- Check understanding of any new vocabulary, e.g., *column*.

Answers

1 Electric wire chart

2 Wire length (in feet)

3 Current in amps

4 Curve 1 = readings (continuous rating amps) for cables in conduit or bundles

   Curve 2 = readings for single cables

5 Four different voltage types

6 Wire sizes

B Intensive reading: understand an explanation of a graph

- Give students time to read the explanation and example by themselves.

- If students are uncertain about the calculation process, go through the example again, pointing out the appropriate sections on the OHT.

- Now have the students do the Workbook exercises for Unit 9, Lesson 2.

Workbook answers

Spelling

A Correct the spellings

No answers at this point.

B Check your answers

1 resistance
2 voltage
3 current
4 formulate
5 power
6 conductor
7 chart
8 wire
9 relationship
10 supply

Language: comparisons

A Complete the sentences

1 The higher the temperature, the lower the breakdown voltage.
2 The larger the surface area, the better the heat dissipation.
3 The higher the wire size number, the smaller the conductor.
4 The more dense the gas, the more the temperature rises.
5 The less energy used, the more efficient the machine.

B Complete the sentences

1 as
2 including
3 In other words,
4 i.e.
5 like
6 such as
7 e.g.
8 including

Lesson 3   Electrical maintenance check

Objectives

To practise listening for specific information and fill in an electrical maintenance check form.

To use language from the unit to discuss electrical problems.

Language

Vocabulary for electrical components, e.g., *conduit, terminal block*.

Language for discussing problems with cables/parts, e.g., *It's too* + adj. *It could* + verb.

Vocabulary and pronunciation

A Match vocabulary to pictures

- Refer students to the diagrams. Elicit what the first picture shows.

- Set for pair or group work and monitor for problems with meaning and pronunciation.

- Elicit what the different items are used for in feedback. Encourage students to guess the function of items they do not know and check in a dictionary or the Glossary.

Answers

1 terminal block   5 cable

2 conduit   6 cable tie

3 cable clamp   7 lug

4 grommet

B Understand descriptions of the electrical items

- Set for pairwork. Emphasise that there may be more than one possible answer for some questions. The object of the task is to check that students understand the underlined words.

- Monitor and assist. Check concepts of any problem vocabulary in feedback, e.g., *What's the difference between a loop and a ring?* (A ring is usually fixed, but a loop is flexible.)

Answers

1 a cable tie   5 a terminal block

2 a lug   6 conduit

3 a cable clamp   7 a grommet

4 a cable

C Preparation for listening: identify the main stress in phrases

- Remind students of the importance of identifying stressed words when they are listening. Elicit what sorts of words are normally stressed (words giving new information, mainly content words such as nouns, verbs and adjectives).

- Do the first phrase together with the class. Point out that there is no single correct answer; the compound noun *identification number* is very likely to be stressed and, depending on the context, *see* could also be stressed.

- Set the task for pair-work. Encourage students to discuss their ideas and practise saying the phrases aloud.

- For feedback, model the phrases in the same way that they are said on the recording.

Suggested answers

1 you can see from the <u>identification number</u>

2 <u>nobody's</u> likely to walk <u>into</u> them

3 they haven't left a <u>loop</u>

4 I can move it <u>up</u> and <u>down</u> at <u>least</u> a couple of inches

5 I'm afraid it'll <u>have</u> to be <u>redone</u>

Reading

A Orientation to an electrical maintenance form

- Encourage students to scan quickly for the answers. It is not necessary to spend too long on the reading activities, as they are primarily preparation for the listening.

- Conduct brief feedback. Discuss/compare conventions of form-filling, e.g., why it is important to have a form code and signature, the use of note form to write in additional details.

Answers

1 G. Townsend

2 E1

3 Conductor and conduit inspection

4 [left to right] ticks and crosses to indicate pass/fail inspection; notes to describe faults; summary of recommendations

B Predict the content of the listening text: discuss possible electrical problems

- Go over or ask students to check unfamiliar vocabulary, e.g., *deformation*, *radii* (plural of *radius*).

- Model the task by eliciting example problems for some of the categories before setting for pairwork.

- Go through their ideas briefly. Stress that they will be able to add to them after listening to the recording of a maintenance technician explaining an inspection.

Suggested answers

Conductor check

Identification number: unclear or missing

Conductor size: too big or small

Conductor type: inappropriate material or shape

Loop: absent, insufficient or too big

Route: may cause obstructions, or be in an unsafe place, e.g., in an area where there is leaking of fluids or excess heat

Hole size: chafing if they are too small

Grommets: chafing if they are absent

Clamping: may pinch the cable if too tight

Deflection (slack): not enough (cable will be pulled too tight) or too much (cable will hang down)

Cable ties: insufficient quantity or too loose/tight

Terminals and lugs: incorrect type used

Conduit check

Position: obstruction of aisles, etc., could be a safety issue

Material: breakages if material is not strong enough

Size: could obstruct if too big

Bending radii: may not allow sufficient bending of cable, e.g., around corners

Drainage holes: insufficient, badly made

Deformation: may go out of shape too easily

Listening

A 🔊 **Listen for the main ideas: identify problems (CD2 Track 11)**

- Establish that Mr Townsend is reporting back to the owner of a small plane after an electrical inspection.

- Play the recording. Pairs compare their ideas.

- Feed back orally. Elicit what students were able to pick up about the nature of the problems, i.e., there is no loop, a grommet is missing and the drainage holes are too rough (they have burrs).

Answers

The following all have problems: *b*, *c* and *d*.

B 🔊 **Intensive listening: fill in a form (CD2 Track 11)**

- Set for individual work and pair-checking. Emphasise that they should not immediately try to complete all the notes and the *recommendations* section because there will be time afterwards to do this.

- Monitor as the students listen and complete the form, checking for problems.

Answers

No answers at this point.

C **Discuss answers**

- Allow time for students to add to their notes and write in the recommendations after they have discussed their answers.

- When students have fed back their answers, let them look at the model answer below (use an OHT if possible) to compare with their ideas. Clarify vocabulary as necessary.

Suggested answer

Ace Aircraft Ltd
Electrical Maintenance Check

Aircraft and I.D.		Date	Inspector (print name)		Usual signature
ACE 306 – P72		23/5/––	G. TOWNSEND		G A Twnsd
Conductor check		Pass ✓ Fail ✗	Nature of fault		Recommendations
Identification No.	✓				REMOVE, REPOSITION AND SMOOTH DRAINAGE HOLES (REMOVE BURRS)
Conductor size	✓				
Conductor type	✓				
Loop	✗		NO LOOP		
Route	✓				
Hole size	✓				
Grommets	✗		MISSING GROMMET		
Clamping	✗		TOO TIGHT		
Deflection (slack)	✗		TOO MUCH (2 in.)		
Cable ties	✓				
Terminals and lugs	✗		WRONG TYPE		
Conduit check		Pass ✓ Fail ✗	Nature of fault		
Position	✗		PROTRUDES TOO FAR		
Material	✓				
Size	✓				
Bending radii	✓				
Drainage holes	✗		ROUGHLY FINISHED		
Deformation	✓				
Form E1					

Speaking

A Practise discussing possible problems caused by electrical installation faults

- Direct attention to the prompts in the book and elicit possible endings for each one.

- Reactivate the language from the lesson. Encourage students to discuss the problems in pairs using their notes on the form.

- Monitor and note problems while students do the task.

- Feed back on impeding errors.

Answers

Answers depend on the students.

- Now have the students do the Workbook exercises for Unit 9, Lesson 3.

Workbook answers

Language: multi-word verbs

A Match the common multi-word verbs to their meanings

stick out – protrude

put in – install or fit

point out – show

run out – exhaust supply of something

get back – return

break down – stop working

B Complete the sentences

to allow for	I can see
could eventually	is likely to be
just in case	can … if
can … if	to allow for

Language: probability

A Find the expressions

See the tapescript of CD2 Track 11. The expressions are found in the following order: *is not likely to, just in case, to allow for, x can … if, could eventually, I can see.*

B Complete the sentences

1	to allow for	5	I can see
2	could eventually	6	is not likely to
3	Just in case	7	can … if
4	can … if	8	to allow for

Language: *adj + to do something*

A Make sentences using the given vocabulary phrases

Answers depend on the students.

Lesson 4 — Long life

Objectives

To discuss properties of and differences between batteries.

To develop skimming skills in a technically dense text.

To practise interpreting and describing graphs.

Language

Vocabulary associated with batteries, e.g., *long life, performance, discharge rate.*

Speaking and vocabulary

A Discuss characteristics of batteries

- Draw attention to the batteries in the pictures (or bring in some real examples) and elicit what they are.

- Set the questions for pair or group work.

- Elicit students' ideas and suggestions for feedback.

Possible answers

1 domestic uses: radios and audio devices, laptop computers, cameras, toys and mechanical devices, mobile phones, watches

   industrial uses: starter motors for vehicles, back-up for electrical equipment, safety devices and alarms, electric tools, e.g., drills, lights and torches [accept any other reasonable suggestions]

2 accept any reasonable suggestions

3 long life: lithium batteries last longer than lead-ion, especially if stored at low temperatures; primary (or disposable batteries) hold their charge for longer than rechargeable batteries, but all batteries self-discharge when they are not being used, e.g., alkaline batteries left on a shelf can lose up to 20 per cent of their charge in a year.

4 Performance is affected by speed of discharge, how fully it is recharged, temperature, chemical degradation of the battery, etc.

5 Batteries are <u>recharged</u> when an electrical current is used to restore their original chemical composition.

6 Lead-acid batteries are found in most cars.

B Focus on key vocabulary

- Direct attention to the underlined words in the previous exercise and elicit the differences between the groups of words in *1* and *2*.

Answers

1 self-discharge: the loss of charge when a battery is not in use through chemical reactions inside the battery

recharge: use of electrical current to restore the original chemical composition of a battery

2 performance: how well the battery works

life: length of time that a battery can be used (including recharging)

capacity: how much energy a battery can produce

Reading I

A Scan the text and answer comprehension questions

- Point out that students should be able to answer these questions very quickly by using the layout of the text (headings and bullet-pointed information).

- Set for individual work and pair-checking.

Answers

1 self-discharge

2 lead-acid, lithium-ion, nickel-cadmium, nickel-metal-hydride

3 temperature

B Mark the sentences true or false

- Tell the students they will need to read more carefully and do some calculations to answer the next set of questions. Set for individual work.

Answers

No answers at this point.

C Compare answers with a partner

- Have students check answers with a partner.

- Feed back with the class. See if students can provide any more information about the different types of batteries.

Answers

1 False (30% more quickly)

2 False (lithium-ion batteries self-discharge more slowly after the first 24 hours)

3 True

4 False (they may have lost 20% but they are not *at* 20% of their capacity)

5 True

- **Note:** Lead-acid batteries are low-cost and have a high power-to-weight ratio, so they are popular for use in road vehicles; Lithium-ion batteries have high energy density and are popular with the defense and aerospace industries; NiCd batteries are used for motorised equipment that runs for a short time; NiMH batteries are generally used for high-tech devices.

- More information about batteries can be found at:

  http://electronics.howstuffworks.com/battery.htm

Reading II and speaking

A Interpret a graph

- Ask students to look at the graphs and comment on the similarities and differences between them, e.g., the horizontal axes: one represents hours and the other represents weeks; both show batteries at three different temperatures.

- Set the task for pairwork discussion. Monitor and encourage stronger students to discuss the graphs (and their implications) in more detail.

- Feed back to the class. Ask further questions about the graphs, e.g., *What happens to a battery's performance in very cold conditions?* (It discharges more quickly.) *What happens if batteries are stored at high temperatures?* (They self-discharge more quickly.)

Answers

1 just under 9 hours
2 approx. 78%

Extension: language

A Focus on the use of the time verbs *take* and *last*

- Write questions on the board with *take* and *last*, e.g., *How long does the battery take to run down? How long does the charge last?* Elicit which verb asks about the time needed to <u>do something</u> (take) and which asks about the <u>duration of something</u> (last).

- Get students to practise asking and answering questions about the graphs using *take* and *last*.

Answers

Answers depend on the students.

- Now have the students do the Workbook exercises for Unit 9, Lesson 4.

Workbook answers

Writing: describing graphs

A Draw a graph

Answers depend on the students.

B Complete the description

Suggested answers

1 the performance of lithium-ion batteries.
2 discharge time
3 voltage
4 the discharge time is approximately 7 hours
5 the discharge time is approximately 9 hours
6 the discharge time is approximately 9 hours

C Write a description

Answers depend on the students.

Language: *take place/occur*

A Referring to something planned or positive (P) and something less definite (LD)

1 LD
2 P

B Complete these sentences with *occur* or *take place*

1 take place
2 occurs
3 takes place
4 occur
5 occur
6 is taking place

Percentages

A The difference between *per cent* and *percentage*

per cent	noun, adj., adv.
percentage	noun

B Circle the examples in Lesson 4

From the text on Course Book pages 168–169: lines 7, 8, 9, 11, 12 and 13.

C Write sentences to express the percentages

Suggested answers

1 Work out 33.3 per cent of _____.

2 The temperature rises by a small percentage every minute.

3 The percentage of production is higher this month than last month.

4 4 out of 6,500 is a low percentage.

5 The required rate is between 10 and 12 per cent.

6 I need this number as a percentage of the total.

Lesson 5 — Battery problems

Objectives

To practise listening for information to complete a form.

To notice the form and usage of the past perfect tense for a series of events in a narrative.

Language

Past perfect simple, e.g., *The battery had self-discharged.*

Verbs connected with battery/engine parts and problems: *charge up, run down.*

Introduction

- Elicit battery problems that occur in cars, e.g., the battery runs down because the lights are left on; the battery goes flat if the car is not used for a long time; the battery does not charge up if the alternator is defective.

- Go over key vocabulary connected with batteries, such as: *run down, charge up, flat.*

Vocabulary

A Raise awareness of multi-word verbs

- Point out/elicit the fact that *run down* is a multi-word verb. Elicit other multi-word verbs from recent lessons (students could look at the tapescript for Unit 9, Lesson 3 to remind themselves), e.g., *take out, put back, hang down, move back.*

- Set the exercise for individual work and pair-checking.

- Go over any problem vocabulary, e.g., *hand-propping.*

Answers

1 Don't leave your headlights on. You'll run *down* the car's battery.

2 Did you leave the battery to charge *up* overnight?

3 Can you hear the engine turning *over*?

4 The engine has cut *out* completely.

5 We could try hand-propping: turning the propeller *round* by hand.

6 It's better if you have an alternator fitted *to* the plane.

7 The plane is starting to taxi *off.*

8 Be careful you don't run *into* anything!

Speaking

A Preparation for listening: discuss possible causes of an accident

- Elicit which parts of the plane are damaged (rear fuselage, starboard wing). Explain that students are going to listen to a radio news report about the accident that caused the damage in the picture. Encourage them to speculate about the cause of the accident.

- If students lack ideas, write (and clarify) some key vocabulary on the board to prompt them, e.g., *starter motor*, *hand prop*, *taxi off*, *hangar*.

- Elicit ideas, but do not confirm or correct. Students will listen to check.

Answers

No answers at this point.

Listening

A Listen to confirm answers (CD2 Track 12)

- Students should listen without looking at the accident form yet.

Answers

No answers at this point.

B Check answers in pairs

- Give students an opportunity to compare what they thought they heard before feeding back.

Answers

Brief summary of accident:

A light aircraft had a flat battery. The pilot left his brother in the cockpit while he tried hand-propping the plane to start it. When the plane started immediately, the brother lost control and it taxied off into the hangar, where it ran into another light aircraft.

C Intensive listening: fill in a form (CD2 Track 12)

- Give students time to read through the form before playing the recording again. Clarify that they do not need to fill in every section.

- Monitor as the students listen and complete the form, checking for problems.

- Students should pair-check and add to their answers as necessary.

- Feed back by either writing up a model answer on the board or showing the model answer on an OHP. Discuss any potentially confusing terms, e.g., the difference between a generator and an alternator: the alternator charges the battery even when the engine is just ticking over.

Suggested answer

Form HIR1/a HANGAR INCIDENT REPORT: Aircraft damage	
Type and ID of aircraft	Two-seater single-engined
Name of owner	Mike Grigson
Other persons involved	Pilot's brother/Tom Grigson
Type of accident (collision, fire, etc.)	collision
Cause (human/mechanical failure, etc.)	pilot lost control
Description of incident Plane started unexpectedly when pilot was hand-propping it. Pilot's brother lost control and it taxied into hangar – collided with twin-engined four-seater	
Damage to aircraft rear fuselage and starboard wing of other aircraft damaged	
Damage to hangar/equipment none	
Personal injury (brief description only; fill in and attach form HIR3 as necessary) none	
Signature	Date
To be signed and handed to the Airfield Office within 24 hours of incident	

D Practise word/phrase recognition in continuous speech (CD2 Track 13)

- Ask students to read the sentences silently and then aloud in pairs.

- Play the recording. Pairs compare answers. Feed back onto the board.

- Drill the pronunciation of the sentences.

- Discuss whether the written sentences that are different convey a different meaning.

Answers

1 D (The engine ~~wouldn't~~ *didn't* start ~~up~~ on the electric starter ~~motor~~.)

2 S

3 D (The electric starter motor ~~becomes an alternator~~ *turns into a generator* and ~~charges~~ *starts charging* the battery up.)

4 S

5 D (~~I went up four weeks ago~~ *I hadn't been up for four weeks* so the battery ~~was charged~~ *had self-discharged* – it was flat.)

6 D (We've got ~~two~~ *a* single-engined thing~~s~~. Light aircraft.)

E Practise reading the sentences aloud

- Have the students practise reading the sentences aloud, focusing on the sentence stress.

Answers

Answers depend on the students.

Language

A Focus on past perfect

- Refer students to the Language Box. Clarify as necessary, highlighting the example sentences with a timeline and concept questions.

Answers

No answers at this point.

B Analyse order of past events in past perfect sentences

- Explain that the past perfect is often used in a narrative to clarify the sequence of events. The grammar exercise is designed to illustrate this.

- Set for pairwork. Monitor and assist.

- For feedback, ask students to identify the use of the past perfect in each of the original sentences. They should read them out using the correct stress and contractions.

Answers

1 *I ran the checks. Then I discovered it was the battery.*

2 *I didn't go up for six weeks. Then I went up.*

3 *The battery self-discharged. As a result, it was flat.*

4 *Someone left the doors open. Then the plane taxied in.*

5 *I hand-propped the plane several times. The last time I hand-propped the plane, it was easy.*

- If you have time, get students to listen again to the recording as they follow the tapescript, underlining examples of the past perfect.

- Now have the students do the Workbook exercises for Unit 9, Lesson 5.

Workbook answers

Language: words and expressions with *charge*

A Mark the words *v*, *n* or *adj*

1	adj	4	v
2	v	5	adj
3	adj	6	n

B Write phrases or sentences

Answers depend on the students.

Language: *have/get something done*

A Suggest a course of action in these situations

1 We'll get the air brake inspected.

2 We'll replace the plane's radio.

3 We'll send off for some more parts for an old engine.

4 We'll check the machine.

5 We'll send help to the aircraft.

6 Can you redo the wiring?

7 I'll send for the drawings from the other office.

8 I will redo the calculations.

Reading and writing: hand-propping

A Fill in the gaps with the phrases

1 by hand

2 battery

3 power

4 procedure

5 taxied off

6 starters

Objectives

To practise describing a graph that shows engine performance.

To raise awareness of syntactic patterns in a text describing maintenance problems.

Language

Language to describe speed and quantity, e.g., *torque reading*.

Vocabulary associated with starter motors, e.g., *brush*, *spark*, *capacitor*.

Introduction

- Introduce the topic of motors and generators. Elicit the difference between a generator and a motor (a motor uses power and a generator generates power, but both functions may be present in a motor/generator).

- Elicit what can go wrong with them.

Vocabulary

A Check understanding of key vocabulary

- Draw attention to the words in the box and elicit which ones are verbs and adjectives and which can be nouns.

- Clarify new vocabulary and elicit words that each can be used with, e.g., *dirt/dust/residue can build up*.

- Set the task for individual work. Feed back orally.

Answers

1 snag

2 worn out

3 build up

4 short circuit

5 pitted

6 burn out

Reading I and speaking

A Identify parts of a graph

- Introduce the multi-graph and clarify that it shows four different things: efficiency, RPM (revolutions per minute), power and torque. Check comprehension of *torque*.

- **Note:** Torque is a measurement of how much force on an object causes it to rotate: power = torque x angular speed.

- Set the task for brief pairwork discussion.

- If possible, feed back by projecting the graph onto the board and highlighting the different scales.

Answers

1 The efficiency and RPM values scales are on the left-hand vertical axis. The power output and torque values scales are on the right.

2 efficiency: as a percentage; RPM in numbers 0–2,000; power in kilowatts; torque in Newton metres

3 0–300

B Interpret a graph

- Set for pairwork as before and feedback by highlighting the points on the graph that give the answers.

Answers

1 80%

2 250 amps

3 39 NM

C Practise asking and answering questions about the graph

- Elicit one or two more questions with the whole class.

- Set the task for pairwork and monitor for problems.

Reading II

A Read quickly to identify topics in a text

- Direct attention to the diagram of a standard brushed motor. Look at the words in the box and elicit/clarify what they are with the help of the diagram and the Glossary.

- Set the task for individual work and pair-checking. Point out that each paragraph deals with one topic.

Answers

Start-up and cooldown period	(paragraph 1)
brush wear	(paragraph 2)
bearings	(paragraph 3)
radio static	(paragraph 4)

Language

A Find expressions for explaining problems in the text

- Explain that the expressions are used when talking about cause and effect, e.g., for explaining problems in a system. Look at *are liable to* as a group and point out that *x is liable to happen* is often followed by an *if* or *when* clause.

- Set for individual work and encourage the students to notice the parts of speech and syntax around the expressions as they circle them.

- Feed back, focusing on correct pronunciation and stress.

Answers

are liable to	(line 1)
causes them to	(line 5)
they can	(line 10)
occurs	(line 12)
resulting in	(line 12)
can have a serious effect on	(lines 17–18)
is often a sign of	(line 22)
is also caused by	(line 23)

B Use the expressions in context

- Set for pairwork discussion. Students can find the answers in the text.
- Elicit several possible answers for each scenario.

Suggested answers

1 Incorrectly positioned brushes *are liable to* snag on the brush holders.

2 Electrical arcing is *often a sign of* air gaps between the brushes and the commutator *resulting in* accelerated brush wear, commutator pitting and burning and radio interference.

3 Vibration *occurs* if the armature is out of balance.

- Now have the students do the Workbook exercises for Unit 9, Lesson 6.

Workbook answers

Language: sentence structure

A Match the sentence halves

1	e	5	f
2	g	6	c
3	h	7	d
4	a	8	b

Language: noun phrases

A Complete the phrases

1	generators	5	thumb
2	manual	6	static
3	inspection	7	carbon
4	holders	8	sparking

B Check your dictionary

Answers depend on the students.

C Find other multi-word nouns

Suggested answers

aircraft starter generators

starting lengths

cooldown periods

routine brush inspection

balance armatures

terminal block

D Write your own sentences

Answers depend on the students.

Writing: adverbs

A Write the ~*ly* form and make a sentence

1	correctly	6	rapidly
2	possibly	7	basically
3	excessively	8	slowly
4	continuously	9	quickly
5	easily	10	mainly

Sentences depend on the students.

B Write a description of the graph in Lesson 6

Answers depend on the students.

Objectives

To expand vocabulary for describing printed circuit boards.

To practise listening to descriptions and taking notes.

To develop the skill of inferencing and using clues in a text to complete it with missing words.

Language

Vocabulary for describing transistor radio parts and stages, e.g., *antenna*, *loudspeaker*, *signal*.

Introduction

- Draw one or two simple circuit diagrams on the board or get students to look at the diagrams in the book. Elicit what they are. Elicit the meaning of the lesson title and explain that diagrams are often used to identify problems with electronic equipment.

Speaking and listening

A Assess knowledge of circuit boards and diagrams

- Set for pairwork. Monitor for problems.
- Feed back, paying attention to the new vocabulary, e.g., *conducting tracks*.

Answers

1	2	**4**	1, 4
2	5	**5**	1, 3, 4
3	1		

B 🔊 Listen and match descriptions to images (CD2 Track 14)

- Set the task for individual work and pair-checking.

- Elicit what students remember about each image by asking questions, e.g., *How is Diagram d used? Who might use Diagram c?*

Answers

1	d	**4**	c
2	a	**5**	e
3	b		

C 🔊 Listen and take notes (CD2 Track 14)

- Tell students to write the name of each diagram as a heading in their notebooks. Tell them to make notes on what each diagram shows and who uses it (if mentioned).

- Play the recording again, pausing after each description for students to complete their notes.

- Students should compare notes in pairs and also add to them during whole-class feedback. Use the model notes on the board or an OHP to compare with students' ideas.

- Elicit further ideas on who might find Diagram 5 useful, e.g., students learning about the system.

Suggested answer

d) PCB track layout diag.

shows: layout of copper tracks + holes for component leads

used: in PCB prod.

a) Component and track layout diag.

shows: components viewed from top + track routing on underside

used: by service technicians to I.D. location of components

b) Pictorial view of electronic board

shows: components in 3D

used: to show what the PCB actually looks like

c) Electronic circuit schematic

shows: elec. connections and component values clearly

used: by technicians for troubleshooting

e) Block diag.

shows: system broken down into blocks or sub-systems

used: to clarify how system operates, e.g., to students

Reading

A Read the text and complete the gaps

- Exploit the diagrams. Explain that they are both used for fault-finding and that one is a flow chart and the other is known as the half split method. Elicit what systems are shown (a lamp and a transistor radio).

- Check that students are familiar with the vocabulary in the box. Clarify that *aerial* is a synonym for *antenna*.

- Set the task for individual work and pair-checking. Students should read the text quickly and then go back and fill the gaps.

- Feed back orally.

Answers

line 7	sequence
line 9	instructions
line 14	several
line 15	antenna
line 16	emitted; signal
line 19	volume; inject
line 20	loudspeaker
line 25	locate

- If you have time, draw attention to the symbols for the radio antenna and loudspeaker. Draw some other symbols of electrical components, e.g., battery, switch, fuse, conductor, etc., on the board and ask students to guess/discuss what they are. Direct attention to the electrical symbols key at the back of the book to check answers.

- Now have the students do the Workbook exercises for Unit 9, Lesson 7.

Workbook answers

Language: electrical symbols and terms

A Match the symbols with the components

1	f	7	c
2	b	8	e
3	h	9	k
4	g	10	l
5	i	11	j
6	d	12	a

Language: word-building

A Complete the table with different forms of the words

verb	noun	adjective
behave	behaviour	*behavioural*
conduct	*conduct*	*conductive*
insulate	insulation	*insulating*
prevent	*prevention*	*preventative*
protect	*protection*	*protecting*
differ	*difference*	different
dissipate	dissipation	*dissipating*
damage	*damage*	*damaging/ damageable*
circle	circle	circular
flex	flexibility	*flexible*
separate	separation	separated
———	*durability*	durable
strengthen	*strength*	strong
corrode	*corrosion*	*corrosive*

Online purchase

Objectives

To extend vocabulary for describing electrical tools and instruments.

To practise listening and identifying advantages of different tools.

To raise awareness of the importance of avoiding intrusive vowels in consonant clusters.

Language

Terms for electrical tools, e.g., *solder gun, wire strippers*.

Introduction

- If possible, bring in a few tools, such as a multimeter, side cutters and a socket set. Check which tools students are familiar with.

- Elicit how it is possible to buy tools without going to a shop (by phone, mail or online).

- Refer students to the title of the lesson and check that the meaning is clear.

Vocabulary and speaking

A Identify tools in the pictures

- Set for brief pairwork discussion.

- Feed back with the class, focusing on the pronunciation of the items.

Answers

multimeter: 1 (digital), 10 (analogue): used for measuring current, voltage and resistance

pliers: 3 (electrical), 6 (multi-grip), 15 (long-nosed): tool with jaws used for gripping, bending and cutting small objects; c/f diagonal pliers or side cutters, which are specifically used for twisting and cutting wire

tape measure (analogue and digital):
5 (steel measuring tape): used for taking measurements

soldering equipment: 8 (soldering iron), 9 (soldering stand), 12 (soldering gun): rod-shaped tool for joining metals

screwdriver: 4 (electrical screwdriver set): tool for turning screws

B Identify remaining items and focus on pronunciation

- Students should work in pairs and groups to write down the names of as many of the tools in the pictures as possible. Go round and monitor and help with vocabulary and spelling.

Answers

Answers depend on the students.

C Mark the stress and practise saying the words aloud

- Feed back onto the board. Point out spelling patterns and use of hyphens. Go over any problems, e.g., the difference between *analogue* and *digital* (see Unit 10, Lesson 3).

- Drill for correct pronunciation and elicit the stressed syllables in each phrase. Point out that there are several ways to mark word and syllable stress. Dictionaries often indicate the word that is normally stressed in a phrase or compound by placing ' before it. In addition, it is useful for students to underline the stressed vowels.

Answers

1 digital 'multimeter

2 reel of 'solder

3 set of electrical 'pliers

4 electrical 'screwdriver set

5 analogue/digital tape measure

6 multi-grip 'pliers

7 small 'hacksaw

8 'soldering iron

9 'soldering stand

10 analogue 'multimeter

11 'side cutters

12 'soldering gun plus accessories

13 'socket set

14 'voltage tester

15 long-nosed 'pliers

16 'wire strippers

17 'crimping tool

18 WD40 'lubricant

Listening

A Listen for specific information: identify the tools the customer buys (CD2 Track 14)

- Set the scene. Make sure that students understand they are going to listen to someone buying some of the tools in the pictures

- After students listen, they should check answers with a partner.

- Play a second time if necessary.

Answers

2, 5, 6, 10, 11, 12, 13, 14, 15, 16, 17

B Listen and take notes (CD2 Track 14)

- Direct attention to the example notes given for the pocket voltage tester. Elicit the meaning of the number and abbreviations.

- Play the recording again.

- New partners compare answers. Feed back onto the board.

Answers

Suggested answers

1 **pocket voltage tester:** v. cheap, 115 V dangerous, easy + quick to use – kept in pocket

2 **analogue multimeter:** v. cheap, v. simple, easy to replace

3 **tape measure:** combi – can read off tape or use digital reading at side, v. damage-resistant

4 **solder gun:** comes with diff. sized bits, more flexible than std. soldering iron + safer for inside plane

5 **socket set:** saves time (for terminal nuts), gives good close contact, small set has all sizes nec.

6 **multi-grip pliers:** allows good grip + frees hands to work – like 2nd pr. hands

7 **single-side cutters and long-nose pliers:** vital – single prs. better qual. than sets

Pronunciation

A Practise saying consonant clusters with *l* and *r*

- Drill the consonant clusters and words in each row of the table. Make sure that students do not add a vowel between the two consonant sounds.

- Get students to practise the words in pairs. Monitor and correct where necessary.

Answers

Answers depend on the students.

B/C Complete the table with other words

- Set for pairwork. Students should look in the Glossary or in their dictionaries if they cannot think of words for all the sections.

Do not insist that they find words for every row. They can write in extra words during feedback.

- As feedback, draw the table on the board or project it onto an OHP. Elicit one or two words for each row of the right-hand column of the table.

Suggested answers

fl	flammable
pl	display, plane
bl	assembly, cable
cl	close, clockwise
cr	screw, aircraft
tr	extra, instrument
pr	compressor, propeller
dr	drill, drogue

Extension

- Set up a role play: students take it in turns to play the customer and the salesman. The customer should ask about different tools, and the salesman should give advice and information.

- Write up some useful expressions on the board, e.g.,

  Which is better quality/cheaper ...? What do I need for ...?

  I'd go for ... I'd recommend ... These are good quality.

- Now have the students do the Workbook exercises for Unit 9, Lesson 8.

Workbook answers

Language: verbs for electrical engineering

A Write correct definitions

1a dismantle an engine for cleaning or maintenance

1b take the insulation sleeve off a wire

1c correct

2 join two pieces of metal by crushing them

3a hold tightly using the hands or a tool

3b correct

4 join two items, usually metal, using a melted copper alloy

5a correct

5b pull a wire so that it does not bend

Language: tools

A Match the two columns from memory

1 wire – strippers
2 socket – set
3 single-side – cutters
4 tape – measure
5 analogue/digital – multimeter
6 pair of long-nose – pliers
7 crimping – tool
8 soldering – gun

Writing: sentence structure

A Expand the sentences

1 One of the biggest jobs is soldering in small spaces.

2 I will probably work alone for most of the time.

3 I was going to use a pair of normal pliers to do the job, but then changed my mind.

4 If you are going to do a lot of soldering, then you need a good kit.

5 The analogue version is not as good as the digital one, but it is cheaper.

6 As far as I can tell from the manual, there is another connection next to this one.

7 Mains supply is 28 V, whereas / but avionics use 115 V.

8 The reading is given both on the display and on the side of the tape measure.

Power systems

Objectives

To extend vocabulary connected with power system.

To practise detailed reading and note-taking.

To practise interpreting notes to convey detailed information in a short talk.

Language

Vocabulary and abbreviations for power systems, e.g., *AC inverter, DC equipment, EPU*.

Vocabulary

A Match the words to their meaning

- Set the task for individual work and pair-checking. Students can use their dictionary and the Glossary.

- Check concepts by eliciting how different items are powered. See if students can put each term into an example sentence, e.g., *An EPU is used when an aircraft is on the ground.*

Answers

1	EPU	4	inverter
2	online	5	AC
3	generator	6	DC

B Discuss aircraft electricity supply

- Set for brief pairwork discussion.

- Feed back to the class. Stress that they will find more information when they read the text.

Answers

No answers at this point.

Reading

A Check answers to the previous exercise

- Have students work and check their answers in groups A, B, C. Monitor closely to assist and correct.

Suggested answers

1 The battery, engine, lights, cockpit instruments, AC, heating, emergency systems

2 By EPU, AC and DC systems

B Scan for numbers in the text

- Have students circle the numbers in the text.

- Check answers with a partner.

- Feed back as a group.

Answers

32 V	(line 13)
400 hertz	(line31)
115 V	(line 31)
700	(line 21)
26 V	(line 31)
40 amp/h	(line 17)
24 V	(line 23)

C What the numbers refer to

- Have students work in pairs to figure out what the numbers refer to.

- Feed back as a group.

Answers

32 V	thirty-two volts
400 hertz	four hundred hertz
115 V	one hundred and fifteen volts
700	seven hundred

26 V	twenty-six volts
40 amp/h	forty amps per hour
24 V	twenty-four volts

D Read carefully and take notes

- Have students complete this exercise individually.

Answers

No answers at this point.

E Compare notes with a partner

- Have students compare their notes with a partner.
- Feedback orally or on the OHP.

Answers

Answers depend on the students.

Speaking

A Give a short talk

- Ask students to look again at the Vocabulary section on page 178. They should choose five words and write an example sentence for each.

- Monitor and assist.

- Ask students to repeat the sentences to each other and memorise them, to reinforce the sentence stress, new vocabulary and engineering concepts.

- The wikipedia website again has a clear general illustration of the mechanical flight control linkage system, with links to further illustrations of <u>ailerons</u>, <u>elevators</u> and <u>rudder</u> at:

  http://en.wikipedia.org/wiki/Aircraft_flight
  _control_systems

- Now have the students do the Workbook exercises for Unit 9, Lesson 9.

Workbook answers

Word-building

A Make a noun, a verb and an adjective from the word roots

1 electricity (*n*) / electrical (*adj*)
2 generator (*n*) / generate (*v*) / generated (*adj*)
3 connector (*n*) /connect (*v*) / connected (*adj*)
4 operation (*n*) / operate (*v*)
5 distribution (*n*) / distribute (*v*)
6 termination (*n*) / terminate (*v*)
7 excess (*n*) / exceed (*v*) / excessive (*adj*)
8 conductor (*n*) / conduct (*v*) / conductive (*adj*)
9 approximation (*n*) / approximate (*v*) / approximately (*adj*)

Pronunciation: word stress

A Find words from Exercise A with the relevant stress patterns

Answers depend on the students.

Language: sentence structure

A Order the sections of these sentences

1 The main consideration for conductors is their current-carrying capacity.

2 To avoid voltage drop, the shorter the cable run the better.

3 The loop of wire makes it easier to take out and put back cabling.

4 Output is an important value as a percentage of maximum output.

5 He had not connected the battery to the charging set the night before.

6 Electrical interference in the radio unit is often a sign of sparking.

7 The internationally recognised symbol for a lamp is a filament bulb.

8 Automatic wire strippers are easier to use than the ordinary ones.

9 The system is automatically disconnected from the generator.

10 The circuit-breaker prevents the circuit being overloaded.

Lesson 10   Spreading it around

Objectives

To review concepts and language from Unit 9.

To practise reading and giving short explanations of electrical components, using vocabulary from the unit.

To practise note-taking and expanding notes into a short text.

Language

Manoeuvres and operation of control surfaces.

Vocabulary

A Match the words with a set of words in the table

- Have the students work in pairs to match the words in the table.
- Monitor and assist.
- Feed back orally.

Answers

1	*current*	**6**	equipment
2	socket	**7**	a fuse
3	electrical	**8**	a circuit
4	a fault	**9**	operated
5	a switch		

Reading and speaking

A Complete the text

- Have students read through the text and fill in the gaps.
- Then have students discuss their answers with a partner.
- Feed back orally.

Answers

line 2	electrical distribution
line 3	common connection
line 6	heavy current
line 8	wide flat strips
line 9	surface area to volume
line 10	maximum amount

B Work with a partner

1 Have students work with a partner. Student A reads Texts A and B. Student B reads Texts C and D.

2 Have students take turns explaining the components without using the text itself. While Student A is speaking, Student B should take notes, and when Student B is speaking, Student A should take notes.

Answers

Answers depend on the students.

Writing

A Write your notes into a brief text

- Have students use their notes from Reading and speaking Exercise B to write a brief text.
- Monitor and assist where necessary.

Answers

Answers depend on the students.

B Compare texts with the originals

- Have students compare their texts with the original texts from Course Book page 181.

Answers

No answers at this point.

C Exchange texts and look for errors

- Have students exchange texts with a partner.
- Students should look for possible sentence structure errors and suggest ways to correct them.
- Monitor and assist.

Answers

Answers depend on the students.

- Now have the students do the Workbook exercises for Unit 9, Lesson 10.

Workbook answers

Spelling and pronunciation

A Correct the incorrectly spelled words

1	correct	7	correct
2	inverter	8	generator
3	housing	9	correct
4	correct	10	malfunction
5	connection	11	correct
6	primary	12	battery

B Write a word from Exercise A that contains the same sound

1 gener*a*tor
2 s*o*cket
3 malf*u*nction
4 isolat*e*d, *i*nverter, hous*i*ng, d*i*stribution, sock*e*t, maxim*u*m
5 distrib*u*tion
6 m*a*ximum, m*a*lfunction, b*a*ttery
7 conn*e*ction, g*e*nerator, t*e*mperature
8 inv*e*rter
9 is*o*lated, invert*e*r, distributi*o*n, connecti*o*n, prim*a*ry, gener*a*tor, maxim*u*m, malfuncti*o*n, temper*a*ture, batt*e*ry
10 primar*y*, batter*y*

C Find other words that have the same sound

Possible answers:

1 b*a*sed, sc*a*le, repl*a*ce
2 dr*o*p, c*o*pper, electr*o*nic
3 n*u*mber, c*o*ver, pl*u*s
4 f*i*ller, pos*i*tion, s*y*stem
5 cond*ui*t, c*oo*l, scr*e*wdriver
6 d*a*mage, v*a*lue, sl*a*ck
7 pr*o*tection, b*e*dding, corr*e*ct
8 t*ur*n, t*er*minal, f*ir*st
9 insulati*o*n, diag*o*nal, f*o*cus, p*e*rformance, original, motor,
10 capacit*y*, tax*i*, easil*y*,

Reading: accident narrative summaries

A Decide what the main cause of each accident was

1 weather conditions
2 mechanical failure
3 pilot error
4 mechanical failure / human error

Unit test

A test for this unit is available at:

http://www.garneteducation.com/reps/documents/
1253/SDT-u9-test.pdf

Contact your local Garnet Education
representative for information about how to
access these resources.

Objectives

To raise awareness of key terms relating to flight information.

To practise reading and categorising items in a text.

To raise awareness of the use of diphthongs in the text.

Language

Word combinations, e.g., *landing gear, weather conditions*.

Vocabulary connected with information systems, e.g., *collision, trim, attitude*.

Speaking

A Activate schemata: important information for taking off, cruising and landing

- Ask students what the three main stages of a flight are (taking off, cruising and landing).

- Ask them to work individually and make a list of factors that are important to a pilot during these flight stages. Refer them to the pictures of cockpit instruments to stimulate ideas.

- Give them one example before they start, e.g., *weather conditions*.

- Ask them to compare their lists with a partner.

- Feed back as a whole class and make a combined list on the board.

Answers

Answers depend on the students.

Reading

A Read for gist: compare the ideas in the text with students' lists

- Set the task for individual work and pairwork checking.

- Tell students to find the information as quickly as possible; set a time limit.

- Feed back as a whole class briefly.

Answers

Answers depend on the students.

B Read for detail and categorise information

- Make sure students understand the task and what the abbreviations stand for, e.g., W = warning.

- Set for individual work and pair-checking. Monitor and check for problem vocabulary. Feed back as a whole class.

Answers

External conditions (E):

wind speed

air temperature

local traffic

weather conditions

Performance (PE):

air speed

landing gear status

trim

engine speed

ground speed

cabin pressure

vertical speed

fuel supply

oil pressure

battery status

cabin temperature

Warning (W):

collision

cabin pressure drop

fire

hot battery

generator voltage low

electrical system failure

fuel low

Position (P):

altitude

height

attitude

distance from departure point

distance to destination

heading

Vocabulary

A Word combinations

- Set the task for individual work and pairwork checking. Point out that students can find unknown combinations in the text above.

- Feed back as a whole class.

- Note the stress on the noun + noun compounds is on the first element, e.g., <u>ground</u> speed, in the adjective + noun compounds on the second element, e.g. local <u>traffic</u>, and in the multiple-word expressions on the penultimate word if that word is a noun, e.g., electrical <u>system</u> failure. There are no examples here, but if the penultimate word in a multiple-word expression is an adjective, then the last word carries the main stress, e.g., unidentified flying <u>object</u>.

Suggested answers

landing – gear

wind – speed

ground – speed

cabin – pressure

pressure – drop

distance from – departure point

fuel – supply

weather – conditions

electrical system – failure

local – traffic

hot – battery

oil – pressure

B Check the meaning

landing gear – the wheels that support an aircraft on the ground

wind speed – the speed at which air is moving

ground speed – the speed at which an aircraft in flight is moving over the ground

cabin pressure – the air pressure inside an aircraft cabin

pressure drop – a decrease in pressure

distance from departure point – how far the aircraft is from where it took off

fuel supply – the system of carrying fuel to an engine

weather conditions – the current state of the atmosphere outside an aircraft

electrical system failure – when an electrical system stops working

local traffic – other aircraft in the air space nearby

hot battery – a battery in which the temperature is higher than normal

oil pressure – the pressure of the oil in a system

Pronunciation

A Underline the long vowel sounds

- Remind students that differences in vowel length are important in English, e.g., students need to differentiate long and short vowels such as *sheep* and *ship*, as well as pure vowels and long vowel sounds (or *diphthongs*), e.g., *let* and *late*.

- Set the task for individual work.

- Elicit or model the five diphthong sounds and ask the class to repeat /eɪ/, /aʊ/, /aɪ/, /əʊ/ and /ɔɪ/.

B/C Find words in the text with long vowel sounds

- Ask students to work with a partner and find more examples in the list in the Vocabulary exercise and in the text in the Reading section. They should choose words with the same sounds as each of the words in the box.

- If you prefer, give each pair or group of students one diphthong sound to look for.

- Feed back onto the board, drilling problem sounds as necessary.

Possible answers

/eɪ/
failure, generator, information, safety, communication, data

/aʊ/
ground, nowadays

/aɪ/
supply, pilot, flight, height, fire,

/əʊ/
local, low

/ɔɪ/
oil, point

D Pronunciation practice: say word combinations with correct stress and vowel sounds

- Set for pairwork. Students should take it in turns to test each other on the pronunciation of the items in the list. Monitor and assist.

- Feedback with the whole class and correct any errors of pronunciation.

Answers

Answers depend on the students.

- Now have the students do the Workbook exercises for Unit 10, Lesson 1.

Workbook answers

Language: word combinations

A Complete the word combinations

1 pressure drop
2 fuel supply
3 weather conditions
4 ground speed
5 cabin pressure
6 distance from departure point

Language: word order in indirect questions

A Correct the word order

1 The pilot needs to know what the position of the aircraft is.

2 It is important to know how external conditions can affect the safety of the aircraft.

3 You can see if the temperature is too high.

B Finish these sentences

1 The fuel gauge tells you how much fuel is in the tanks.

2 To avoid overheating, you need to know what the temperature is.

3 By looking at the instruments, he can see how far away the ground is.

4 The pilot can tell if there is danger of fire.

5 Air traffic controllers tell him when there is bad weather.

Lesson 2   The basic six

Objectives

To practise listening and note-taking.

To raise awareness of language related to six basic cockpit instruments.

To use language to distinguish between facts and possibilities.

Language

Zero conditional and first conditional, e.g., *The lights come on if you press this button.*

Vocabulary for cockpit instruments, e.g., *turn coordinator.*

Speaking

A Match cockpit instrument diagrams with vocabulary

- See if students can guess what the figures on the different instruments represent.
- Set the task for individual work and pairwork checking.
- Feed back with the whole class.

Answers

1 attitude indicator

2 vertical speed indicator

3 altimeter

4 gyro compass

5 air speed indicator

6 turn coordinator

B Discuss the function of the instruments

- See if students can say what each instrument measures, but do not go into too much depth, as the listening text will explain this in more detail.
- Feed back on the board.

Answers

1 The attitude indicator informs the pilot of the orientation of the aircraft in relation to the horizon.

2 The vertical speed indicator informs the pilot of the rate of descent or climb.

3 The altimeter measures the altitude of the aircraft (the height at which it is flying).

4 The gyro compass is used to find true north.

5 The air speed indicator measures and displays the speed of the aircraft through the air.

6 The turn coordinator displays information to the pilot about the rate of turn and the rate of roll.

Listening and writing

A 🔊 **Listen and identify the position of the instruments (CD3 Track 1)**

- Check students understand the task.
- Set the task for individual work and pairwork checking.
- Feed back with the whole class.

Answers

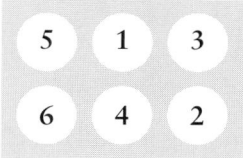

| 5 | 1 | 3 |
| 6 | 4 | 2 |

B–D 🔊 **Listen and take notes about each instrument (CD3 Track 1)**

- Set the task for individual work.
- Note that you may have to play the recording more than once as there is quite a lot of information to note down. Point out that information is not always given for every category in the table, e.g., the instructor does not always mention why an instrument is needed.
- Ask students to compare notes with a partner.
- Feed back with the whole class.
- As an option, give out the tapescript so that students can follow it while they check their answers.

Suggested answers

	what it tells pilot	why needed	how it works
altimeter	height above sea level		weighs air pressure above plane
attitude indicator	shows wing position in relation to horizon and whether nose is pointing up or down	needed if cockpit visibility is poor, e.g., in cloud	shows position of wings relative to the horizon
airspeed indicator	how fast plane is going	ensures pilot maintains enough speed not to stall	measures pressure of moving air against front of plane
turn coordinator	checks plane is turning correctly	prevents plane slipping sideways when banking + back-up if attitude indicator fails	
gyro compass	indicates direction	magnetic compass not always accurate	uses an internal gyroscope
vertical speed indicator	measures rate of climb or descent		measures changes in pressure

Pronunciation

A Mark the stressed words

- Ask students what kind of words are normally stressed (important words, nouns, verbs and adjectives) and which words are not normally stressed (prepositions, articles, auxiliary verbs and so on).
- Set the task for individual work and pairwork checking.

Answers

No answers at this point.

B 🔊 **Listen and check answers (CD3 Track 1)**

- Play the recording through once.
- Feedback with the whole class.
- Ask students to work with a partner and practise saying the sentences aloud.
- Listen to some examples with the whole class.

Answers

1 The <u>lay</u>out of the <u>pan</u>el is <u>like</u>ly to be the <u>same</u> in any small <u>plane</u> you <u>fly</u>.

2 You <u>can't</u> <u>look</u> at an <u>in</u>strument and then <u>switch</u> <u>off</u>.

3 <u>How</u> does it <u>work</u>?

4 You'll <u>shake</u> it and <u>break</u> it.

5 <u>This</u> <u>meas</u>ures the <u>aircraft's</u> <u>rate</u> of <u>climb</u> or <u>des</u><u>cent</u>.

C Practise saying the sentences aloud

- Drill the sentences as necessary.
- Give students an opportunity to take it in turns saying each sentence to a partner.

Answers

Answers depend on the students.

Language

A Identify types of conditional

- Ask students to read through the examples in the Language Box.
- Focus on the different <u>form</u> used in *if* sentences with possible results.
- Set the task for individual work and pairwork checking.
- Feed back with the whole class.

Answers

True

Possible

True

Possible

B Find examples of conditionals in the tapescript

- Set the task for individual work and pairwork checking.
- Feed back with the whole class.

Answers

True

It's easier to adapt if you change to flying a different plane.

Possible

We can run through them again if you like.

If I fly into a different airspace, I might need to recalibrate it.

If you're in cloud, this can be a life-saver.

If you go too slow, you'll stall.

If you go too fast, you'll damage the plane.

I'll give you more explanation, if it's still not clear.

C/D Write and compare sentences using zero and first conditionals

- Set the task for individual work and pairwork checking. Tell students they can look at the tapescript if they wish.
- Students should compare sentences with a partner and check that the grammar is correct.
- Feed back with the whole class. Compare students' ideas with the model below; project it onto an OHT if possible.

Possible answers

1 If you don't calibrate your altimeter before take-off, it may give an incorrect reading.

2 If the battery becomes overheated, the instrument may malfunction.

3 If a magnetic compass is moved around a lot, it may give an incorrect reading.

4 If you can't see clearly out of the cockpit, an attitude indicator can be a life-saver.

5 If you sit in the cockpit of any light aircraft, you will see the same six basic instruments.

- Now have the students do the Workbook exercises for Unit 10, Lesson 2.

Workbook answers

Language: movement and direction

A Complete the sentences with a word or expression from the box

No answers at this point.

B Check answers against the tapescript

1 against
2 through
3 up
4 sideways
5 up or down
6 below; down
7 out of
8 anticlockwise; relative to

Reading and vocabulary: instruments

A Match the correct instrument with each paragraph

1 altimeter
2 attitude indicator
3 vertical speed indicator
4 gyro compass
5 turn indicator
6 air speed indicator

Lesson 3   Digital revolution

Objectives

To practise reading and note-taking.

To practise asking and responding to questions about analogue and digital instruments.

To focus on the language of tendency, e.g., *x is prone to ~ing*.

Language

Frequency expressions.

Vocabulary connected with digital and analogue instruments, e.g., *dial*, *display*.

Speaking and vocabulary

A Activate vocabulary

- Introduce the topic by writing the word *thermometer* on the board.
- Ask students what a thermometer is used for and what it looks like. Use the pictures to establish the idea that there are different kinds of thermometer, e.g., *digital* and *mercury*.
- Set the task as pairwork.
- Feed back with the whole class.

Answers

1 spring balance
2 magnetic compass
3 digital watch
4 digital multimeter
5 mercury thermometer
6 electricity meter
7 digital thermometer

B Discuss differences between analogue and digital instruments

- Set for pair or group work.
- Feedback with the whole class.

Answers

1 They measure different things: time, direction, temperature, current/voltage, weight.

2 analogue: spring balance, magnetic compass, mercury thermometer

   digital: digital watch, digital multimeter, digital thermometer, electricity meter

3 The digital thermometer displays its reading in numbers on a digital display. An analogue thermometer is transparent glass filled with a liquid, e.g., mercury. The level of the mercury is measured on a scale. A digital thermometer can be made of plastic. It has no scale, but has an LCD display showing the temperature as numbers.

4 analogue: a type of physical representation – has a mechanical dial and pointer to show quantity and degree.

   digital: display uses only numbers (digits) to show quantity and degree. Devices are 'solid state', i.e., they have no mechanical parts.

Reading and writing

A Question formation

- Set the task for individual work. If you wish, expand the first sentence together as a group, e.g., *What are analogue instruments used for?*

- Feed back with the whole class and make sure their questions have been formed accurately.

Suggested answers

1 What are they used for?

2 What does analogue/digital mean?

3 What are the advantages?

4 What are the disadvantages?

B Read to answer the questions

- Nominate one student in each pair as A and the other as B.

- Ask students to find the answers to the questions they have written in their text only. Tell them not to look at the other text.

- Tell them to use note form to make notes more quickly.

Suggested answers

Text A

1 older instruments, e.g., electronic equipment in aircraft

2 a type of copy – has a dial and pointer to show quantity and degree

3 work in a continuous way; show a complete range of readings; direct relationship between input to meter and output on display

4 likely to be affected by extremes of temperature and pressure; prone to malfunction; wear out and may require lubrication; difficult to read accurately

Text B

1 a lot of modern control, measuring and communications equipment

2 display shows only digits or numbers – solid state devices

3 have no moving parts; less fragile; do not wear out or require lubrication; less affected by extremes of temperature and pressure, and mechanical shocks; easy to read quickly; cheaper and simpler to make and repair

4 do not always show the operator maximum and minimum on the scale being measured; do not show gradual changes in same obvious way as analogue devices

Speaking

A Ask and answer questions

- Set the task as pairwork. Monitor for problems.

- Feed back with the whole class. Use the suggested answer from Reading and writing Exercise B to summarise on an OHP.

Answers

Answers depend on the students.

B Personalisation: discuss preferences

- Look back at the list of advantages and disadvantages and have a short discussion about individual preferences. Encourage students to expand on their preferences.

Answers

Answers depend on the students.

Language

A/B Focus on language for expressing tendencies and patterns

- Set this up in the usual way. Students should circle the expressions when they find them.

- Feed back with the whole class. Point out that *not always* can be used in a positive sense and *usually* and *tend to* in a negative sense, depending on the context.

Answers

not always: Text B, line 12 (negative – can be used in positive and negative sentences)

be prone to: Text A, line 13 (negative – normally used in negative sentences)

tend to: Text B, line 6 (positive – can be used in positive and negative sentences)

be affected by: Text A, line 12 (negative – normally used in negative sentences)

(be less affected by): Text B, line 6 (positive)

usually: Text B, lines 2 and 5 (positive – can be used in positive and negative sentences)

Extension

- Elicit other sentences using the expressions from the previous exercises by writing the following prompts on the board:

  Digital watches tend to [bare infinitive of verb]

  Analogue watches are affected by [noun]

  Digital meters usually [present or past simple verb]

  Analogue meters are prone [verb + ~*ing*]

- Focus on the grammar used after each one and elicit possible endings for each sentence.

- Now have the students do the Workbook exercises for Unit 10, Lesson 3.

Workbook answers

Language: prepositions

A Put the words on the right in the correct place in the sentences

1 The pointer moves an equivalent distance <u>on</u> the scale.

2 The weather can change a lot <u>from</u> one day to another.

3 It is always <u>between</u> a maximum and a minimum reading.

4 This design is more prone <u>to</u> malfunction than the other.

5 This feature makes it more suitable <u>for</u> use in aircraft.

6 These are less affected <u>by</u> ambient temperature and pressures.

7 It works <u>in</u> a different way from other devices.

8 It is the same <u>as</u> driving a car.

Language: tendency

A Find the expressions in the Reading texts

not always	Text B, line 12
be prone to	Text A, line 14
tend to	Text B, line 7
be affected by	Text A, line 13
usually	Text B, line 3

B Negative and positive usages

not always negative

… they do not always show the operator the maximum and minimum …

be prone to negative

They are also more prone to malfunction …

tend to positive

They also tend to be less affected by extremes of temperature …

be affected by negative

… analogue devices are more likely to be affected by extremes of temperature …

usually neutral / positive

… and usually has a display which shows only digits or whole numbers.

… they are usually less fragile …

C Complete the following sentences using the expressions

1 usually
2 is prone to
3 not always
4 be affected by
5 usually
6 are prone to
7 tend to
8 is affected by

Writing: advantages and disadvantages

A The greatest advantage or disadvantage

Answers depend on the students.

Lesson 4   Know-how

Objectives

To introduce terminology and concepts associated with avionics.

To review verbs connected with avionics operations and activities, e.g., *fine-tune*, *coordinate*.

To practise scan-reading.

Language

Common verb + noun collocations, e.g., *diagnose a fault*.

Speaking

A Discuss what avionics technicians need to know

- Elicit what students understand by the term *avionics*. Ask students to read the brief paragraph about avionics to check.

- Set the task for brief whole-class or group discussion. Draw attention to the examples given.

- Feed back with the whole class. Elicit and write a few ideas on the board, e.g., *installation and maintenance of electronic systems, how to calibrate instruments*, etc.

Answers

Answers depend on the students.

Reading I

A Scan reading

- Set the task for individual work and pairwork checking.

- Feed back with the whole class. Most of the vocabulary should be familiar to students, but clarify any problems, e.g., with *calculus*, *chips*, etc.

Answers

Answers depend on the students.

B Matching exercise

- Set the task for individual work and pairwork checking.

- Feed back with the whole class.

Answers

1	c	4	a
2	c	5	a
3	b		

Vocabulary

A Make verb/noun collocations

- Highlight the example that is given on the board with the whole class: *interpret drawings – data*. Highlight the fact that only a restricted number of nouns can go with certain verbs. Elicit other nouns that can go with *interpret*, e.g., *figures, handwriting, document*.

- Set the matching task for individual work and pairwork checking.

- Feed back with the whole class. Elicit any alternative answers, e.g., *install – set up equipment*.

Answers

1	**interpret** drawings	data
2	**coordinate** work	a schedule

3	**install** wiring	cables
4	**set up** a support system	an account
5	**fine-tune** an engine	equipment
6	**carry out** an inspection	a test
7	**adjust** equipment	components
8	**diagnose** faults	malfunctions

B Extend collocations

- Clarify that students should think of another noun for each verb and add it to Column 3. Set for individual work and pair-checking.

- If students cannot think of a noun for each row, they should leave a gap and add one after they have scanned the text in the Reading exercise.

Answers

Answers depend on the students.

Reading II

A Scan reading to find collocations

- Ask students to underline the verbs from Vocabulary Exercise A in the text and see what nouns they go with.

- Feed back with the whole class. Encourage students to make a note of any new collocations.

- Get students to read the text again.

Answers

interpret flight test data	line 5
coordinate work	line 12
install electrical and electronic components	line 14
set up ground support equipment	line 1
fine-tune avionics equipment	line 4
carry out functional flight tests	line 2
adjust malfunctioning components	line 10
diagnose malfunctions and systemic performance problems	line 5

B Read and discuss duties of an avionics technician

- Set the task. Students should look at the task instructions and then read the text again carefully, using a dictionary to check unfamiliar vocabulary.

- Each student or pair of students should choose the activities they think are most challenging/interesting.

- Feed back by having a whole-class discussion comparing ideas.

- Encourage students to discuss the activities from memory, without reading from the text.

Answers

Answers depend on the students.

- Now have the students do the Workbook exercises for Unit 10, Lesson 4.

Workbook answers

Language: spelling

A Fill in the vowels in these words

1	layout	7	troubleshoot
2	recalibrate	8	avionics
3	relative	9	diagnose
4	airborne	10	equipment
5	magnetic	11	performance
6	analogue	12	instruments

Pronunciation: syllables and stress

A Group together the words which have the same number of syllables

2-syllable words

systems	records
adjust	circuit
repair	science
install	radio

3-syllable words

technician	malfunctions
equipment	component
regulate	statistics
troubleshoot	instrument
diagnose	

4-syllable words

avionics	electronic
systematic	technology

B Match the parts of compound nouns and mark the main stress

1 c (flight test proc<u>e</u>dures)

2 a (<u>ci</u>rcuit board)

3 b (<u>e</u>lectromechanical drawings)

4 f (<u>main</u>tenance systems)

5 h (<u>e</u>lectronic data)

6 e (<u>junc</u>tion box)

7 d (malfunctioning comp<u>o</u>nents)

8 g (rep<u>air</u> work)

Language: giving information about jobs and activities

A Give more information

Answers depend on the students.

Writing: systems

A Write sentences to describe the systems

Answers depend on the students.

Objectives

To raise awareness of conventions and language used in technical drawing.

To review vocabulary, symbols and descriptions for mechanical, electrical and electronic parts, e.g., *amplifier*, *gearbox*.

Language

The verb *make* + *it* + adjective, e.g., *This makes it more complex.*

Introduction

- Review types of drawings and schematics that students are already familiar with.

Speaking

A   Discuss circuit diagram layout

- The diagrams show the fact that schematics can be drawn in different ways to show the same thing. Both diagrams are representations of the same circuit (a switch for dimming a lamp).

- Elicit what the symbols are in the diagram. It might be helpful to draw the symbols for a battery, a lamp and joining wires on the board, or to refer students to the Electrical Symbols section on page 263.

- Set the task for pairwork. Students should not worry about describing the diagrams in too much detail, but should be able to notice that the two diagrams show exactly the same things, e.g., the wire junctions are on either side of the battery in each case.

- Feed back to the class. Elicit what makes a diagram clear and easy to understand.

Answers

1 circuit diagrams

2 A is laid out in a clearer, more logical way than B.

3 Answer depends on the students.

- For further information about these diagrams, go to:

  www.22142.zen.co.uk/Prac/readdiag.htm

Reading

A   Match texts with diagrams

- Draw attention to diagrams c and d. Elicit what students think each one shows (c shows a DC motor and d shows a gyro compass).

- Set the task for individual work and pairwork checking.

- Feed back with the whole class. Elicit what each diagram is useful for.

- Ask students to name some of the parts shown in each of the diagrams. Write some parts on the board, e.g., *power source, gearwheels, resistor, dials, oscillator, amplifier, cable, switch, fuse, capacitor*, and get students to work in pairs to locate where they are and in which diagram. Alternatively, they could test each other by pointing to various images and symbols and eliciting names from their partner.

Answers

1 Text 1 – Diagram d; Text 2 – Diagram c

2 Diagram c – layout diagram; Diagram d – electromechanical diagram

Extension

- See if students can identify components in Diagram d which are mechanical, e.g., gears, dial; *electrical*, e.g., fuse, cable, switch; and *electronic*, e.g., amplifier, oscillator.

Language

A Focus on *make* + *sth* + adjective

- Write the following sentence from the text onto the board: *An overview like this makes it easier for an avionics technician to understand how the assembly works.* Highlight the verb *make* and elicit which part of speech follows it.

- Ask students to read the information in the Language Box carefully. Model a few more example sentences.

- Highlight the difference in form between *makes it easier/more difficult to do something* and *makes it likely that something will happen*.

- Set the task for individual work and pairwork checking.

- Look at the first prompt together with the class. Ask what other word(s) could replace *hard* (*difficult*).

- Feed back with the whole class.

Suggested answers

1 Poor lighting makes it hard (for the operator) to see the job properly.

2 Standard instrument layout makes it possible for pilots to fly different types of aircraft.

3 Multi-grip pliers make it easier to grip objects firmly.

4 Digital displays make it easier to read information quickly.

5 Bad weather makes it more difficult to fly aircraft.

6 A cockpit turn coordinator makes it possible to check if the wings are level or banked.

7 Workplace safety regulations make it less likely that accidents will happen.

8 Internationally recognised symbols make it easier to follow diagrams.

9 Exploded drawings make it possible to see details that are normally hidden.

- Now have the students do the Workbook exercises for Unit 10, Lesson 5.

Workbook answers

Language: word-building

A Make words from the roots

Answers depend on the students.

B Write sentences

Answers depend on the students.

Language: differences in meaning

A Explain how the pairs of words are different

1 no difference

2 Test data is the information which a test provides, whereas test equipment includes all the devices used in doing a test.

3 A circuit board is a physical object, whereas a circuit diagram is only a representation on paper/screen.

4 no difference

5 Repair work refers to work carried out to correct a specific fault, while maintenance work may also be routine cleaning/checks.

6 The installation procedure is the series of steps that a technician takes in doing an installation; the installation manual is the book where the procedure is written/described.

7 A component is one element, e.g., a diode, switch or capacitor, of the circuit that electricity flows round.

8 A gyroscope is a wheel which spins inside a freely movable frame; a gyrocompass is a type of compass (for measuring direction) that uses a gyroscope as a major component.

9 no difference (in electronics)

10 Communications systems use radio to transmit messages. Radar is a system used to identify the position or speed of objects, e.g., aircraft or ships.

Language: passive forms

A Complete the sentences with the correct passive form of the verb in brackets

1 If a component is faulty, it <u>should be</u> replaced.

2 There's no reading because these connections <u>have not been soldered</u> properly.

3 The system is malfunctioning. It has got <u>to be overhauled</u>.

4 There's a crack in the board. It seems <u>it was put under</u> bending force.

5 Your unit isn't ready. It is <u>being repaired</u> at the moment.

6 It was dropped by an operator during assembly and <u>thrown away</u>.

7 All of these assemblies <u>will be tested</u> tomorrow.

8 Interference can result if a PCB <u>is not arranged</u> properly.

Lesson 6 Time is money – troubleshooting

Objectives

To extend vocabulary for describing electrical faults and solutions (in the context of PCBs).

To practise listening for detail and completing outline notes.

To raise awareness of how words are linked in a stream of speech.

Language

Pronunciation – linking, e.g., *throw_it_in*.

Vocabulary related to troubleshooting, e.g., *faulty*, *cracked*.

Speaking

A Discuss the layout of PCB diagrams

- Set the task for individual work and pairwork checking.
- Feed back with the whole class.

Answers

The key difference between the PCBs is that the first one is neatly laid out with components either parallel or at right angles to each other, while the second one is a mess. This is referred to in the recording later in the lesson.

Vocabulary

A Focus on verbs connected with PCB faults

- Set the task as pairwork.
- Feed back with the whole class.

Suggested answers

1 a circuit board (or any piece of equipment) can develop a fault

2 a technician can repair a fault

3 a repair tool or device such as a multimeter can trace a fault

B Categorise troubleshooting vocabulary

- Set the task for individual work and pairwork checking. Monitor for problems with comprehension and pronunciation.
- Feed back with the whole class. Deal with any vocabulary problems that might arise.
- Elicit personalised examples for the 'bad news' terms.

Answers

good news: repairable, OK, fitted neatly, tidy

bad news: fault comes and goes, cracked, faulty, badly arranged, badly made, broken away, intermittent fault, loose, mishandled, put under stress, not soldered properly, insecure connection

Listening

A1 ⊙ Listen for specific information (CD3 Track 3)

- Tell students to look again at the list of words and expressions in Vocabulary Exercise B.
- Tell them they will hear some (but not all) of these words and expressions on the recording and ask them to tick the ones they hear.
- Play the recording once and ask them to check with a partner.
- Feed back with the whole class.

Answers

(in the order in which they are mentioned on the recording)

tidy

put under (a lot of) stress

not soldered properly

insecure connection(s)

intermittent fault

it (the fault) comes and goes

OK

faulty

cracked

badly made

broken away

badly arranged

A2 ⊙ Listen and complete a table (CD3 Track 3)

- Ask students to look at the problems and solutions table. Tell them to think about the problems and possible solutions that were mentioned and make brief notes.
- Play the recording again and encourage students to check and add to their notes.
- Pair-check and then feed back with the whole class.

problems	solutions
1 faulty component: resistor	replace the resistor
2 cracked copper track: connection	replace the whole PCB (or repair the connection if PCB cannot be replaced)

B ⊙ Listen and complete outline notes on PCBs (CD3 Track 3)

- Ask students to read through the notes on PCBs carefully first.
- Ask them to listen to the recording and fill the gaps in the notes.

- Play the recording once and check progress.
- If necessary, play the recording again.
- Feed back with the whole class.

Answers

PCBs

Layout

v. imp to keep board <u>neat</u> and <u>tidy</u>.

Direction of comps. should be <u>parallel or at</u> <u>right angles</u> because <u>easier to work with</u>.

2 basic kinds of fault:

<u>Component not working</u> and <u>copper path</u> <u>broken</u>.

Specific faults, e.g., <u>cracks</u>, insecure connections

Causes, e.g., <u>board mishandled</u> or <u>put under</u> <u>stress or vibration</u>.

Faults sometimes intermittent, i.e., <u>comes</u> <u>and goes</u>.

Repair

1st check visual – use <u>strong light</u> and magnifying glass – check for visible <u>cracks</u> & bad connections.

2nd check PCB: connect to a multimeter and press on component side.

3rd check indiv. <u>components</u>.

If comp. faulty, always <u>replace it</u>.

Pronunciation

A 💿 Raise awareness of linking (CD3 Track 4)

- Ask students to read the information in the Skills Box carefully.
- Highlight the fact that this phenomenon applies to all words beginning with a vowel sound.
- Write a simple example on the board, e.g., *pick it up*, and focus on the way the sounds run together. Ask students to repeat *pi ki tup*.

- Now ask them to look at the six sentences from the tapescript and mark the places where they think the links will be. This should be quite straightforward if they bear in mind it will be before words beginning (or sometimes ending) with a vowel sound.
- Play the recording so they can check their predictions.

Answers

1 If the board_is put_under_a lot_of stress,_ it might crack.

2 One_of the problems with_insecure connections _is that the fault_is often_intermittent.

3 Throw_it_in the rubbish.

4 Press down with_a pencil_in different places.

5 If the board_itself seems_OK, put very gentle pressure_on_each component.

6 Here_is the circuit diagram_and_a list_of the readings.

B Practise saying the sentences aloud

- Ask students to practise saying the sentences aloud. This can either be done in pairs or as choral drilling with the whole class.

- If you have time, revise the passive by asking students to read the tapescript and underline the passive structures. Discuss why they are used.

- Now have the students do the Workbook exercises for Unit 10, Lesson 6.

Workbook answers

Writing: workshop session notes

A Write notes

Possible answers:

Workshop Training

Date:

Tutor: **Mr Patel**

Area (delete where applicable):

~~machine shop / lab~~ / electronics / ~~clean room / mechanics~~

Job: PCB troubleshooting

Work done (inc. materials, techniques, faults, repair, etc.):

Checked PCB and found fault.

(Rasheed) Checked resistor with multimeter. No reading – replace resistor.

(Carlos) broken copper path – remove component, repair path and replace component.

Important points to remember:

Keep components tidy – parallel or at right angles

Check PCB for cracking / good connections

Use strong light to inspect

Usually throw away faulty PCBs/components (time is money!)

Language: corrections

A Choose the correct sentences

1 It's quite difficult to find faults in a badly arranged board.

2 A strong light and a magnifying glass is the best way to find cracks.

3 One of the components is probably loose.

4 The PCB should not be untidy.

5 They should be at right angles to each other.

Reading: avionics equipment

A Read the text and complete the gaps

1 avionics

2 systems

3 chips

4 components

5 fault

6 checks

7 multimeter

8 oscilloscope

Lesson 7 — Procedure

Objectives

To extend vocabulary connected with PCB repair.

To practise skimming and scanning instructions.

To practise speculating about the purpose of individual requirements.

Language

Modal verbs for speculation – *might, can, could, may.*

Verbs and nouns connected with PCB overhaul, e.g., *solder, trim, lacquer.*

Adverbial constructions for instructions, e.g., *whenever possible, thoroughly.*

Vocabulary

A/B Review terms connected with cleaning and repair

- Set the task for individual work and pairwork checking.

- Encourage students to use monolingual English dictionaries to carry out this task.

- Feed back orally.

- Make sure students understand the meaning of all the terms.

- Check students can pronounce all these terms with the correct word stress, e.g., noun + noun: <u>solvent</u> cleaner, <u>heat</u> sink and <u>solder</u> sucker; adjective + noun: short <u>circuit</u>, foreign <u>matter</u>, protective <u>lacquer</u>, short <u>gap</u>.

Answers

solvent cleaner
    a solution used to clean different equipment

short circuit
    a short circuit allows a charge to flow along a different path from the one intended

foreign matter
    anything that does not belong in a particular place

solder sucker
    a device used to remove solder from a PCB

grease
    industrial lubricant

a heat sink
    a device that is placed in contact with a component's hot surface to stop the heat affecting other components

reinstallation
    the process of installing something again

protective lacquer
    a coating with a hard finish that protects the surface of something

a small gap
    a small space between two different things

Reading

A Scan reading

- Ask students to look in the text and find and underline the nine expressions from the Vocabulary exercise. (They are all in the text.)

- Feed back with the whole class.

Answers

a solvent cleaner	3
a short circuited	6
foreign matter	2
a solder sucker	6
grease	3

a heat sink	5
reinstallation	9
protective lacquer	9
a small gap	7

B Read for gist

- Ask students to read the text quickly and answer the question.

- Feed back with the whole class. Point out that each sentence is an imperative.

Answer

1 General instructions for repairing PCBs.

C Read and fill the gaps with the correct verb

- Ask students to read the vocabulary items carefully before they begin the task.

- Check all the words are understood, especially *trim*. Use gestures and examples to clarify.

- Note that *remove* is used twice.

- Set the task as individual work and pairwork checking.

- Feedback with the whole class.

Answers

1	Avoid	6	Remove
2	Brush or blow	7	Ensure
3	Remove	8	Trim
4	Solder	9	Spray
5	Clip on		

D Match instructions to pictures

- Exploit the pictures. See if students can say what each one is.

- **Note:** There are two sprays: discuss which one is an aerosol.

- Set the task as pairwork.

- Feedback with the whole class.

Answers

1	(solder sucker)	6
2	(heat sink)	5
3	(pliers for trimming wires)	8
4	(gap between components on a PCB)	7
5	(lacquer spray)	9
6	(brush)	2
7	(suitable stand for a PCB)	1
8	(aerosol solvent cleaner)	3
9	(soldering iron)	4

Language

A Practise using modal verbs to speculate

- Set the task for pairwork.

- Feed back with the whole class.

- **Note:** There are a number of possible answers for each one, and *may*, *might*, *can* and *could* are more or less interchangeable, although *could*, *may* and *might* suggest something is less likely than *can*. For example, *Excess solder can cause short circuits* is more likely than *Excess solder might cause short circuits*.

Answers

Answers depend on the students.

B Discussion

- Set the task for pairwork.

- Feed back with the whole class.

- **Note:** There are a large number of possible answers for each of these. Accept any answers that are logical.

Answers

Answers depend on the students.

- Now have the students do the Workbook exercises for Unit 10, Lesson 7.

Workbook answers

Language: spelling

A Correct the spelling

1	correct	**9**	correct
2	spray	**10**	correct
3	circuit	**11**	correct
4	conductor	**12**	aerosol
5	components	**13**	grease
6	particle	**14**	correct
7	correct	**15**	cracked
8	excess		

B Copy ten other words

Answers depend on the students.

Language: quantities with *of*

A Write the words from the box

1	solder	**5**	aviation fuel
2	heat	**6**	protective lacquer
3	dust	**7**	aluminium
4	oil	**8**	pliers

B Think of expression with *of*

Answers depend on the students.

C Write sentences

Answers depend on the students.

D Write what's being described

Answers depend on the students.

Language: *allow, prevent* and *avoid*

A Verbs with similar meanings

1	stop something	prevent
2	permit something	allow
3	not do something	avoid

B Circle the correct option

1	avoid	**4**	prevent
2	prevent	**5**	allows
3	forces	**6**	allow

Objectives

To extend vocabulary to describe navigation systems and aids.

To practise listening to a lecture for detail.

To raise awareness of the use of linking words in a text.

Language

Vocabulary and abbreviations connected with navigation, e.g., *ATC*, *VHF*.

Linking words, e.g., *because, still, although, and, but, so*.

Vocabulary and speaking

A Communications devices

- Start the lesson by asking students if they know any means of communication that helps with navigation.

- Write the abbreviation *ATC* (air traffic control) on the board and ask the class what it stands for.

- Then ask them to look at the drawing of the Boeing 737 and ask if they know any other abbreviations.

- Set tasks *1* and *2* for pairwork. Point out that even if they do not know the words, they will be able to do tasks *1* and *2*.

- Feed back with the whole class. Elicit ideas for what each device is for, but do not confirm, as students will listen to a lecture explaining navigation aids.

Answers

1

ADF   automatic direction finder

GPS   global positioning system

TCAS   traffic alert and collision avoidance system

VHF   very high frequency

HF   high frequency

ELT   emergency locator transmitter

DME   distance measuring equipment

ATC   air traffic control

2 (Suggested answers)

automatic direction finder
> a navigation device used on aircraft

global positioning system
> a navigation system which uses the position of satellites in orbit round the earth (the globe) to detect the exact position of an object on the surface of the earth

traffic alert and collision avoidance system
> detects the presence of other aircraft

very high frequency
> the radio frequency range from 30 MHz to 300 MHz

high frequency
> the radio frequency range from 3 MHz to 30 MHz

emergency locator transmitter
> sends out distress signals in an emergency

distance measuring equipment
> measures distance by timing the delay of VHF radio signals

air traffic control
> the system used to communicate with ground control

B Look at the picture

- Have students look at the picture and say from memory what each acronym stands for.

Answers

Same as above in Exercise A.

Listening

A Check understanding of vocabulary

- Encourage students to use monolingual English dictionaries to carry out this task.
- Monitor the groups to check progress.
- Feed back with the whole class. Check concepts of any problem vocabulary, e.g., *quadrant = a quarter of a circular area.* Explain Morse code by clarifying that it is a system of communication using short and long telegraphic sound signals (represented by dots and dashes, e.g., SOS = … /---/…).

Answers

transmitter: a device that sends a radio signal

receiver: a device that receives a radio signal

quadrant: one of the areas created by dividing a circle equally into four

dashes and dots: long and short marks on paper, or long and short sounds used in Morse code

tower: the place in an airport where air traffic controllers are positioned

controller: air traffic controller; the person on the ground communicating with pilots in flight

path: the route followed by an aircraft; the flight path

navigation: planning and controlling the flight from A to B

radio signal sent: the signal that goes out from a transmitter

echo received: repeated sound reflected from an object, e.g., a radio signal in a radar system

frequencies: in radio, rates of oscillation of radio waves; different frequencies are used for different types of communication

chart: map or graph which display information; in air traffic control, it shows the position and speed of aircraft

B Pronunciation

- Model the pronunciation of each word.
- Highlight the stressed syllables, e.g., *trans<u>mit</u>ter.*

Answers

re<u>cei</u>ver	path
<u>qua</u>drant	navi<u>ga</u>tion
<u>dash</u>es and <u>dots</u>	radio signal <u>sent</u>
<u>tow</u>er	echo re<u>cei</u>ved
con<u>trol</u>ler	<u>fre</u>quencies

C 🎧 Listen for specific information (CD3 Track 5)

- Ask students to read through the five items carefully before they listen to the recording.
- Play the recording once and check progress.
- If necessary, play the recording again.
- Feed back with the whole class. If possible, project the diagrams onto the board and label them.

Answers

The items are mentioned in the following order:

two-way air-to-ground radio

four-course radio range

airport control towers

Very High Frequency Omni-directional Radio

radar

D Listen and label the diagrams

- Ask students to look carefully at the three diagrams on Course Book page 250. Elicit the names of items in the pictures.

- Play the recording again. Students should write in the titles and any other labels they can, e.g., *tower, echo, transmitter, controller.*

Answers

A ATC (Air Traffic Control)

B Radar

C Four-course radio

Language and writing

A Gap-fill exercise

- Introduce the activity by writing two short sentences on the board that can be linked with a linking word (for example, *It was cold* and *I stayed at home*). Ask the class to make one sentence linking these two ideas. Possible linking words here are *so, or, and.*

- Tell students to skim the text to establish that it is a summary of the main points of the lecture that they have just heard.

- Set the task for individual work and pairwork checking Feedback with the whole class. Focus on punctuation; elicit which connectors can start a sentence (*because, although*) and that commas are needed before *but* and *so.*

Answers

1 because	6 so
2 but	7 Because
3 so	8 Although
4 and	9 and
5 still	

B Linking ideas

- Set the task as pairwork.

- Ask students to use the linking words from Exercise A.

- Tell them there may be more than one possibility in some cases.

- Feedback with the whole class.

Answers

1 Workshops can be dangerous, <u>so</u> safety procedures are essential.

2 Digital displays are very useful, <u>but</u> analogue displays have advantages.

3 Radial engines are easier to cool than in-line engines, <u>and</u> they are smoother.

4 Humidity is important for ATC <u>because</u> it affects weather conditions, <u>so</u> it is important for pilots.

- **Note:** The comma in *3* is optional.

- If you have time, ask a few questions about the different navigation systems, e.g., *How precise are they?*

- Now have the students do the Workbook exercises for Unit 10, Lesson 8.

Workbook answers

Language: dependent events with *as*

A Answer the questions

1 The two events happen at the same time.

2 The situation is a long process.

3 The dashes and dots got louder because he flew towards the transmitter.

B Write sentences about dependent events

1 The temperature of gas rises as the pressure increases.

2 As the project gets longer, the overall cost goes up.

3 As you do the job, it becomes easier.

4 As control towers become more common, navigation becomes less dangerous.

5 As the speed increases, the pointer moves clockwise.

Language: word-building

A Adjectives that end in –based

Answers depend on the students.

B Adjectives that end in –free

Answers depend on the students.

C Write sentences using the adjectives

Answers depend on the students.

Language: expressions with *not*

A Answer the questions

Answers depend on the students.

Lesson 9 · · · · · GPS/ELT

Objectives

To practise word-building.

To practise reading and taking notes to summarise information in a text.

To practise asking/answering questions about navigation equipment.

Language

Question forms.

Vocabulary connected with satellite navigation, e.g., *acquisition, transmitter*.

Introduction

- Review the early navigation systems from the previous lessons, using the diagrams.

Vocabulary

A Word formation

- Do one or two examples on the board with the whole class, e.g., *locate – location, navigate – navigator – navigation – navigational*.

- Set the task as individual work with pairwork checking.

- Feedback with the whole class.

Suggested answers

acquire	acquisition
transmit	transmitter; transmission
receive	reception
track	track; tracking
inform	informer; information
navigate	navigator; navigation; navigational
activate	activator; activation
locate	location
measure	measurement
equip	equipment; equipped

B Write sentences using vocabulary

- Elicit one or two possibilities, e.g., *Higher frequency <u>transmitters</u> were built after the war.*

- Encourage students to look back at the tapescript for the previous lesson for ideas.

- Set the task for pairwork. Monitor closely and assist. Feed back onto the board.

Answers

Answers depend on the students.

Language

A Expand skeleton prompts into full questions

- Do the first one with the whole class:
  What kind of clocks are used?

- Point out that some of the sentences are active and some are passive, and that this exercise is a revision of question forms.

- Set the task as individual work with pairwork checking.

- Feedback with the whole class.

Answers

1 What kind of clocks are used?

2 How is the unit activated?

3 What does the GPS unit measure?

4 How are ELT units identified?

5 How does information reach the RCC?

6 What does the unit look like?

7 What is 'acquisition'? / What does 'acquisition' mean?

8 What do MCC and RCC stand for?

9 Are there any problems with the system?

10 How will the system be developed in the future?

Reading

A1 Read to find answers to five of the questions

- Divide the class into two groups, A and B.

- Ask students in Group A to read the text about GPS and Group B to read the text about ELT.

- They should identify five questions that are relevant to their text and answer them by making notes.

A3 Ask questions about the other text and take notes

- Students should be paired A and B to swap information about their texts. Tell them to take turns to ask their partner the questions they have <u>not</u> answered themselves. They can choose the order in which they ask them. Monitor and check for problems.

- Students use notes to answer their partner's questions. As students listen to the answers, they should make brief notes, so that they have made notes on all ten questions.

- Feed back briefly on any problems with the whole class.

Answers

Text A: GPS

1 atomic clocks

3 it measures the time delay between transmission and reception of each satellite's radio signal

7 finding the nearest three satellite signals

9 it is affected by atmospheric conditions

10 they will be made more precise and include ground transmitters at ends of runways

Text B: ELT

2 by the pilot selecting an emergency radio frequency manually or by automatic self-activation

4 they send out a signal, so can be identified to within 100 m

5 satellite transmits info. to a receiver system at MCC and is passed on to RCC

6 brightly coloured, waterproof and about 30 cm square

8 MCC = mission control centre
  RCC = rescue co-ordination centre

Speaking

A Give a short description of a navigation system

- Ask students to work in pairs and describe how their partner's navigation system works.

- Students should listen to their partner and correct where necessary.

Answers

Answers depend on the students.

B Read to check information

- Students read their partner's text and check the information.

- Feed back with the whole class. Discuss the differences in function between the two systems.

Answers

Answers depend on the students.

- More information about GPS can be found at:

  http://www.pilotfriend.com/training/flight_training/nav/global.htm

- Now have the students do the Workbook exercises for Unit 10, Lesson 9.

Workbook answers

Language: verbs with prepositions

A Find and circle the words

See Course Book page 199.

B Complete the following

1 communicate <u>with</u> the ground
2 connect something <u>to</u> something
3 know <u>about</u> external conditions
4 it increased <u>by</u> 30%
5 depend <u>on</u>
6 make a difference <u>to</u> the result
7 a fault <u>with</u> the system
8 remove something <u>from</u> an assembly
9 the gap <u>between</u> two components
10 responsible <u>to</u>
11 the distance <u>from</u> A to B

Language: verb forms

A Circle the correct form

Pupils should circle:

1	to be located	5	be made
2	sends	6	transmits
3	are required	7	is written
4	is affected	8	interpret

Writing: note form

A Rewrite the sentences in note form

1 Signal allows unit located quickly & accurately.
2 Transmitter sends location to receiver on ground.
3 GPS required to transmit on 406 MHz.
4 GPS accuracy affected by atmospheric conditions.
5 GPS more accurate in future.
6 Satellite transmits message every 12.5 mins.
7 Zulu time written in four digits.
8 On-board GPS interprets satellite signal.

Objectives

To practise listening and reading for very specific detail.

To interpret and correct notes.

To raise awareness of consonant minimal pairs.

Language

Vocabulary connected with metal inspection, e.g., *oscillator*, *eddy*, *current*.

Pronunciation, e.g., voiced and voiceless sounds.

Vocabulary

A Discuss differences in meaning

- Do the first example with the whole class.
- Set the task for pairwork.
- Feed back orally.

Answers

- *Alternating current* (AC) is a current the direction of which varies cyclically, while *direct current* (DC) flows in the same direction.

- A *crack* is a split in something, whereas a *void* is an empty space.

- An *eddy current* is created when a conductor moves across a magnetic field; a *magnetic field* is an area affected by magnetic power.

- *Permeability* is the property of something that allows a liquid or gas to pass through it; *durability* is the strength or toughness of a material.

- *Corrosion* is damage caused to metal by a chemical reaction; *dirt* is foreign matter.

- A *conductor* is a material that allows an electric charge to pass through it; a coil is a copper wire wound around a core to form an electromagnet.

- *Something is sound* if it has no faults or problems, the opposite of *something that is faulty*.

B Pronunciation

- Model the pronunciation of the words in Exercise A.
- Ask students to repeat chorally and individually.
- **Note:** The stress patterns in *magnetic field*, *eddy current*, *permeability*, *durability*, *corrosion* and *conductor*.

Listening

A Discuss the pictures

- Ask students to work with a partner and discuss what the pictures show and what the devices are for. They should check unknown vocabulary in a dictionary or in the Glossary.
- Feed back orally. Elicit whether any students have used a metal detector.

Answers

No answers at this point.

B Listen for gist: number the items as they are mentioned (CD3 Track 6)

- Ask students to look at the pictures again.
- Tell them to write the numbers *1* to *5* next to the pictures as they hear them on the recording.
- Play the recording once and check progress.

- If necessary, play the recording again.

- Feed back with the whole class.

Answers

1 metal detector

2 oscillator circuit

3 headphones

4 eddy current

5 cracks

C 💿 Listen and correct notes (CD3 Track 6)

- Ask students to read through the notes carefully. Point out that there are some mistakes in them and they need to be improved.

- Elicit the meaning of the abbreviations, e.g., *elec. prob. obj.*

- Play the recording once.

- Ask students to compare their notes with a partner. Monitor and check for problems.

- If necessary, play the recording again.

- Feed back with the whole class.

Answers

Metal-detecting instruments

- Use oscillator circuit to create ~~DC~~ AC voltage in coil.

- ~~Coil~~ Oscillator connected to headphones. Coil has elec. mag. field.

- Elec. mag. field in detector induces current in any ~~object~~ metal object on ground – eddy current in obj creates another mag. field.

- Instrument detects ~~shape~~ sound of other object.

- ~~Not exactly~~ Similar detectors used for airport security.

Eddy current inspection instruments

- Used to check for faulty, cracked metal parts ~~instruments~~ of planes.

- ~~Large~~ Hairline cracks ~~always~~ often under paint or dirty surfaces.

- Eddy current flows through damage ~~easily~~ less easily.

- First take reading (reference) from sound piece of metal.

- Compare test reading with first reading. If same, no prob., if different, ~~fix~~ replace part.

Pronunciation

A Group the words

- Ask students to read through the information in the Skills Box.

- Model the difference between voiced /z/ and voiceless /s/. Demonstrate by asking students to say each sound with their hand on their throat and notice the difference (no vibration of vocal chords for /s/).

- Ask students how the /s/ can be heard if there is no voice (a lot of air is used).

- Set the sorting task as individual work and pair-checking.

- Point out that there are two words in each of the ten groups.

- Feed back with the whole class.

Answers

/g/	signal; gear
/k/	direction; current
/f/	frequency; physics
/v/	excessive; various
/b/	bit; mobile
/p/	operate; piece
/tʃ/	check; change
/dʒ/	engine; dangerous
/θ/	something; through
/ð/	that; there

B Add more words

- Set the task as pairwork.
- Feed back with the whole class.

Suggested answers

get	good; go
cool	can; cold
affect	from; field
very	vary; verb
bolt	big; ball
open	play; part
watch	choose; child
large	bridge; hinge
three	thing; thin
this	though; then

C Practise saying the words aloud

- Model items that students have particular problems with, e.g., p/b, t/d.
- Ask students to repeat chorally and individually.

Answers

Answers depend on the students.

- To learn more about aircraft environmental control systems, go to:

  http://en.wikipedia.org/wiki/Environmental _Control_System.

- Now have the students do the Workbook exercises for Unit 10, Lesson 10.

Workbook answers

Language: word-building

A The meaning of the underlined words

1	auto-shutoff	automatically
2	microscope	extremely small
3	de-energise	deprive
4	anti-static	inhibit
5	pre-flight	before
6	self-regulating	functioning automatically

B Use your dictionary to find other words

Answers depend on the students.

C Write six words to remember

Answers depend on the students.

Reading: eddy current inspection

A Decide if the statements are true or false

1	T	4	F
2	T	5	T
3	F		

Unit test

A test for this unit is available at:

http://www.garneteducation.com/reps/ documents/1254/SDT-u10-test.pdf

Contact your local Garnet Education representative for information about how to access these resources.

UNIT 11 Lesson 1 — Doing the paperwork

Objectives

To extend vocabulary for aircraft certification.

To become more familiar with and practise describing different types of maintenance forms.

Language

Vocabulary for certification, e.g., *mandatory*, *regulate*.

Expressions with so: *I think so, do so*, etc.

Vocabulary

A Complete the definitions

- Put the words from the box on the board. Elicit ideas as to the meanings of some of them.

- Give students time to read through all the definitions *1–8* and fill in the words they are sure about. Check in pairs.

- Students use dictionaries to complete the rest of the definitions. Monitor and assist. Feed back orally.

Answers

1	factor	5	certify
2	mandatory	6	appropriate
3	authority	7	relationship
4	regulate	8	overhaul

Reading

A Activate ideas

- Ask students to cover the text on page 203 and look only at the diagram.

- Set the task for pairwork. As feedback, elicit some ideas from the group.

Answers

No answers at this point.

B Scan to confirm answers

- Set a time limit for students to read to check their ideas and label the icons in the diagram.

- Feed back orally.

Answers

1 the external environment

2 people/personnel who work with aircraft

3 machinery: airframes, systems and all component parts

C Read for detailed understanding

- Refer students to the questions *1–3*. Give time to read them.

- Set the task for individual work and pairwork checking. Feed back orally.

Answers

1 environment: avoid extreme conditions; personnel: correct certification; machinery: checks for airworthiness

2 they all control airworthiness

3 maintenance personnel

D Vocabulary in context: complete the gaps

- Refer students to the vocabulary on page 202. Elicit plural forms of nouns, e.g., *factors*, and past participles of verbs, e.g., *regulated*.

- Set the task for individual work and pairwork checking. Stress that students will need to use a plural or past participle in some cases. Monitor and assist.

- Feed back orally.

Answers

line 1   factors

line 2   relationship

line 4   regulated

line 6   appropriate; authority

line 8   certified

line 13   mandatory

line 14   overhaul

Speaking

A Skim aircraft safety forms

- Tell students they are going to look at some examples of the aircraft safety forms described in the Reading text. Clarify the meaning of *1–4*.

- Refer students to the forms on Course Book pages 251–252. Set the task for individual work and pairwork checking. Set a time limit.
  Feed back orally.

Answers

1   3 IAA preliminary inspection report

2   1 FAA malfunction or defect report

3   4 FAA major repair and alteration

4   2 FAA suspected unapproved parts report

B/C Study a form and explain it to a partner

- Set the task. This could be done in groups of four, with each student looking at a different form. Point out that students should look closely at the forms and infer the answers to the questions. Monitor and assist.

- Students then work with their partners and explain the forms to each other.

- If you wish to do further work on them, project each form onto an OHP and go through the sections.

Suggested answers

FAA MALFUNCTION OR DEFECT REPORT: used to register details of a fault and how it happened, and to suggest how to stop it happening again

Sections 1–4: details of the aircraft

Section 5: details of the faulty part

Section 6: details of the assembly that includes the faulty part

Section 7: the date this form is being submitted

Section 8: longer comments on the fault. Below Section 8, a box should be ticked and the date of the accident or incident entered

FAA SUSPECTED UNAPPROVED PARTS REPORT: used to register the discovery of a part, previously fitted to the aircraft, that the operator suspects should not have been fitted as it is not the correct part

Section 1: the date on which the suspect part was found

Sections 2–4: details of the part

Section 3: how many of the same part there are

Section 6: the assembly in which the part is included

Section 7: details of the aircraft

Section 8: details of the company who supplied or installed the suspect part (US *ZIP Code* equates to UK *Post Code*)

IAA PRELIMINARY INSPECTION REPORT: used when applying for a new or renewed Certificate of Airworthiness

Sections 1–3: details of the aircraft, such as its registration and serial numbers

Section 4: number of hours' flying that the aircraft has had, and number of separate flights the aircraft has completed – since manufacture/since its last Certificate of Airworthiness/in the last 12 months

FAA MAJOR REPAIRS AND ALTERATION: used to register details of large-scale work carried out

Section 1: details of the aircraft

Section 2: details of the owner

Section 3: only Federal Aviation Administration staff can write here

Section 4: details of the engine and propeller

Section 5: what kind of work was carried out on the items in 4 – repair or alteration

- Now have the students do the Workbook exercises for Unit 11, Lesson 1.

Workbook answers

Language: spelling

A Correct the spelling as necessary

1	machine	7	defect
2	correct	8	correct
3	appropriate	9	environment
4	correct	10	propeller
5	overhaul	11	correct
6	inspection	12	control

Language: word endings (suffixes)

A Make words with the suffixes

Suggested answers

correc<u>tion</u>   environment<u>al</u>

airworth<u>iness</u>   inspec<u>tion</u>

require<u>ment</u>   regular<u>ly</u>

replace<u>ment</u>   authorisation

B Check the words

Answers depend on the students.

C Find examples of new words

Answers depend on the students.

Writing: giving reasons

A Underline the part of the sentence that gives a reason

All parts must be certified <u>in order for work to be considered acceptable</u>.

The motor is stripped <u>so that maintenance is easier</u>.

The regulator is there <u>to ensure a constant electricity supply</u>.

B Complete the sentences

Answers depend on the students.

Language: *so*

A Complete the sentences

1	I'm afraid so	4	do so
2	I hope so	5	If so
3	says so	6	I think so

We all make mistakes

Objectives

To practise discussion of maintenance procedures and possible effects of incorrect procedures.

To review the language of hypothetical situations.

Language

Conditionals zero, 1 and 2.

Introduction

A Personalisation: discuss servicing problems

- Exploit the photo. Elicit what is happening. Clarify the word *service* if necessary.

- Set the task for pairwork discussion and feed back experiences to the class.

Vocabulary

A Establish meaning of new vocabulary

- Put on the board some examples of pairs of words with different and similar meanings, e.g., *large/big*; *inspect/overhaul*. Elicit and mark *S* and *D* respectively.

- Set the task for pairwork. Tell students to first mark those items which they feel sure of. They should then use dictionaries to complete the rest. Point out that some answers may be subjective, e.g., *unsuitable* and *wrong* can be used interchangeably, e.g., *That screwdriver looks unsuitable/wrong for this job*, but not always, e.g., *You can't afford to get it wrong.*

Answers

No answers at this point.

B Discuss differences in meaning

- Students discuss answers with a new partner. Feed back orally.

- **Note:** It does not matter that there are grey areas around whether meanings are, strictly speaking, different or similar, e.g., *wrong* and *unsuitable* can mean the same in some situations, but not others. The aim is to check comprehension and stimulate discussion.

Answers

1 S

2 D insufficient = not enough; inappropriate = unsuitable

3 S

4 S

5 D spare = extra or reserve; useless = of no use

6 S

7 S

8 S

9 D bolt = fastener used with a nut or washer (see diag.); screw = fastener that does not need a nut or washer

10 D unsuitable = not appropriate for a particular situation; wrong = incorrect (usually stronger than unsuitable)

Language and speaking

A Discuss maintenance scenarios

- Ensure that students understand the two scenarios, A and B.

- Work through the first item as a group. Set the task for individual work and pairwork checking. Feed back orally.

Answers

1 wrong kind of hydraulic fluid put into brake system — A

2 insufficient lubrication of motor shaft — B

3 equipment cover panel not refastened — A

4 5-amp capacity cable put into 10-amp circuit — A

5 screw hammered into position — A

6 tools left in work area — A

7 connectors not securely done up — B

8 oil-based grease used on rubber components — A

9 maintenance record documents not completed — B

10 spanner used instead of torque wrench — A

11 spare solder left on PCB repair — A

12 reassembly instructions ignored — B

13 live cable cut with insulated sidecutters — A

14 no warning tags on unfinished repair — B

15 bolt refitted without locking washer — B

B Focus on language and use the language in context

- Direct students to the Language Box. Review the grammar of hypothetical situations – conditionals zero, one and two. Ensure that students understand the differences between, e.g., *What happens if you (leave tools in the work area)?* and *What will/might/could happen if you (leave tools in the work area)?* and *What would/might/could happen if you (left tools in the work area)?*

  Establish which scenarios are more likely/more hypothetical.

- Use one or two of the scenarios in Exercise A to elicit and write on the board the possible results of, for example, leaving tools in a work area or not completing maintenance documents.

- Set the sentence completion task for individual work and pairwork checking. Point out that sometimes more than one answer is possible. Monitor and assist. Feed back orally.

Answers

No answers at this point.

C Talk about the scenarios from Exercise A using conditionals

- Look at the example given and elicit the fact that the second conditional is used because it is unlikely that someone would do this. Elicit a few more sentences about other errors, ensuring that students use conditional structures.

- Set for pair or group work. Feed back by eliciting sentences from various groups.

Suggested answers

1 That's very hot – if you <u>touch</u> it, you <u>will/'ll</u> burn yourself.

2 What <u>happens</u> if you <u>increase</u> the volume of a gas?

3 The project is well advanced now. If we <u>stop/stopped</u> now, we <u>will/would</u> lose a lot of time and money.

4 This is an electric fire. What <u>would happen</u> if I <u>used</u> a water extinguisher on it?

5 You should avoid dropping an electric motor – the drive shaft <u>will/would become</u> misaligned.

6 If I <u>put</u> this copper wire into the acid, you <u>will see</u> a reaction begin to take place. Watch carefully please.

D Explain how the mistakes could affect an aircraft system

- Look at the example given.

- Elicit a few more explanations, using the list of maintenance errors from Exercise A.

- Set for pair or group work. Feed back by eliciting explanations from various groups.

Answers

Answers depend on the students.

- Now have the students do the Workbook exercises for Unit 11, Lesson 2.

Workbook answers

Verb forms: *get/be*

A Think of situations

Answers depend on the students.

B Choose best option to complete the sentences

1	gets	**5**	will get
2	will be	**6**	is
3	has been	**7**	got
4	got	**8**	have been

Writing: notes

A Rewrite the notes to make them shorter

1 Mandatory maintenance not carried out.
2 Metal surface & cabling abraded.
3 Power supply left on.
4 Screws not tightened.

B Rewrite the ideas as notes

Answers depend on the students.

Language: verb-noun-adverb combinations

A Write sentences about the items

Answers depend on the students.

Lesson 3 — Getting it the right way round

Objectives

To extend vocabulary through word-building.

To practise discussion of (re)assembly procedures.

Language

Expressions for position, e.g., *upside down, the right way round*.

Vocabulary for (dis)assembly: *installation, uninstall, reinstall*.

Introduction

- As an introduction, ask students to underline all the nouns in the text. They are: *majority, mistakes, reassembly, reinstallation, assemblies, subassemblies, components* (as well as *maintenance*, used here as an adjective; *it* and *something*).

- Elicit which are plurals, what *re-* means and what *sub-* could mean.

Vocabulary

A Word-building

- Copy the table onto the board. Clarify the meaning of the headings, particularly *undo* and *process*. Work through the first line as a group.

- Set the task. Monitor and assist. Encourage the use of dictionaries.

- Feed back onto the board.

Answers

verb	undo it	do it again	process
maintain	–	–	maintenance
assemble	disassemble	*reassemble*	*assembly*
install	*uninstall*	reinstall	*installation*
put together	*take apart*	put back together	–

B Complete the table

- Set the task. Monitor and assist. Encourage the use of dictionaries.

- Feed back onto the board.

Answers

mount	*dismount*	*remount*	*mounting*
fit	–	*refit*	fitting

C Identify expressions for position

1

- Draw attention to the illustration of the car battery. Elicit what the problem is (the wires are connected to the terminals the wrong way round).

- Set the task for individual work and pairwork checking. Stress that students are looking for opposite meanings.

- Do the first item as a group. Monitor and assist.

- Feed back orally. Demonstrate the positions if possible, using convenient objects.

Answers

a upside down

b leave out

c on the wrong side

d the other way round

e the wrong way round

2

- Set for pairwork. Point out that students should look at both the expressions in the box and in items *a–e* in Exercise 1. Feed back orally.

Answers

correct orientation: the right way up; the right way round

incorrect orientation: upside down; the wrong way round

vertical orientation: the right way up; upside down

horizontal orientation: the right way round; the wrong way round

Reading and speaking

A Use vocabulary in context: discuss possible problems with assembling a food mixer

- Give students time to study the illustration. Set the task for pairwork. Feed back orally.

Answers

The washer could go under the cutter, instead of on top, or be left out altogether.

B Extension: possible problems with assembling a power saw

- Give students time to study the second illustration. Set the task for pairwork. Feed back orally, using an OHP if possible.

Answers

The inner and/or outer clamp washer could be left out.

The inner clamp washer could be installed on the wrong side, i.e., in the place of the outer clamp washer, and vice versa.

Either washer could be installed the wrong way round.

The blade could be installed the wrong way round.

C Discuss a more complex assembly

- Form new pairs; set the task for pairwork. Monitor and assist. Feed back orally, using an OHP if possible.

Answers

1 the assembly is an aircraft wheel.

2 **Suggested answer:** see Writing below.

3 You could leave out/forget to put in: O-rings, nuts, bolts, washers, balance weights, balance weight screws, inflation valve, valve cap, grommet.

   You could put items the wrong way round: o-ring/grommet.

   You could put items on the wrong side: nut/tie bolt; balance weights/balance weight screws.

Extension: writing

A Write instructions

- Give students time to study the expressions in the box in Exercise C on page 206. Elicit examples of their use.

- Write on the board the first instruction: *1. Position O-ring on inner wheel half.*

- Encourage students to include warnings such as, *Ensure that components are installed the right way round.*

- Set the task for individual work. Monitor and assist.

- Feed back onto the board.

Answers

1 Position O-ring on inner wheel half.

2 Insert balance weight screws into holes.

3 Fit wheel halves together.

4 Pass tie bolt through washer and wheel assembly. Fit washer and nut on other side.

5 Put balance weights and nuts on balance weight screws

6 Tighten all nuts.

7 Fit inflation valve in the order: O-ring, valve, grommet, cap.

- If you have time, get students to practise describing the disassembly and reassembly of objects they have to hand, e.g., mobile phones, pens, etc., OR review the expressions for position in Vocabulary Exercise B.

- Now have the students do the Workbook exercises for Unit 11, Lesson 3.

Workbook answers

Writing: comparing actions

A Write seven sentences using the prompts

Answers depend on the students.

Language: sentence structure

A Complete the sentences

1 Once it has been thoroughly cleaned, it can be put <u>back</u> together.

2 Someone has put the wires here instead <u>of</u> here.

3 If it's the right way <u>up</u>, you'll see a red marker.

4 Fill a discrepancy form <u>in</u> and show it to me before you leave.

5 The bolts have been taken out and replaced with the wrong size.

6 CRACKS AND ABRASION DAMAGE <u>ON</u> DISC SURFACE

7 I've installed a new circuit <u>in</u> the selector.

8 The cable needs to pass <u>through</u> this hole and connect to the block.

Pronunciation: syllables and stress

A Count and write the number of syllables

1 1		**7** 2		**13** 2	
2 3		**8** 1		**14** 3	
3 4		**9** 4		**15** 2	
4 3		**10** 2		**16** 3	
5 2		**11** 3		**17** 1	
6 3		**12** 1		**18** 2	

B Say the words aloud

Answers depend on the students.

Lesson 4 — Go by the book

Objectives

To practise listening for general ideas and for detail.

To practise reading for general ideas and detail.

To raise awareness of the features of maintenance documentation text by rewriting poorly written text.

Language

Features of maintenance instructions, e.g., use of mixed lower- and upper-case letters, imperatives, the passive voice.

Listening and speaking

A Activate ideas

- Elicit one or two ideas. Set the questions for pairwork discussion.

- Feed back orally. If possible, circulate some examples of different types of text.

Answers

letters, forms, reports, manuals, catalogues, memos, etc.; headings; subheadings; section numbering; items bullet-pointed/numbered; illustrations; clear layout; boxes around certain information; coloured text/background; clear font/handwriting; use of lower case, not block capitals; capitalisation for important words, especially warnings; clear referencing to footnotes/other pages and documents/parts lists for diagrams

B Listen for main ideas (CD3 Track 7)

- Set for individual work. Play the recording once and give time for pairs to compare ideas.

- Feed back orally.

Answers

2 maintenance documentation style

C 🔊 **Listen for detail and make notes on problems and solutions (CD3 Track 7)**

- Set the task. Stress that students do not need to find a single solution to each problem, but to note problems in general and the type of solutions applied.

- Play the recording again. Feed back in pairs, and then orally as a whole group.

Answers

Problems: mistakes can be caused by badly designed maintenance documents, including texts that are ambiguous, wordy or repetitive; some maintenance engineers use English as a second language

Solutions: aviation documents now written in simplified English, which is simple, concise and clear for maximum accessibility; improvements in layout, diagrams and warnings; non-use of block capitals, which are difficult to read

Reading and writing

A **Skim the text for gist**

- Exploit the photos and elicit what they are: a desktop fan, a fridge, a photocopier and a wall-mounted AC unit.

- Set questions *1* and *2* for individual work and pairwork checking.

Answers

The instructions are for servicing a wall-mounted AC unit, including dismounting, cleaning, lubricating and remounting.

B **Read and identify problems with the text**

- Elicit one or two ideas. Set the task for individual work and pairwork discussion. Feed back orally.

Answers

1 The instructions are in block capitals.

2 The vocabulary is too complex; words such as *facilitate* and *obviate* are an unnecessary strain on engineers using English as a second language.

3 The language is too complex and sentences are over-long, especially Instruction 4, which starts with a very long adverbial clause.

4 Instruction 3 is repetitive (*remove* only needs to be used once).

5 There are a lot of unnecessary words, e.g., constructions with *so as to*.

6 The style is not consistent: some instructions use the imperative, some use the passive voice and some use subject + modal, e.g., *you must* ...

C **Rewrite the instructions more appropriately**

- Elicit which instructions use the passive voice (1, 4, 6) and active voice (3, 5, 6) and which use the imperative (2, 5).

- Focus on Item 1 of the instructions; elicit suggestions for an improved version and write it on the board, e.g., *The heater should be isolated from the mains /disconnected from the power supply.* Elicit alternative versions using the active voice and imperative, e.g., *You should/must disconnect the heater from the power supply; Disconnect the heater from the power supply.*

- Point out that there is no single correct way to write the instructions, but that the aim is clarity and concision.

- Divide the class into three groups. Each group should write a shorter, simpler version of the instructions using either the passive, active or imperative. This final version could be written as a poster or onto an OHT.

Suggested answers

1

1 Disconnect the heater from the power supply.

2 Remove it from the wall.

3 Undo 4 base screws and remove cover.

4 Clean the appliance thoroughly, avoiding damage to components.

5 Put a few drops of A36 light machine oil on both motor bearings.

6 Reassemble and remount the heater.

2

1 First you must disconnect the heater from the power supply.

2 Then you should remove it from the wall.

3 Next you have to undo 4 base screws and remove the cover.

4 Next you must clean the appliance thoroughly, avoiding damage to components.

5 Then you should put a few drops of A36 light machine oil on both motor bearings.

6 Finally you need to reassemble and remount the heater.

3

1 The heater must be disconnected from from the power supply.

2 It should be removed from the wall.

3 The 4 base screws must be undone, and the cover must be removed.

4 The appliance should be cleaned thoroughly, avoiding damage to components.

5 A few drops of A36 light machine oil should be applied to both motor bearings.

6 The unit should be reassembled and remounted on the wall.

D Compare texts

- Groups swap their drafts and check for accuracy and consistency.

- Feed back with the class. Discuss the differences in style, e.g., the passive is more formal; the use of *you* + modal makes the instructions stronger and more personalised; the imperative is short and clear.

Answers

Answers depend on the students.

- Now have the students do the Workbook exercises for Unit 11, Lesson 4.

Workbook answers

Language: word forms

A Circle the false words

Pupils should circle:

1	careness	5	resafe
2	lubricately	6	preceetion
3	suiter	7	accumulatory
4	contamine	8	reducement

B Mark the correct words *n, v, adj* or *adv*

1 carefully (*adv*)
  care (*v*)
  careful (*adj*)

2 lubricant (*n*)
  lubricate (*v*)
  lubrication (*n*)

3 suitably (*adv*)
  unsuitable (*adj*)
  suitability (*n*)

4 contaminant (*n*)
  contamination (*n*)
  decontaminate (*n*)

5 unsafe (*adj*)
  safely (*adv*)
  safety (*n*)

6 proceed (*v*)

procedural (*adj*)

procedure (*n*)

7 accumulate (*v*)

accumulator (*n*)

accumulation (*n*)

8 reduce (*v*)

reduction (*n*)

reducing (*v*, *adj*)

C Check a dictionary

Answers depend on the students.

D Write sentences

Answers depend on the students.

Language: verbs in context

A Complete the sentences with the correct verbs

1 isolate   **4** disconnect

2 decontaminate   **5** apply

3 remove

Language: ~*ing* form or infinitive verb forms

A Choose the correct sentence

2, 4, 6, 7

B Complete the table

verbs that take ~*ing*	verbs that take *to* + infinitive
avoid (damaging)	take care (to avoid)
	will help (to keep)
	makes things (easier to see)
	attempt (to bend)

Language: modal verbs

A Circle the correct modal verb in each sentence

Pupils should circle:

1 may   **4** may

2 should   **5** can

3 would   **6** cannot

Lesson 5   NDI and NDT

Objectives

To practise scanning a text for specified items.

To practise cross-referring data in a table with written information.

Language

NDI and NDT.

Vocabulary for damage to and maintenance of brakes, e.g., *scratches*, *wire brushing*, *coning*.

Speaking

A Discuss the questions

- Refer students to the text at the top of page 211. Clarify vocabulary problems. Elicit a few ideas as to the meaning of NDI and NDT.

- Set the three questions for pairwork. Monitor and assist. Feed back orally.

Answers

1 NDI: non-destructive inspection; NDT: non-destructive testing

2 Personnel involved in NDI/NDT *should*: carry out disassembly and reassembly with care; follow manuals carefully; use the correct inspection/testing instruments and tools; fill in the appropriate report documentation; *should not*: damage, remove or break any assemblies or components while inspecting or testing them

3 *advantages*: aircraft are not taken out of service for long; no replacement parts are required; *disadvantages*: thorough inspection/testing may be difficult without damaging a component or assembly

Vocabulary

A Categorise vocabulary

- Give students time to read through the items in the box. Copy the table onto the board.

- Elicit and write in one or two of the items that students are sure of.

- Set the task for pairwork. Encourage use of dictionaries and monitor to assist with this.

- Clarify and check concepts of new vocabulary in feedback. Elicit other types of damage and clarify the difference between *scratches* (surface damage) and *grooves* (deeper cut). Clarify that *face* = the surface of the disc and *brake lining* = surface of the brake pads. Use the diagram to help clarify.

Answers

damage to assembly	environmental factors	assembly parts	maintenance actions
rust	industrial pollution	disc faces	wire brushing
wear	rough ground	brake lining	standing
deep scratches	standing water	brake pad	wiping

B Pronunciation

- Drill and correct pronunciation of the vocabulary. Students will need to use the words in the next exercise.

Answers

Answers depend on the students.

C Discuss the braking system

- Give the students time to study the diagram. Elicit the answer to question *1*. Set questions *2* and *3* for pairwork. Feed back orally.

Answers

1 The diagram shows a brake assembly.

2 Hydraulic pistons push the brake pads onto the metal disc, which is part of the wheel assembly, causing the wheel to stop rotating.

3 The brake pad lining (surface) might be affected by wear; the brake disc by scratches and grooves, and rust caused by standing water, infrequent use and possibly airborne industrial pollution.

Reading

A Scan to find key vocabulary

- Get students to look back at the vocabulary list in Exercise A. Set the task for individual work and pairwork checking. Set a time limit.

Answers

wear	line 5
deep scratches	line 6
sanding	line 14
wiping (wipe)	line 12
brake pad	line 5
standing water	line 2
wire brushing	line 13
rough ground	–
disc faces	line 5
brake lining	line 4
rust	lines 10, 11, and 12
industrial pollution	line 3

B Cross-check and correct data in a table

- Allow students time to read carefully through the table.

- Set the task for individual work and pairwork checking. Feed back orally, if possible using an OHP.

Answers

brake discs – maintenance procedure	
maintenance schedule	
frequent use	*less* frequent inspection
dry conditions	normal inspection schedule
airborne contamination	*more* frequent inspection
tolerances	
coning	max. *0.015*" *0.381* mm) in either direction
grooves	*max.* depth 0.030" (0.76 *mm*)
wear	see manual Fig. "*A2*" dim. *A* for *min.* thickness
actions	
light rusting	clean disc *once or twice during* taxiing to wipe
extensive rusting	remove *disc*; apply wire brush, then *220* sandpaper

Extension: speaking

- Ask students to cover the text in Exercise A.

- Get students to work in pairs to ask about and explain the maintenance procedure to each other in their own words.

- As feedback, elicit explanations of some of the items in the table from the group.

- If you have time, use the text as a basis for work on modals: *may/will*. There are also useful examples of *can*, *would* and *should* that you can point out to students OR quickly review the pronunciation of decimals, such as those in the text on page 211.

- Now have the students do the Workbook exercises for Unit 11, Lesson 5.

Workbook answers

Pronunciation: single and two-part vowel sounds

A Match up the vowel sounds

long single-vowel sounds:

1	part	hard
2	clean	wheel
3	more	install
4	groove	screw

two-part vowel sounds:

1	straight	cable
2	wipe	lining
3	cone	motor
4	ground	power

B Write words in the third column

Answers depend on the students.

Language: ~ing forms

A Make ~ing words

1	rusting	5	wiring
2	pitting	6	conditioning
3	armouring	7	scratching
4	coning	8	wiring

Language: verbs with prepositions

A Delete the unnecessary prepositions

1	–	4	–
2	delete *for*	5	–
3	–	6	–

Reading and language: using *of*

A Complete the text with *of* where necessary

1	of	8	of
2	unnecessary	9	of
3	of	10	unnecessary
4	of	11	unnecessary
5	of	12	unnecessary
6	unnecessary	13	of
7	unnecessary		

B Check your answers

Answers depend on the students.

Lesson 6 — Going through the checks

Objectives

To discuss pre-flight checks and their rationale.

To practise listening for key ideas.

Language

Regular and irregular verb forms.

Past conditions: 3rd and mixed conditionals.

Sentence stress.

Reading and speaking

A Introduction: compare verbs

- Students will probably already know two or three of these words. Ask them to look up those that they do not know and check with a partner as to the meaning they all share.

- With a strong group, simply elicit the shared meaning.

Answer

The verbs all share the meaning of *check/make certain that*.

B Activate ideas

- Give students time to read all the questions. Clarify vocabulary problems.

- Set the task for pairwork. As feedback, elicit one or two ideas. Do not confirm or correct answers yet, as students will read to check their ideas in Exercise C.

Answers

No answers at this point.

C Read to check ideas

- Set a time limit. They are only reading to check their ideas.

- As feedback, elicit answers to the questions in Exercise B.

Suggested answers

1 They are carrying out pre-flight checks.

2 To ensure safe optimal performance of the aircraft.

3 Human factors: fatigue, rushing or carelessness. External factors: a fault may not be visible/picked up by instruments.

4 Airframe, access doors, landing gear, control surfaces, fuel, oil, engine ducts, tyres, cockpit instruments.

Listening

A Revise the verb forms

- Many of these verbs will be well known to students. Choose two that you are sure they know as examples: one with –ed endings and one irregular verb.

- Set the task for pairwork. Encourage the use of dictionaries. Monitor and assist.

- Feed back orally. Focus on correct pronunciation.

Answers

All verbs end in ~ed in the past and past participle forms except:

take off	(took off; taken off)
have to	(had to, had to)
say	(said, said)
be	(was/were; been)
give	(gave, given)
fly	(flew, flown)
do	(did, done)
tell	(told, told)

B Review pronunciation

- Refer students to the Skills Box.

- Elicit the correct pronunciation of the past tense forms of the regular verbs and check that students pronounce the ~ed endings correctly [/d/, /t/ and /ɪd/].

- Get students to practise the pronunciation in pairs.

Answers

/d/	/t/	/ɪd/
called	checked	accepted
discovered	pushed	suspected
tried	missed	indicated
said	noticed	
did		
moved		
failed		
seemed		
arrived		
had to		
handed over		
told		

C Listen for key information (CD3 Track 8)

- Ask students to reread quickly the list of pre-flight inspection checks in Reading and speaking Exercise C.

- Set the task for individual work and pairwork checking. Play the recording once. Feed back orally.

Answers

Report A	3
Report B	9
Report C	1
Report D	8

Writing and language

A Focus on language

- Review the grammar of conditions in the past and their results.

- Refer students to the Language Box. Clarify as necessary, highlighting the tenses in the example sentences on the board using colour.

- Set the question for pairwork discussion. Check students understand the difference between structures with hypothetical past and present results (3rd and mixed conditionals).

Answers

In *1*, the result is in the past; in *2*, the result is in the present.

B Apply the language

- Do the first item as a group.

- Set the task for individual work and pairwork checking. Monitor and assist.

Suggested answers

1 for the fuel delivery document; have known how much fuel there was in the tanks

2 be so annoyed; done a good job

3 have landed at another airport; been so heavy

4 had spotted the damage; be in trouble

5 We would have been able to bring up the landing gear if the locking pin hadn't been in position.

- Now have the students do the Workbook exercises for Unit 11, Lesson 6.

Workbook answers

Reading: finding faults

A Find out what a PASD device does

The PASD can be used to efficiently locate and repair intermittent faults.

B Circle the correct meaning

1	a)	5	b)
2	a)	6	b)
3	b)	7	a)
4	a)	8	a)

Language: multi-word verbs

A Find the verbs in the tapescript

The verbs are found in the following order:

bring up	(Report A)
come up	(Report A)
take (it) back	(Report B)
hand over	(Report B)
turn out	(Report B)
hook up	(Report B)
take off	(Report C)
put in	(Report D)
come on	(Report D)

B Discuss what the words refer to

Answers depend on the students.

C Complete these sentences

1	turned out	5	hook up
2	take it up	6	hand over
3	bringing up	7	put in
4	comes on	8	take it back

Language: past forms

A Distinguishing between active and passive forms

active form	they found
passive form	the aircraft was being inspected

B Expand the sentences

1 I was forced to make an emergency landing.

2 Maintenance had checked it and not seen that it was there.

3 All locks and locking pins were removed before the aircraft was taxied out.

4 We went through the usual list and made sure that everything was OK.

5 The wing tanks were not loaded with fuel.

6 Our tail section was damaged when we hit another aircraft on the ground.

Lesson 7   In safe hands

Objectives

To practise reading copy advertising equipment, for general ideas and detailed information.

To practise note-taking and referring to notes.

To develop speaking skills: present a piece of equipment and respond to questions about it.

Language

Question forms: *What are the dimensions?*

Collocation, e.g., *operational safety, aircraft jacks.*

Noun/verb/adjective forms for access equipment, e.g., *handle.*

Vocabulary

A Complete the verbs

- Do the first item on the board with the whole group. Elicit or point out that the vowels are missing.

- Set the task for pairwork. Feed back onto the board.

- Check students remember the meanings of the verbs by eliciting common collocations and uses, e.g., *something may protrude from a tyre, you cancel an appointment.*

Answers

1 ope<u>r</u>ate

2 pr<u>o</u>v<u>i</u>de

3 pr<u>e</u>vent

4 l<u>o</u>c<u>a</u>te

5 ext<u>e</u>nd

6 pr<u>o</u>tr<u>u</u>de

7 c<u>a</u>nc<u>e</u>l

8 r<u>o</u>t<u>a</u>te

B Word-building

- Again, do the first item with the group. Set the task for pairwork.

- Feed back onto the board. Point out that *~tion* and *~sion* are common noun endings.

Answers

operate:	operation; operator
provide:	provision; provider
prevent:	prevention
locate:	location; locator
extend:	extension
protrude:	protrusion
cancel:	cancellation
rotate:	rotation

C Preview vocabulary

- Several of the items in this exercise can be both verb and noun or verb and adjective. Encourage discussion of these cases when monitoring pairwork and in feedback.

- Copy the table onto the board. Elicit the first item and write it in.

- Set the task for pairwork. Encourage use of dictionaries; monitor and assist.

Answers

noun	verb	adjective
caster	control	hydraulic
chassis	access	secure
clearance	secure	operational
mechanism	jack	excessive
control	handle	accidental
access	sway	
jack		
handle		

Reading and writing

Activate schemata

- Get students to read the introductory text and, in pairs, to consider some general questions, e.g., *Which parts of the aircraft might be 'inaccessible' or 'high up'? Why might it be necessary to lift the aircraft? What kind of tools and personnel might it be necessary to move high up?*

- Exploit the pictures. Elicit how students think each piece of equipment would be useful to maintenance personnel.

- As feedback, have a general discussion.

Answers

Answers depend on the students.

A Read for general idea

- Divide the group in half: Students A and B.

- Set the task. Students A read about the Rotozoom access platform; Students B read about the AJJ series lift system. They should read once through their texts to get an initial idea of the content.

Answers

No answers at this point.

B Read for specific information and take notes

1

- Set the task for individual work. Students A should check answers together; likewise Students B.

2

- Students should first underline the phrases which contain words from the Vocabulary section. They then copy each phrase onto their notes next to the appropriate category from Exercise 2, e.g.,

Text B:

Safety Features

Low height leg design (maximum gear door clearance)

Answers

Answers depend on the students.

Speaking

A Write questions for your partner

- Write the categories from Reading and Writing Exercise B1 on the board and check students understand each one.

- Write up the question for the first category: *What type of equipment is it?* Elicit and add a question for the Operation category, e.g., *How does it operate/work?*

- Set the task for pairwork. Stress that there may be more than one possibility for each question. Monitor and assist. Feed back onto the board.

Suggested answers

What type of equipment is it?

How does it work?/How is it operated?

How big is it?/What are the dimensions?

What is the maximum load?/How much weight can it carry?

What are the safety features?/Does it have any special safety features?

How is it transported/can it be transported?

B Ask and answer questions about your partner's text

- Set the task. Mix the students so that each Student A sits with a Student B.

- Stress that students should only make notes during the conversation. They should not try to copy down everything their partner says verbatim (or look at their partner's notes).

Answers

Answers depend on the students.

C Read back information from notes

- Students now read back their notes to their partner. Point out that they can use their own words but should cover all of the information that their partner gave.

- Each time, the student in the role of listener should wait until their partner has finished to correct him.

- As feed back, discuss with the group how accurate they were, and therefore how effective their note-taking was.

Answers

Answers depend on the students.

D Application of concepts

- Ask students to read questions *1–4*. Clarify vocabulary problems.

- Set the task for pairwork. Feed back orally.

Answers

1 AJJ Jet & Turbo-Prop Aircraft Lift System
2 Rotozoom Access Platform
3 AJJ Jet & Turbo-Prop Aircraft Lift System
4 AJJ Jet & Turbo-Prop Aircraft Lift System

- Now have the students do the Workbook exercises for Unit 11, Lesson 7.

Workbook answers

Language: adverbs

A Match the words with the definitions

No answers at this point.

B Check answers in the dictionary

first; initially
wrongly; incorrectly
especially; particularly
properly; correctly
potentially; possibly

C Complete the sentences

1	potentially	4	properly
2	especially	5	incorrectly
3	Initially		

Language: time, place, manner

A Match expressions with similar meanings

1	c)	4	d)
2	e)	5	a)
3	b)		

B Complete the sentences

1	next to	4	near
2	under	5	at the same time as
3	by using		

Writing: word-building

A Make necessary changes

1 Easy transportation is provided by large spring-loaded wheels.

2 Operational safety is provided by a work cage.

3 Locking pins prevent accidental lowering of the jack.

4 Dangerous loads are signalled by a siren.

5 The chassis carries a rotating extension platform.

6 This equipment allows access to areas that are normally inaccessible.

7 A height control allows for adjustment of clearance under the aircraft.

8 Excessive movement of the platform is prevented by an anti-sway mechanism.

Lesson 8   Taken out of service

Objectives

To practise listening to a seminar for main ideas and for detail.

To discuss and present a short talk on features of maintenance hangars, e.g., *lighting*, *fire protection*.

Language

Expressions for describing the running of a maintenance hangar, e.g., X *is used to* (verb), X *provides Y.*

Listening

A 🔘 **Generate ideas (CD3 Track 9)**

- As an introduction, exploit the picture on page 216. *What does it show?*

- Give students time to read questions *1–3* and clarify vocabulary problems.

- Elicit a few ideas for question *1*. Set the rest for pairwork. Monitor and assist.

- Elicit ideas for questions *2* and *3*, but do not confirm or correct them yet. Students will listen to check their ideas.

- Play the recording once. Pairs compare answers. Feed back orally.

Answers

- **Note:** These are answers according to the recording, but other ideas the students suggest may be equally valid.

1 heavy maintenance, overhaul

2 jobs which require the aircraft to be taken out if service for a long time

3 <u>facilities</u>: heating, ventilation, and air conditioning systems; lighting; electrical supply; fire detection/protection systems and alarms; water supply

<u>tools and equipment</u>: specialist access and lifting equipment (access docking); undercarriage lifting platforms; jacks

B 🔘 **Listen for key ideas (CD3 Track 9)**

- Allow students time to read through the text. In pairs, ask them to speculate as to what might fit in each gap. Elicit a few ideas.

- Play the recording. Students compare answers. Stress that they should focus on accuracy of grammar and information, not on a verbatim reproduction of the recording.

Suggested answers

line 1   the specialist equipment it contains

line 2   access docking

line 3   maintenance can begin

line 4   removal and testing of; jacking of the aircraft

line 5   along the fuselage

line 6   maintenance of the undercarriage

C ⊙ Listen for detail (CD3 Track 9)

- Set the task. Be prepared to pause the recording where necessary, e.g., students could be asked to say 'stop' when they hear a key word. You then play that section of text again so that they can write down the whole phrase.

- Work through item *1* first on the board with the group.

- Play the recording. Students check answers in pairs. They must agree on a grammatically correct answer in each case.

- Feed back orally. Do not confirm or correct yet.

Answers

No answers at this point.

D ⊙ Listen and read (CD3 Track 9)

Replay the whole text as students follow the tapescript. Tell them to cover the phrases they wrote for the moment and focus on the text.

1 A well-designed maintenance hangar should have the facilities you can see here.

2 Heating, ventilation and air conditioning systems.

3 Overhaul, heavy maintenance or aircraft paint spraying.

4 Works to all parts of the aircraft can continue to be carried out uninterrupted.

E ⊙ Pronunciation practice (CD3 Track 9)

- Replay the relevant sections of text for students to repeat.

- Get students to practise saying the phrases in pairs.

- **Note:** Students should aim to copy the word- and phrase-level stress used by the speaker, both to improve their pronunciation of the items in the phrases and to continue to train their ear in perceiving it.

Vocabulary

A Focus on vocabulary related to maintenance in context

- Set the task for individual work and pairwork checking.

- Feed back orally, using an OHP if possible.

Answers

The expressions are in the following order:

depends on

provided within the hangar

so as to enable

immediately

The provision of X allows Y

It is also used to

The platforms can then be lowered

can continue to be carried out uninterrupted

B Activate vocabulary for speaking.

- Set the task for individual work and pairwork checking. Encourage students to find examples of each pattern in the text and make their own sentences using similar patterns. Monitor and assist.

- Feed back onto the board.

Answers

noun	depends on	noun
noun	allows	noun *to* + verb
noun	is used to	verb
noun	requires	noun
noun	provides	noun
noun	enables	noun *to* + verb
noun	can be	past participle

Speaking

A Apply vocabulary and concepts

- This exercise is an opportunity for students to put into a discussion the vocabulary and concepts they have learned, and to prepare at the same time to give a short talk in Exercise 2.

- Set the task. Encourage students to use the expressions and patterns from the vocabulary exercises. Monitor and assist.

- Any answer which is grammatically correct and conforms to the general outline below is acceptable.

Suggested answers

heating, ventilation, and air conditioning systems
  provide clean air at a temperature suitable for equipment, procedures and operators

lighting, including emergency lighting
  ensure safe and accurate work

main, sub-main, and small power supply
  are used to power systems, equipment and tools

fire detection and alarm systems
  are used to give early warning of fire

fire protection systems
  prevent fire and possible explosion, and to extinguish fire quickly

domestic and process water services
  provide water for human consumption and waste treatment, and cleaning

process ventilation
  ensures clean air around work spaces where aircraft manufacture is being carried out

compressed air
  allows manufacturing processes such as paint spraying to be carried out

lightning protection and main earth
  ensures that the supply of electricity to the workspaces is not affected by adverse weather

energy management
  requires employees to cooperate

B Give a short talk

- Ask students to choose a feature of maintenance hangar design and prepare to speak to a partner for approximately one or two minutes about it, explaining its importance to aircraft maintenance.

- As usual, students rehearse with one partner who suggests changes to the talk, and then present their talk to a second partner.

- To avoid repetition of certain features, assign a feature to each pair of students, who can rehearse a talk together before presenting it to colleagues. Monitor and assist with pronunciation and the use of vocabulary from the lesson.

Answers

Answers depend on the students.

- Now have the students do the Workbook exercises for Unit 11, Lesson 8.

Workbook answers

Language: long nouns

A Match the words to make long nouns

1 main power/sub-main power
2 fire protection/fire detection
3 pre-flight testing/non-destructive testing
4 maintenance schedule/maintenance operations

5 emergency lighting/emergency situations

6 alarm systems/air conditioning systems

B Find a word in the word lists/Course Book

Answers depend on the students.

Spelling: double letters

A Complete the double letters

1 operation
2 compressed
3 cabling
4 appropriate
5 incorrectly
6 suitable
7 assembly
8 damage
9 connection
10 indicator

11 refitted
12 circuit
13 horizontal
14 manual
15 access
16 visual
17 loading
18 rubber

Writing: word-building and definitions

A The meaning of *sub-* and *post-*

sub- below, underneath, beneath

post- after, later

B Write short explanations

Answers depend on the students.

Lesson 9 — Equipment MRO

Objectives

To practise reading manual text for MOR equipment.

To practise cross-checking the accuracy of data.

Language

Vocabulary to describe parts of a trolley jack, e.g., *saddle*, *castor*.

Introduction

- As an introduction, ask students what they think MRO might stand for in the light of the issues and concepts they have so far studied in this unit. (It stands for *Maintenance, Repair and Overhaul.*)

Vocabulary and speaking

A Discuss use of the trolley jack

- Refer students to the statement at the top of the page, stressing the dual problem of damage and injury. Set task *1* for pairwork. Feed back orally.

- Tell students that they are going to study the trolley jack in the illustration with reference to maintenance of maintenance equipment. Elicit or explain that a trolley jack is used for lifting vehicles, e.g., cars and buses, and that is hydraulically operated: it uses oil as a hydraulic fluid.

- Set task *2* for pairwork. Encourage the use of dictionaries and students' general knowledge.

Answers

1 handle

2 release valve (handle knob)

3 saddle

4 foot pedal

5 front wheel

6 rear castor

- Set task *3* for pairwork. Again, stress that students are not expected to know correct answers but that they should use the diagram and their own knowledge to propose answers.

- As feedback, elicit ideas from the students. Accept all reasonable suggestions, and tell the students that they will read to check their ideas later.

Answers

Answers depend on the students.

Reading

A Read and match headings to sections

- Set the task for reading without dictionaries, to a time limit. Pairs check answers. Feed back orally.

- Ask students to check their answers to the questions from Vocabulary and speaking Exercise 3 with the text. (They may need to read the text again more carefully.)

Answers

1 Choosing the correct oil

2 Adding oil

3 Replacing the oil

4 Lubrication

5 Inspecting and cleaning

6 Storage

B Reading to check data

- Give students time to read *1–8* and clarify vocabulary problems.

- Set the task for individual work and pairwork checking.

Answers

1	I	5	C
2	C	6	I
3	I	7	C
4	C	8	I

- If you have time, exploit the Reading text for vocabulary OR build up the main points from the text on the board.

- Now have the students do the Workbook exercises for Unit 11, Lesson 9.

Workbook answers

Reading: hydraulic jacks

A Complete the table

1	d	6	c
2	f	7	i
3	a	8	e
4	h	9	g
5	b		

Writing: reducing sentences

A Write the sentences in short form

No answers at this point.

B Compare your answers

1 release valve not tightly closed

2 contact Omega Tech. Service

3 ensure proper fluid level

4 fluid level low

5 remedy overload condition

6 with ram fully retracted, remove oil filler screw, reinstall oil filler screw

7 air supply inadequate

8 overload condition

9 remove load, drain fluid to proper level

Language: *will (not)*

A Underline expressions with *will*

will lift

will not extend

will not lower

will not lift load

B What does *will (not)* refer to?

2 In all cases, *will* refers to readiness to do something (connected to *willingness*).

C Write actions for problems

Answers depend on the students.

Language: expressing quantity

A Expressions

Answers depend on the students.

B Complete the sentences

1 half full

2 almost complete

3 fully extended

4 on its side

5 at an angle

Lesson 10   Who's done what

Vocabulary and pronunciation

- As an introduction, elicit ideas as to what an aircraft's 'job file' might be. (It is the record containing details of all maintenance work carried out on that aircraft.)

A Discuss the differences

- Work through the first item with the group. The difference (as in all of *1–5*) between the words lies in their meaning, rather than their grammar or pronunciation.

- Set the task for pairwork. Students may use dictionaries if you wish, although this will make the task much slower.

- Feed back orally.

Answers

1 personnel (the others are people with specific authority over work)

2 write (the others mean keep a written record)

3 renewed (the others mean seen and certified by an authorised person or body)

4 airworthiness certificate (the others are documentation for MRO activities themselves)

5 remedy (the others mean issues in need of inspection/repair)

B ⊙ Analyse pronunciation of key vocabulary (CD3 Track 10)

- Work through the first item on the board. Set the rest of the task for pairwork. Encourage students to say the word to themselves and each other to help them.

- Play the recording once. Pairs check their answers.

Answers

1	'supervisor	6	e'lectrics
2	'damaged	7	'structural
3	'ordered	8	'generator
4	signed 'off	9	'battery
5	con'nectors	10	re'placed

C ⊙ Listen for specific items (CD3 Track 11)

1

- Refer students to the Skills Box. Review notions of stressed and unstressed words in a spoken sentence, and of linking.

2

- Work through the first item with the group. Set the task for individual work and pairwork checking.

3

- Refer students to the tapescript on page 316. Go through the sentences correcting stress, and linking as well as the pronunciation of the target vocabulary items.

- Set the task for pairwork. Monitor and assist.

Answers

1 So I've _ordered_ a new pump. – (c)

2 I just _replaced_ it. Put a new one in. – (j)

3 One of the O-ring seals was _damaged_ and there was a leak. – (b)

4 It wasn't charging the _battery_ properly. – (i)

5 No corrosion, no sign of any _structural_ weakness. – (g)

6 I refitted it with new _connectors_. – (e)

7 The boss has _signed_ that _off_, too, so that's OK. – (d)

8 Oh yes, problem with the starter _generator_. – (h)

Listening

A Pre-listening: discuss the job form

- This exercise is designed to familiarise students with a fairly complete document before they listen to a conversation about it.

- In spite of its title of _Maintenance hangar report_, the recording in this Lesson refers to this document as a _job form_. This is intended as a generic term for maintenance forms, of which there is a large number in MRO, and whose titles also vary from company to company.

- Give students time to read _1–4_. Set the task for individual work and pairwork checking. Feed back orally.

Answers

1 It is the 100-hour overhaul for this aircraft.

2 Four (Tahoni, Higgins, Armstrong and Maddox)

3 Five (engine and fuel; airframe; hydraulics/controls and landing gear; electrics; avionics and instruments)

4 In the *Discrepancy* column, any faults which students can reasonably predict, and under Action, their solutions.

B 💿 Listen for main ideas (CD3 Track 12)

- Ensure students understand the questions *1–3*. Set the task.
- Play the recording once.
- Feed back orally.

Answers

1 3 (Mr Greenhill, the owner of the aircraft; Mike Armstrong, technician on Electrics and Avionics & Instruments; Lufti Tarhoni, technician on Engine and Fuel.)

2 by next Thursday (Tarhoni says: *If we finish it before then, we'll call you.*)

3 yes (His responses are very positive: *Good; That's good news; that's all right, I'm not in a hurry; thanks very much … Cheers.*)

C 💿 Complete the Maintenance hangar form (CD3 Track 12)

- Review the use of reduced grammar, including passive structures, in the technicians' notes. Students must copy this style, as well as the use of block capitals.

- Ask students to read over the form again to check that they know what should go in each space. Point out that each of the signatures on the lower left must be okayed by John Maddox, who signs on the right against each category of work.

- Play the recording. Students compare answers.

- Feed back orally, or if possible onto an OHP.

- If you wish, replay the conversation as students follow the tapescript.

Answers

1 SKYBIRD 406

2 overhaul completed as per service manual waiting for new fuel pump

3 inspection completed; no faults found

4 slight nose wheel vibration on landing

5 low charge from starter/generator

6 new compass fitted

7 –

8 –

9 –

10 BOB HIGGINS

11 JOHN MADDOX

12 –

13 BOB HIGGINS

14 –

15 –

16 JOHN MADDOX

17 JOHN MADDOX

Extension: speaking

- Elicit onto the board other possible discrepancies for each of the five categories, and appropriate action to remedy them (students discussed some of these in Listening A4).

- Set up a pairwork role-play in which the owner of an aircraft and a technician discuss the progress of maintenance work on the aircraft: students should semi-script their conversation along the lines of the one on the recording, and use as much of the language from the lesson as possible.

- As usual, students rehearse their role-play before presenting it to colleagues.

- As colleagues listen, they can fill in a job form copied into their notebooks.

- If you have time, elicit the three items which were the same in Vocabulary and listening Exercise A, e.g., in *3*, *signed off*, *okayed*, *authorised* OR work on pronunciation issues arising from the role-play.

- Now have the students do the Workbook exercises for Unit 11, Lesson 10.

Workbook answers

Language: *worth*

A Meaning of the action

1 worth a try

2 is not worth

B Expand the sentences

1 Let's go. It's not worth waiting for.

2 Perhaps we can use this. It's worth asking.

3 Look at this. Is it worth replacing?

4 Throw it away. It's not worth the time it is taking to fix it.

5 We may have made a mistake with the calculations. It's worth taking another look.

Should have/must have

A Expressions about fulfilling obligations in the past

I should have put it in for the service earlier really.

B Write *should(n't) have* or *must have*

1 should have

2 must have

3 should have

4 shouldn't have

5 must have

Reading

A Circle and write out in full

Report 1

L/H	left-hand
AW	according with/in accordance with
PARA	paragraph
M/hrs	man hours

Report 2

DATE DISCD	date discovered
NON-ICE	not icy
CONDS	conditions
lubed	lubricated

B Write what work was done

Answers depend on the students.

Unit test

A test for this unit is available at:

http://www.garneteducation.com/reps/documents/1255/SDT-u11-test.pdf

Contact your local Garnet Education representative for information about how to access these resources.

Take-Off

Objectives

To review vocabulary for emergency systems.

To practise dealing with dense technical text.

To write warnings using appropriate language.

Language

Vocabulary for Emergency Oxygen System:
trip lever, delivery tube.

Expressions for warnings: *It is essential
that ..., X may result in Y.*

Speaking and vocabulary

A Activate ideas

- Set the two questions for pairwork.

- Feed back orally.

Suggested answers

1 Because they fly at very high altitudes where
the air is too thin to be breathable.

2 When the normal supply system breaks
down; if he is forced to leave the aircraft at
high altitude; if his supply runs out, e.g.,
because he is obliged to spend more time
than anticipated at high altitude.

B Study vocabulary for reading

- Look at the compounds in the box and
point out that the words before the final
noun function adjectivally, to describe it.

- Highlight the form of the example
definitions in the rubric on the board,
pointing out that they use a relative
pronoun *that.*

- **Note:** *Which* can also be used.

- Set the task for pairwork. Stress that
students will need to apply their knowledge
of word relationships as well as engineering
concepts. Monitor and assist.

Answers

a contents gauge: a gauge that measures
(gauges) level of contents

a trip lever: a (safety) lever that trips, i.e., starts
or stops, a system

a charging connection: a connection for
charging a system with fuel/gas/electricity

an operating cable: a cable that operates a system

a manual operating handle: a handle that
operates a system manually

a bolted clamp strap: a clamp strap that is
bolted (in position)

a rigid supply tube: a supply tube that is rigid

a tell-tale wire: a wire that is a 'tell-tale', i.e.,
when found broken, it is evidence that the
system it is attached to has been operated

an operating mechanism: a mechanism that
operates a system

C Practise the stress of multi-word nouns

- Work through the first item on the board
with the whole group. Remind students that
stress is marked conventionally using a
vertical mark before the stressed syllable.
Other systems are acceptable; get them to
check the one used by their own dictionaries.

- Set the rest of the task. Students compare
answers in pairs. Feed back onto the board.

- Drill all the items. Get students to practise
them in pairs.

Answers

a 'contents gauge

a 'trip lever

a 'charging connection

an 'operating cable

a manual 'operating handle

a bolted 'clamp strap

a rigid 'supply tube

a 'tell-tale wire

an 'operating mechanism

Reading

A Scan a text and circle the compounds

- Set the task. Remind students to use initial letters, word shape and features such as hyphens to find words in a text. Set a time limit.

- Feed back orally, if possible onto an OHP.

Answers

a contents gauge	lines 10, 13 and 15
a trip lever	lines 4, 9, 11 and 20
a charging connection	line 12
an operating cable	line 8
a manual operating handle	lines 5 and 22
a bolted clamp strap	line 7
a rigid supply tube	line 17
a tell-tale wire	line 10
an operating mechanism	line 11

B Close reading: underline what you understand

- This task and the following one are intended to lessen the reading load of what is a fairly dense text, first by showing students that they do understand a large percentage of the text, and then by getting students to collaborate in extracting the information in it.

- Set Exercise B for individual work. Stress that the task is to mark the parts of the text they do understand and ignore the rest, which will probably seem unusual to students.

Answers

Answers depend on the students.

C Compare answers

- Set Exercise C for pairwork discussion. Point out that even if students still do not feel they understand all of the text, it does not matter. The next task requires only that they label the diagram.

Answers

Answers depend on the students.

D Apply the new vocabulary

- Ask students to look at the diagram and tell you what numbers 1 and 7–10 represent. Clarify that they only need to write numbers 2–6 on the diagram to complete the labelling task.

- Set the task for individual work and pairwork checking. Monitor and assist.

- Feed back orally, using an OHP if possible. Elicit what each numbered part does. Encourage students to use the same language structures as in Speaking and vocabulary Exercise B.

Answers

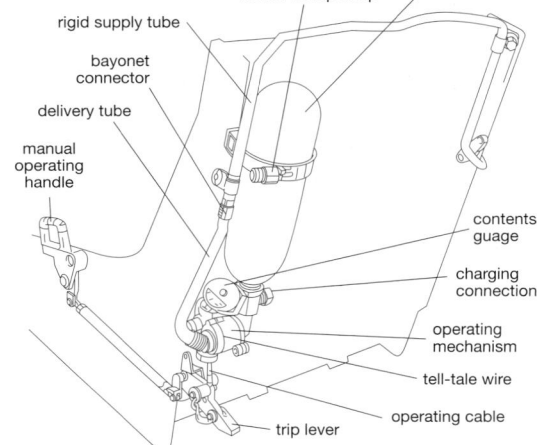

Writing

A Focus on warning language in the text and write a similar warning

- Ask students to reread the warning in the Reading text (the final paragraph). If possible, write or project it onto the board and highlight the warning language, e.g., *It is essential …* and the use of the imperative.

- As a group, look at the examples of other warning language in the box. Elicit warnings for the other topics orally.

- Set the task. Monitor and assist.

- **Note:** Students may prefer to write warnings for other topics, e.g., *hand-propping a plane*.

- As feedback, get students to compare and comment on each other's work.

Answers

Answers depend on the students.

- If you have time, elicit the operating mechanism of the Emergency Oxygen System in the lesson OR review with students the lessons in Units 7–8 that deal with concepts of warnings and of oxygen supply, Unit 7, Lessons 1, 2, 7 and 8.

- Now have the students do the Workbook exercises for Unit 12, Lesson 1.

Workbook answers

Language: long nouns

A Match a word from the first column with a word from the second column

1	a	**6**	e
2	g	**7**	i
3	j	**8**	h
4	f	**9**	c
5	b	**10**	d

B Write sentences

Answers depend on the students.

Language: combinations with *to, across, from, of*

A Circle the verbs

fitted to	line 1
secured to	line 6
fitted with	line 7
consists of	line 7
mounted on	line 12
removed from	line 14
routed across	line 17
connected to	line 9

B Complete the sentences

1	mounted on	**5**	removed from
2	secured to	**6**	connected to
3	consists of	**7**	routed across
4	fitted with	**8**	fitted to

Sentence structure

A Put the sections of the sentences in the correct order

1 Seat belts are designed to withstand extremely high loads.

2 The main parachute opens and the ejector seat is automatically discarded.

3 If there is any damage to a seat belt, its load-bearing ability will be reduced.

4 CO_2 and halon disperse almost as soon as they are sprayed into the open air.

5 Foam can be used for Classes A & B fires.

6 An ambulance should be called in case of personal injury.

7 The accumulator is filled by introducing nitrogen through a charging valve.

8 All checks must be carried out in accordance with a set procedure.

Fire extinguishers: Review

Objectives

To review concepts relating to features of fire extinguishers.

To practise listening to check data on a test sheet.

Language

Vocabulary for fire testing: *ignite*, *smother*, *backing foam*.

Speaking

A Discuss existing knowledge

- Check that students recognise what the icons show.

- Set tasks *1–3* for pairwork. Stress that students should try to remember the different classes of fire from memory.

Answers

No answers at this point.

B Review information on fire extinguishers from Unit 7

- Students should look at Unit 7, Lessons 3 and 6 to check their answers.

Answers

From Unit 7:

1

Water	Class A
Halon	Class A, B and C
CO_2	Class C (also B)
Dry powder	Class A, B and C

(Class D fires require specialist extinguishers)

2 Class A fires involve natural solid materials, e.g., paper, wood, cloth, rubber, certain plastics which leave a residue when they burn.

Class B fires involve flammable liquids and gases, e.g., gasoline, paint thinner, kitchen grease, propane and acetylene.

Class C fires involve energised electrical wiring or equipment, e.g., motors, computers, panel boxes, wiring and cabling.

Class D fires involve the combustion of unusual metals, e.g., magnesium, sodium and titanium.

3 See Unit 7, Lesson 6 for the diagram of the relationship between the three necessary elements to create and maintain fire: fuel, oxygen and heat.

C Review details of extinguishers

- Elicit some ideas from the group.

- Ask students to continue using Unit 7 to complete the table.

- Feed back by drawing a table on the board and eliciting key advantages and disadvantages from students. You may wish to compare the finished table that they come up with to the model answer below (ideally show it on an OHP).

Suggested answer

	advantages	disadvantages
dry powder	multi-purpose (Class A, B and C fires)	messy; possible damage to equipment
CO_2	does not conduct electricity; effective in closed areas; does not support combustion; ideal for Class C; can be used for Class B; no mess/damage to equipment	no cooling effect; stored at high pressure; damages ozone layer
halon	versatile (Class A, B and C fires); effective in closed areas; very small amount of gas required to extinguish fire; no mess/damage to equipment	expensive; gas stored at low pressure; small, light extinguisher; risk of suffocation to personnel; damages to ozone layer
water	stored at low pressure; cooling effect on hot fires; no damage to environment	limited use: only appropriate for Class A fires; messy: possible damage to equipment

Vocabulary

A Study vocabulary for Listening

- Set the task for individual work and pairwork checking.

Answers

1 upholstery fabric
2 ignite
3 glue
4 smother
5 dust
6 backing foam
7 mess

B Practise pronunciation

- Drill pronunciation with the whole group. Get pairs to practise saying the words to each other.

Answers

Answers depend on the students.

Listening

A Listen for advantages and disadvantages of different extinguishers (CD3 Track 13)

- Set the task. Play the recording once. Pairs compare ideas.

- Feedback orally. Refer back to the table on the board and discuss which ideas were mentioned. Add any additional ideas to the table.

Answers

No answers at this point.

B Listen and correct information in a table (CD3 Track 13)

- Give students time to read through the information in the table. Clarify any unknown vocabulary, such as *squirt*.

- Set the task. Play the recording again.

- Students check answers in pairs. Feed back orally, using an OHP if possible.

Answers

Fire test			
fuel: upholstery, foam, gasoline			
	halon	CO$_2$	dry chemical
extinguished Y/N	Y	<u>Y</u>	Y
no. of squirts required	<u>1-2</u>	4	<u>2</u>
time	minimum: 2-3 secs	10 secs	4 secs
reignition Y/N	<u>N</u>	<u>Y</u>	N
approx. amount used (%)	<u>50%</u>	100%	<u>66%</u>
immediate environmental effects of use	none	<u>none</u>	extensive layer of chemical powder on all surfaces in the area; <u>airborne particles unpleasant</u>

- Now have the students do the Workbook exercises for Unit 12, Lesson 2.

Workbook answers

Writing: explanations

A Label the parts

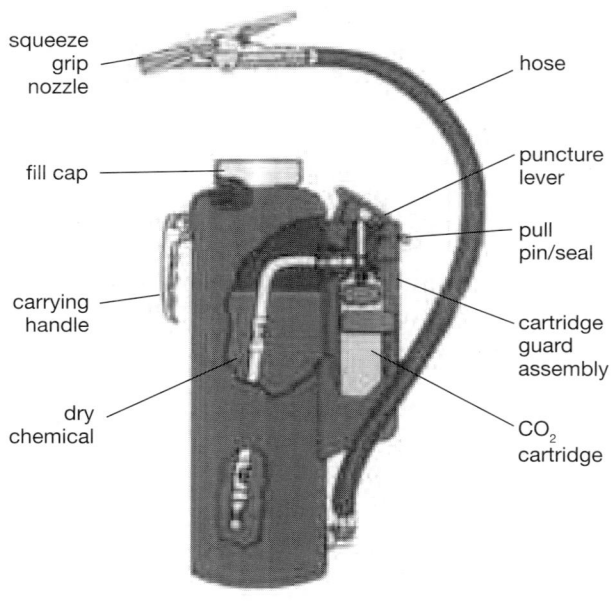

squeeze grip nozzle

hose

fill cap

puncture lever

pull pin/seal

carrying handle

cartridge guard assembly

dry chemical

CO$_2$ cartridge

B Use expressions to explain the functions

Answers depend on the students.

C Write notes

Answers depend on the students.

Language: quantity

A Complete the sentences

1 a couple

2 nearly empty

3 two-thirds of

4 in minimum time

5 not particularly large

6 overall

7 about half of

8 one or two

Language: verb forms

A Complete the table

read	read	read
cut	*cut*	cut
break	*broke*	broken
find	*found*	*found*
see	saw	*seen*
go	*went*	gone
do	*did*	*done*
fly	flew	*flown*
become	*became*	become
put	put	*put*
have	*had*	*had*
write	*wrote*	written

Objectives

To review concepts in air conditioning, including evaporation and humidity.

To practise listening to label a diagram.

To practise notes on a diagram.

Language

Vocabulary around air conditioning, e.g., *moisture*, *intake*.

Speaking

A Review of concepts

- Set the task. Refer Students A to Unit 8, Lesson 5 to find information about evaporation; Students B look at Unit 8, Lesson 6 to find information about humidity.

Answers

No answers at this point.

B Exchange review information

- Set for pairwork. Encourage pairs to ask each other questions. As feedback, clarify any outstanding problems students have with concepts.

Answers

Evaporation

- refers to the change of matter from a liquid to a gaseous state
- heat is absorbed during the evaporation process
- tends to proceed more quickly where there is air movement

Humidity

- refers to the amount of water vapour in the air
- measured in three ways: absolute humidity, relative humidity (used in weather forecasts) and specific humidity

Listening

A Activate ideas

- Set for pairwork.
- Feed back orally; hold a general discussion on the features of the buildings. Present/ check understanding of key vocabulary, e.g., *tower*, *tunnel*, *roof*, *dome*.

Answers

Answers depend on the students.

B Listen for information to discuss questions (CD3 Track 14)

- Give students time to read through the three questions. Play the recording once.
- Set the questions for discussion in pairs. Feed back orally.

Answers

1 because the group are near the sea

2 air movement increases evaporation of perspiration, which cools the skin

3 accept all reasonable suggestions, e.g., hanging or laying out of wet mats; building living areas on upper floors

C Listen and label the diagram (CD3 Track 15)

- Give students time to study the diagram. Set the task.

- Play the recording once. Students compare answers in pairs. Feed back using a drawing on the board or an OHP.

Answers

1 wind tower
2 air intake
3 tunnel
4 house

D Listen and take notes on how the system works (CD3 Track 15)

- Tell students to take brief notes on the diagram. Replay the recording.

Answers

No answers at this point.

E Compare notes with a partner

- Students compare answers in pairs. Encourage them to comment on and amend their notes. Feed back onto the board.

Suggested answers

1 tower as tall as possible: high wind force + low temp.

2 cool air down tower + through tunnel

3 evap. cools tunnel = cool air behind. cool air into house + out of windows. cycle

F Listen to the last part of the recording (CD3 Track 16)

1 Set the task. Reassure students that a high standard of drawing is not required. They need only represent the appropriate ideas.

- Play the recording. Students compare their drawings in pairs. Feed back onto the board.

Suggested answer

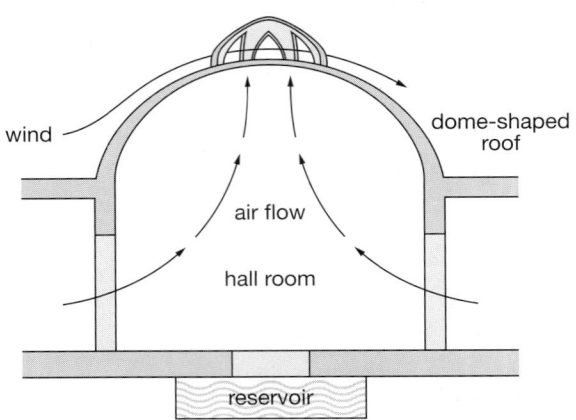

http://www.cais-soas.com/CAIS/ Architecture/wind.htm

2 Get students to read the text and, in pairs, to discuss their ideas for the gaps. Do not feed back yet. Students will listen to check.

3 Play all of the recording again while students complete the gaps.

- They should check and confirm answers in pairs. Feed back orally.

Answers

line 2	pressure
lines 3–4	warm air
line 5	airflow
line 7	reservoir
line 8	ambient temperature
line 9	evaporation

4 Have students discuss which AC system is which in the photos in Exercise A.

Answer

The photographs show the wind tower system.

Speaking and pronunciation

A Find the odd one out (CD3 Track 17)

- Set the task. As they listen, students circle the word in the line with a different number of syllables from the other three items. Allow them to read all of the words in each line before they hear them.

- Students compare answers in pairs.

Answers

1	average	**5**	cool
2	falls	**6**	constant
3	distribution	**7**	saturate
4	ducting		

B Check concepts

- Refer students to the example on page 227. Demonstrate the task with a strong student. Point out that correct pronunciation is important in this task, as well as giving the part of speech, the meaning and a typical example phrase.

- Set the task. Students can choose any words from the exercise. Students have the option of saying *I'm sorry, I don't know.*

- At the end of the task, students should look up any words they did not know in the dictionary.

- As feedback, elicit a few examples.

Answers

Answers depend on the students.

- Now have the students do the Workbook exercises for Unit 12, Lesson 3.

Workbook answers

Reading and writing: AC systems

A Decide which AC system is used

The wind tower system

B Circle the expressions in the tapescript

The words can be found in the following order:

and

because …

so

but

as

then

because of

This means

C Look at how the expressions are used

Answers depend on the students.

D Describe how the Environmental Control System works

Answers depend on the students.

E The number of forms used in the text

build	1
ventilate	2
control	1
circulate	1
keep	1
sustain	1
comfort	1

F Correct the phrases

1 large vents in the towers collect desert winds

2 fresh air is circulated around the building

3 this energy-saving system drastically reduces the need for air conditioning during large parts of the year

4 a comfortable temperature-controlled environment all year round

Objectives

To review concepts connected with air pressure.

To practise saying numbers correctly.

To practise transferring data from a graph to a table.

Language

Numbers: decimals, fractions, thousands.

Question forms, including passive forms.

- **Note:** Pneumatics, strictly speaking, is the use of pressurised gas to do work such as transmit signals or energy. The topic of pressurised aircraft cabins does not fit into this category. However, it gives useful scope for revision of the vocabulary and concepts around gas pressure covered in Unit 8.

Speaking and vocabulary

A Numbers review

- Use the board to review a few fractions, decimals and longer numbers as a group.

- Set the task for pairwork. Feed back orally.

- **Note:** American English typically uses, e.g., *four hundred fifty,* where UK English uses *four hundred and fifty*, and in decimals, *zero* where UK English uses *oh* or *nought*.

Answers

1/3	one-third
7,450	seven thousand, four hundred and fifty
3.14	three point one four
4½	four and a half
159	one hundred and fifty-nine

1,450,000	one million, four hundred and fifty thousand
+/– 0.3	plus or minus nought point three
7.08	seven point oh eight
8/91	eight over ninety-one
1/6	one-sixth
308	three hundred and eight

B Pre-check vocabulary

- This exercise is intended to divide the vocabulary load and provide opportunity for practice of explanation of terminology.

- Groups of Students A and Students B should check their words. Monitor and assist. Encourage use of dictionaries and discussion. Students can make notes.

Answers

No answers at this point.

C Explanation of terms

- Put pairs of Students A and B together. Set the task. Students should explain rather than read definitions for their terms.

- Feed back orally.

Possible answers

A: pressurised cabin – aircraft cabin where the air pressure is kept at a level comfortable for crew/passengers even when the aircraft flies at high altitudes where the outside (atmospheric) pressure may not provide enough oxygen for people to survive.

climb rate – how quickly an aircraft gains height

fuselage structure – the central part of the airframe, made from bulkheads, longerons, stringers and skin

outward forces – forces inside an aircraft pushing the skin outwards

B: cruising altitude – the altitude where an aircraft flies during the longest part of its journey

sea level – the average level of the surface of the sea. Aircraft altitude is given as X meters above sea level

cabin pressure – the air pressure inside the aircraft cabin

flight engineer – a member of the aircraft crew who checks and controls aircraft systems during flight

Reading

A Scan for numbers in the text

- Set the task. Remind students that they should not look at the text from top left to bottom right line by line, but look for numbers only, in most cases here, zeros. Set a time limit, e.g., ten seconds.

- Feed back orally. Elicit the fact that one reason why numbers are easily identified is that they are the size of capital letters.

Answers

777	line 10 (a type of plane)
11,000	line 12 (altitude in feet)
43,000	line 11 (altitude in feet)
7,200	line 11 (altitude in feet)
40,000	line 18 (cruising altitude in feet)

B/C Form questions and find answers in the text

- Tell students to cover the text on page 229.
- Work through the first item on the board. Set the task for individual work and pairwork checking.
- Feed back onto the board.
- Set the reading task. As feedback, pairs ask and answer the questions from memory.

Answers

Why is it important to pressurise aircraft cabins? (The cabin pressure reinforces the fuselage and enables it to stand up to the pressure placed upon it from the outside air.)

How are aircraft pressure systems controlled? (Modern electronic systems are automatically controlled. Older systems are controlled manually.)

At what altitude is it necessary to pressurise a (the) cabin? (Normally above 10,000 ft.)

Who is responsible for controlling the cabin pressure? (The flight engineer.)

D Focus on vocabulary in the text

- Set the task for individual work and pairwork checking. Feed back orally.

Answers

1 pressure differential
2 climb schedule
3 flying the cabin

E Relate the text to graphical data

- Give pairs time to discuss the question. Feed back orally.

Answers

The graph shows details of the climb schedule: the numerical relationship between increasing aircraft height on the horizontal axis, and effective altitude inside the cabin on the vertical axis. That is, the exact pressure differential for increasing altitude.

F Data transfer

- Set the task for individual work and pairwork checking. Do the first item as a group, using an OHP if possible. Elicit the meaning of *effective altitude* and *actual altitude*. Point out to students that it is possible to be accurate only to a limited degree because of the scale of the graph.

- Encourage students to read out the figures they have added to the table to each other. Monitor and check pronunciation.
- Feed back onto the board. Allow disagreements of 1–2,000 ft.

Answers

effective altitude	actual altitude
6,000 ft	*36,000 ft*
4,000 ft	32,000 ft
3,500 ft	*28,000 ft*
2,000 ft	20,000 ft
1,000 ft	*12,000 ft*
-200 ft	sea level
500 ft below sea level approx.	*-3,000 ft*

Extension: speaking

- This exercise requires further practice of numbers as well as of the graph.
- In pairs, students will ask and answer their own questions, e.g.,

   Student A: *How high is the plane flying if the cabin altitude is 5.5 thousand feet?*

   Student B: *Slightly more than 35,000 feet. About 36 or 37 thousand.*

- Elicit the questions onto the board. Set the task. Monitor and assist.

Answers

Answers depend on the students.

- If you have time, work on prepositions in the text: *above*, *inside*, *up*, *outside*, *up to*, *around*, *below*.

- Now have the students do the Workbook exercises for Unit 12, Lesson 4.

Workbook answers

Language: word-building

A Label the words

pressurisation	noun
pressurised	adjective/verb third form
unpressurised	adjective/verb third form
pressurise	verb

B Complete the table

1	-ion	combination	compression	location		
2	-ly	slowly	damply	respectively	randomly	automatically
3	-ment	alignment	requirement			
4	-ing					
5	-ure	pressure	moisture			
6	-ate	evaporate	contaminate			
7	-ity	humidity	quantity	electricity		
8	-ic	pneumatic	hydraulic			
9	-ent	equivalent	dependent			

C Mark the endings

1 n (noun)
2 adv (adverb)
3 n (noun)
4 v, adj, n (verb, adjective, noun)
5 n (noun)
6 v (verb)
7 n (noun)
8 adj (adjective)
9 adj (adjective)

Language: sentence structure

A Mark subject, verb and object

1 Maintenance crews (S) check (V) aircraft tyres (O) for signs of wear.

2 Where is (V) the pre-flight documentation (O)?

3 Ground crews (S) do not usually fly (V) aircraft (O).

B Tick the correct sentences

1 incorrect (Is the supervisor in his office?)

2 correct

3 incorrect (I'm sorry. I do not have the report form.)

4 incorrect (My partner came with me.)

5 incorrect (How did he break this?)

C Order the parts of the sentences

1 Planes operating below 10,000 ft are not usually pressurised.

2 An airliner cruising at around 40,000 ft has a cabin altitude of about 7,200 ft.

3 Pressurising a cabin exerts stresses on the fuselage.

4 The flight engineer's job is known as 'flying the cabin'.

5 I can give you an idea of the schedule.

6 Too much reinforcement would make the plane heavy.

7 The pressure inside is different from the external pressure.

8 This relationship is called the pressure differential.

 **Lesson 5** — Electrical systems: Review I

Objectives

To review terms and symbols for electrical components and aircraft electricity supply.

To practise scan-reading, reading for detailed information.

Language

Abbreviations in electrical supply, e.g., *AC*, *INV*.

Speaking

- As an introduction, you may wish to briefly review electrical symbols using the chart on pages 263–265.

A Review aircraft electricity supply

- Refer students to Unit 9, but tell them they must scan the unit and find the relevant lesson/text themselves (it is Unit 9, Lesson 9).

- Give students time to study it individually in silence and remind themselves of the concepts it contains.

- Feed back onto the board, build up a mind map and have a general discussion on the systems: *EPU*, *Battery*, *SG*, etc.

Answers

Answers depend on the students.

Reading

A Review electrical components and symbols on a schematic

- Elicit or clarify that the diagram is an inverter. Students may wish to check the definition in the Glossary.

- Set the task for pairwork discussion. Encourage students to look at the electrical symbols at the back of the book and compare them with the ones in this diagram (there are some differences).

- Feed back by eliciting ideas and writing them on the board, e.g., *switch*, *circuit breaker*, *battery*, *generator*. Talk about what an inverter does.

Answers

Answers depend on the students.

B Review abbreviations

1 Set the task for pairwork. Point out to students that they have met all but two of these abbreviations before in the course.

2 Stress that students must find all the examples in the text.

- **Note:** Sometimes the abbreviation follows the full versions of the words, e.g., *Central Warning System*. Set a challenging time limit as usual.

3 Set the task for individual work and pairwork checking. Feed back onto the board.

Answers

Hz	*Hertz*	(lines 6 and 24)
CB	*circuit breaker*	(line 17)
BAT	*battery*	(lines 14, 17 and 23)
DC	*direct current*	(lines 6, 10, 12, 19 and 21)
GEN	*generator*	(lines 10, 14, 17 and 23)
CWS	*central warning system*	(lines 5 and 14)

AC	*alternating current*	(lines 6, 12, 14, 16, 18, 21 and 24)
INV	*inverter*	(lines 15 and 17)
V	*volts*	(lines 6, 10, 14 and 24)

C Reading for key information

- As with other texts in the course, this manual text is quite dense in terms of vocabulary and concepts. The focus is on developing the skills required to deal with such text, rather than understanding every word of this particular example.

- Set the task for individual work and pairwork checking.

Answers

1	a	**4**	a
2	a	**5**	a
3	b	**6**	a

- If you have time, work on the difference between *by* and *with/using* in such cases as:

  Supply is controlled <u>by</u> an automatic regulator. / The system is protected <u>by</u> circuit breakers.

  and

  Components must be removed <u>with/using</u> the appropriate tools. / PCBs are checked <u>with/using</u> a multimeter.

- Now have the students do the Workbook exercises for Unit 12, Lesson 5.

Workbook answers

Language: the passive voice

A Put the verbs in the sentences into active/passive voice

1 The system consists of two inverters and a switch.

2 It is protected by circuit breakers.

3 The system is monitored by the CWS.

4 The two inverters supply 115-V AC and 26-V AC to the system.

5 A switch is used to select one inverter at a time.

6 The output is monitored by the CWS. The circuit breaks if the voltage falls too low.

7 Both inverters are supplied with DC input.

8 They are prevented from changing DC into AC by inhibitors.

9 A remote ON/OFF loop controls the inverter inhibiting circuits.

Pronunciation: stress

A Divide the words into groups

Oo	Ooo	ooOo	oOoo
circuit	actuates	automatic	emergency
current	dissipate	situation	mechanical
busbar	multiple		electrical
	actuate		

B Complete the text

1	electrical	5	mechanical
2	automatic	6	actuates
3	circuit	7	current
4	emergency		

Word-building: *anti-*, *semi-*, *self-*

A Write the words next to the prefixes

1 anti- static/oxidant

2 semi- circular/conductor

3 self- discharge

B Find other words

Answers depend on the students.

C Write sentences

Answers depend on the students.

Lesson 6 — Electrical systems: Review II

Objectives

To review concepts connected with batteries and electric motors.

To practise interpreting data in a table.

Language

Comparatives: *far more than*; *as ... as*.

Vocabulary for starter generators.

Speaking I

- As an introduction, elicit different types of engine, e.g., *radial*, *rotary*, *turboprop*, *gas turbine*, and a few advantages of each over its predecessors.

- Set the Speaking task for pairwork. Feed back orally, but do not confirm ideas yet, as students will hear someone talking about advantages in the recording.

Answers

No answers at this point.

Listening

A 🔊 Listen for advantages of aero engines (CD3 Track 18)

- Exploit the visuals. Elicit what the photos show, and whether or not students think that these aircraft could use electric engines.

- Set the task. Play the recording once.

Answers

No answers at this point.

B Check ideas

- Pairs check answers to Exercise A. Feed back orally.

Suggested answers

1 Electric engines are quieter than other engines.
2 They produce no dangerous sparks.
3 They are cleaner.
4 They have a longer life and are far more reliable than combustion engines.
5 They are efficient: they can convert more of the electrical energy into mechanical energy than conventional engines.

C 🔊 Listen to label a diagram (CD3 Track 18)

- Give students time to study the diagrams. Tell students that the recording uses the term *BLDC*, or *brushless DC motor*, for this equipment. Elicit how it is different from the brushed motor on page 173.

- Get students to look back at page 173 and establish that the brushless motor is like a reversed version of the brushed motor.

- Play the recording once. Students compare answers. Feed back onto a drawing on the board or an OHP.

Answers

1 armature
2 electromagnets
3 magnets
4 rotor
5 shaft

Reading and speaking

A Read for main ideas

- Explain to students that the same lecture continued and that they are going to read a student's notes from it.

- Set the task. Tell students to read the three titles, and remind them of what *BLDC* stands for. Pairs check answers. Feed back orally.

Answers

Disadvantages of electric engines

B Interpret a table

- Briefly review Unit 9, Lesson 4. Elicit the types of battery in the table: lead-acid; nickel-cadmium; nickel-metal-hydride; lithium-ion; zinc-air, gasoline.

- Set the task for pairwork. Feed back orally.

Suggested answer

The lecturer compared six types of electric battery against each other and against kerosene in terms of the watt hours of energy produced per kilogram that they weigh. Kerosene is by far the most effective.

Writing and language

A Focus on comparatives

- Review comparatives briefly if you feel it is necessary for the group.

- Set the task for individual work and pairwork checking. Feed back orally.

Answers

See tapescripts 57–58.

B Expand student notes

- Elicit an example sentence onto the board. Draw attention to the use of the comparative structure *as … as*.

- Set the task for individual work and pairwork checking. Monitor and assist.

- Feed back onto the board.

Suggested answer

Disadvantages of electric engines

Improved motors, e.g., BLDC, are not the whole story. There is still the problem of the power supply:

1 Even modern batteries cannot convert chemical energy into electric energy as effectively as kerosene.

2 Batteries are heavy. (The Sunseeker can carry its pilot because it gets power from solar cells and can convert this sunlight into electricity, so it has no need to carry a heavy battery.)

3 The plane does not get lighter as it flies, unlike with kerosene. The only way to get round the problem would be to throw batteries out of the plane when they were used up. This is not practical.

- If you have time, work on note-taking forms in the text in Reading and speaking 1: numbered points; abbreviations such as *bats, e.g.*; shorthand such as *cos/w.*; reduced grammar such as *Only way to get round prob*.

- Students could take turns to be the lecturer and say exactly what he must have said OR students could take notes on the recorded parts of the lecturer's talk.

- Now have the students do the Workbook exercises for Unit 12, Lesson 6.

Workbook answers

Language: electrical terms

A Complete the words and phrases

1	busbar	8	malfunction
2	insulation	9	loop
3	current	10	transverter
4	voltage	11	connect
5	conductor	12	lubricant
6	restarter motor	13	short circuit
7	switch log		

B Write connected words

Answers depend on the students.

C Check your spelling

Answers depend on the students.

Language: the same or different?

A Use these expressions

1 the same as

2 different from

3 equivalent to

4 approximately

5 equal to

Reading and writing: brushless motors

A Look at the notes

No answers at this point.

B Write notes

Answers depend on the students.

Objectives

To review concepts of flight instrumentation.

To practise transferring information from a diagram to a written text.

To focus on prepositions in a text.

Language

Prepositions for position and direction, e.g., *inside, at, to.*

Speaking I

A Revise what students know about the 'basic six'

- Focus attention on the diagram. Check that students are familiar with the vocabulary in the labels.

- Set the task for pairwork. Stress that you do not expect students to know exactly how the instrument works, but that they should speculate and apply their knowledge to discussing it.

- As feedback, elicit some ideas from the group. Do not confirm or correct them yet. Students will read to check their ideas. See what other information students can remember about the 'basic six'.

Answers

1 It is an altimeter.

2 Answer depends on the students.

Reading

A Transfer information

- Tell students to read through the text without completing the blanks. Point out that it is a description of the instrument in Speaking I.

- Elicit the fact that it is an altimeter. Elicit any ideas the students have formed on first reading as to how it works.

- Set the task for individual work and pairwork checking. Feed back orally.

Answers

line 1	capsules
line 11	gear wheels
line 12	pointers
line 17	static port
line 20	altimeter setting adjustment knob

B Scan for collocation

- Stress that several answers are possible to this exercise. Do one item with the group and elicit possible combinations, e.g., *pressure: the pressure inside the plane/the pressure outside the plane/the pressure inside the instrument.*

- Pairs discuss possibilities for other phrases.

- Feed back onto the board. Accept any ideas that collocate and are logically possible and grammatically correct.

- Students read and check their ideas against the actual phrases in the text.

Answers

fastened	to	the body of the instrument
the pressure	inside	the instrument
the atmosphere	outside	the plane
the ports are	at	the rear of the fuselage
attached	to	the altitude display
set a display	to	zero

C Transfer language

- Set the task for individual work and pairwork checking. Do the first item with the group.
- Feed back orally.

Answers

1	to	**4**	inside
2	to	**5**	outside
3	at	**6**	to

Speaking II

A Apply new language

- Give students time to study the diagram. Elicit what the air speed indicator does.
- Point out the presence of gear wheels and a shaft connecting the components to the display. Elicit what *ram air* is, i.e., air entering the aircraft as it moves forward through the air.
- Set the task for pairwork. Monitor and assist. Feed back orally. Encourage the correct use of the prepositions, and of vocabulary and collocations, from the reading text.

Suggested answer

Ram air enters the instrument via the pilot tube, transmitting the outside air speed to the diaphragm. The diaphragm is free to move: when the pressure of the ram air increases, the diaphragm will move forward, and when the plane slows down and the pressure of the moving air decreases, it will move backwards to its original position. The moving assembly is connected to the display via shafts and gear wheels. The display has a pointer that gives the air speed.

- Now have the students do the Workbook exercises for Unit 12, Lesson 7.

Workbook answers

Reading and writing: the airspeed indicator

A Reread the text

No answers at this point.

B Write notes

Answers depend on the students.

Language: *the other, another, each other, both and each*

A Complete the sentences

1	each other	**6**	both
2	both	**7**	another
3	the other	**8**	the other
4	each other	**9**	both
5	another	**10**	another

B Complete the sentences

1	both	**5**	each other
2	each other	**6**	both
3	other	**7**	other
4	another	**8**	another

Language: similar or different?

A Mark the pairs of words S or D

No answers at this point.

B Check your answers

1	S	**11**	S
2	S	**12**	D
3	S	**13**	S
4	D	**14**	D
5	D	**15**	S
6	D	**16**	D
7	S	**17**	S
8	D	**18**	S
9	D	**19**	D
10	D	**20**	S

C Replace words

4	fail	**10**	environmental
5	adjust	**12**	inspect
6	suitable	**14**	extinguish
8	fault	**16**	fall
9	okayed	**19**	forbidden

D Write an expression

Answers depend on the students.

E Mark the stress

ad'just	fall
in'spect	for'bidden
environ'mental	fail
ex'tinguish	'suitable
o'kayed	fault

Lesson 8   Instruments: Review II

Objectives

To review concepts around 'glass cockpit' instruments.

To practise listening for main ideas and details.

To practise writing and reading notes.

Language

Sentences with *if*, e.g., *If he didn't have an altimeter, he wouldn't know how high he was flying.*

Speaking

A Discuss current knowledge of flight instruments

- Elicit the name of the flight instrument in the previous lesson (*altimeter*). Set the task for pair or small-group work.

- As feedback, elicit some ideas from the group. Do not confirm or correct yet. Students will check their ideas in the next exercise.

Answers

No answers at this point.

B Review flight instruments

- Set the task for individual work. Point out that students will also need to look at the tapescript for Unit 10, Lesson 2.
- Students check their ideas in pairs.
- Feed back orally.

Answers

The 'basic six' instruments and their functions are:

air speed indicator
> forward speed of aircraft through the air

attitude indicator
> position of wings relative to the (artificial) horizon

vertical speed indicator (VSI)
> rate of ascent or descent of aircraft

altimeter
> height of aircraft above sea level

turn coordinator
> position of aircraft during a turn

gyro compass
> direction (bearing) of flight

C Activate ideas

- Set the task for pairwork. Feed back orally.

Answers

The picture on the left shows a more traditional cockpit. The picture on the right is a more modern cockpit in which the information given by the 'basic six' instruments appears on LCD screens.

Listening

A Discuss topic before listening

- Set the task for pair or small-group discussion. As feedback, elicit a few ideas from the group. Do not confirm or correct yet.

Answers

No answers at this point.

B 🔊 Listen for main ideas (CD3 Track 20)

- Give students time to read the two questions. Elicit a few ideas as to the answers.
- Play the recording. Pairs compare their answers. Feed back orally.

Answers

1 The term 'glass' comes from the fact that the screens of the first PFDs (Primary Flight Displays) were made of glass. The PFD contains all the information contained in the 'basic six' instruments.

2 Extra information given by the PFD is: a visual indication of the runway; the ILS signal beams (Instrument Landing System); Navigation Display.

C 🔊 Listen for detailed information (CD3 Track 20)

- Set the task. Play the recording. Pairs compare answers.
- Feed back onto the board using a drawing or an OHP.

Answers

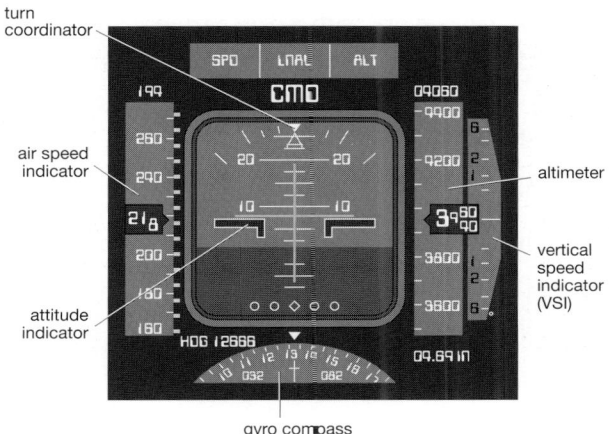

Writing and speaking

A Read the notes

- Give students time to read the notes silently.

B Correct the information

- Set the task for pairwork. Elicit the first correction from the group.
- Feed back onto the board.

Suggested answers

Glass cockpit

'glass' refers to the PFD – Primary Flight _Displays_

basic six info _same as usual_

All info on screen to _make flying safer_

Instr: centre att. indic ~~/ air speed indic~~

L/H side – _air speed indicator_

R/H side – altitude

C Note-taking (CD3 Track 21)

- As usual, draw students' attention to the note-taking techniques in the notes in Exercise A.
- Set the task. Play the recording.

Answers

No answers at this point.

D Check notes

- Students work with a new partner and practise describing the systems using their notes.

Suggested answers

1 Instrument Landing System = ILS

horiz. + vert. signal beams: PFD shows position

Pilot can land using only this instr.

2 Navigation Display (ND)

imp. instr. = combining of radio navigatn instr., e.g., ADF indic) + moving map

Can show: big/small map, shape of land, landing maps, weather maps, 3D images

Pilot can change map, or no map

E Listen and check (CD3 Track 21)

- Play the recording again.
- Students can use the tapescript and listen to the recording to check their answers.

F Pronunciation

- This task is designed both to improve listening skills and to improve students' pronunciation. Choose a short section of the recording where at least two voices appear.
- Play it phrase by phrase and get students to repeat it. Drill the pronunciation until it is an accurate rendering of the recorded speech.

Language

A Practise using second conditional structures

- This task is an opportunity for students to review Unit 10, Lessons 8 and 9.
- Write the example from the Course Book on the board. Elicit further examples of what would happen if the pilot did not have flight instruments, e.g., _He might fly in the wrong direction if he didn't have a gyrocompass._
- Highlight and review the grammar structure if necessary.
- Make two groups, A and B. Refer Students A to Unit 10, Lesson 8, and Students B to Unit 10, Lesson 9. They should study the material there and write four or five sentences following the pattern.

Answers

Answers depend on the students.

B Compare with a partner

- When they are ready, pairs of Students A and B should take turns to tell their partner one of their sentences. Their partner should elicit clarification of the concepts with the question *Why?* For example:

  Student A: *Flying would be much more dangerous if planes did not carry ELT.*

  Student B: *Why? What is ELT?*

- Monitor and assist. As feedback, elicit one or two exchanges.

Answers

Answers depend on the students.

- Now have the students do the Workbook exercises for Unit 12, Lesson 8.

Workbook answers

Language: place and movement

A Choose the correct word from the three options given

1 over	**4** up and down
2 in the centre	**5** on the right
3 At the bottom	**6** on either side of

Language: *whether* and *if*

A Decide if the rules of usage are true or false

Pupils should circle

1 True	**3** False
2 False	**4** True

B Complete the sentences

Answers depend on the students.

Spelling and pronunciation: consonant clusters

A Complete the missing letters

line 1	sp
line 2	st
line 3	ed
line 4	pr
line 5	fr
line 6	l
line 7	ct

B Practise saying the words aloud

specific	spray	inspector	especially
static	installation	adjust	instruments
captured	crimped	charged	arranged
pressure	provide	price	compression
from	friction	free	fridge
sleeve	block	clamp	clear
disconnect	defects	protects	electrical

Objectives

To practise interpreting a troubleshooting flow chart.

To discuss and design a flow chart for troubleshooting equipment failure.

Language

Vocabulary for a simple electrical appliance and faults, e.g., *sensor, fuses, blockage, sparking.*

Question forms, e.g., *Is the fan turning? Can the 'stat be reset?*

- **Note:** The initials *MRO*, as explained in Unit 11, Lesson 9, stand for *Maintenance, Repair and Overhaul*, the term covering all work done on aircraft and equipment.

Vocabulary

A Categorise vocabulary

- Give students time to read through the items in the box.

- Draw a simple table on the board with two columns: *equipment parts/components* and *problems*. Elicit the correct column for, and write up, two or three items from the box.

- Set the task for pairwork. Students check their ideas with a new partner.

- Monitor and assist. Feed back onto the board.

Answers

equipment parts/components	problems
fuses	overheat
thermostat	bad connection
socket	fault
PCB	sparking
fan	blockage
sensor	
terminals	
pressure switch	
lead	
gas valve	
burner	

B Identify equipment

- Elicit ideas from the group. Accept all reasonable answers. In the case of the flow chart the students will look at in this lesson, the words are all connected with a domestic heating boiler.

- Elicit the meaning of the term *flow chart*. Refer students to Unit 9, Lesson 7, to review the faultfinding flow chart for a lamp. Explain that students are going to read a more complex flow chart.

Answer

Accept any justified answers. Students will see later that the equipment is a boiler.

Reading

A Complete the questions

- Set the task for individual work and pairwork checking. Do a few examples with the group, e.g., *1* and *3*.

- Feed back orally.

Answers

Is the fan turning?

Is mains on neon '13' illuminated?

Is there a live supply to both terminals of overheat 'stat?

Can the overheat 'stat be reset when the system is cold?

Is there a supply to fan connector on PCB?

Is the fuse blown?

Is the ignition electrode sparking?

Is the boiler casing correctly fitted?

Is there a live supply on the lead to pressure switch?

Is the pilot injector blocked?

B Use the flow chart

- Refer students to the flow chart on page 253. Give them time to read it. In pairs, they check that they understand it. Monitor and assist; it may take some time for students to feel comfortable with the chart.

- Divide the group into Students A and B. Demonstrate the task with a strong student, if possible using an OHP of the flow chart for the group to follow. If no projection is available, they can follow in their books.

- Set the task. Clarify that students A and B each follow the chart in their own book; they should not look at their partner's book, and only Student A asks questions and Student B answers.

- Monitor closely and assist.

Answers

Answers depend on the students.

C Check your answers

- Have students compare the place on the chart where they have each reached.

- Monitor closely and assist.

Writing

A Prepare questions for a faultfinding flow chart

- Give students time to read the text. Clarify any vocabulary problems.

- Set the task. Elicit onto the board the first question: *Is there a power supply? /Is the power supply switched on? /Is the equipment connected to a power supply?*

- Point out that more than one set of questions for the text may be possible.

- Point out also that questions may begin *Does ...?/Has ...?/Is ...?/Are ...?/ Is there ...?/Are there ...?/Can ...?*

- Monitor and assist. Feed back onto the board.

Suggested answers

Is the equipment connected to a power supply?

Does the equipment work when power is switched on?

Is there a fuse?

Is the circuit breaker correctly set?

Does the appliance work now?

Is the cable damaged?

Does the equipment work now?

Are there signs of burning around the fuse or circuit breaker?

Has the cable been checked?

B Prepare to write: familiarisation with layout conventions

- Check students understand instruction and decision. *Instruction* is taken here to mean a course of action to be taken as part of the troubleshooting process; *decision* is a course of action to be taken based on the results of troubleshooting.

- Elicit the first answer and draw on the board a 'start or finish' box with the text inside.

- Set the task for pairwork. Students should draw the appropriate boxes in their notebooks. Students compare answers with a new partner.

- Feed back onto the board.

Answers

1	start or finish	**4**	decision
2	instruction	**5**	instruction
3	decision		

C Transfer text into a flow chart

- Refer students back to the text in Writing Exercise 1. Elicit the first box – the 'start situation' – in this case, and draw it on the board.

- Set the task for pairwork. Stress that, again, there is more than one possible design for the chart. Monitor and assist.

- Students compare their ideas with a new partner.

- Feed back onto the board.

Suggested answers

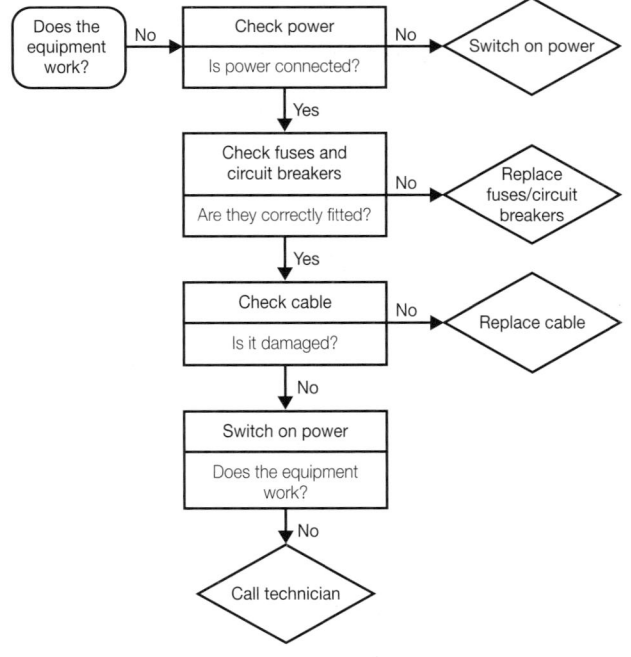

D Design a flow chart

- Draw the first box on the board. Elicit a few reasons why a mobile phone might not work and what should be done to remedy the problem, e.g., *It may be turned off – turn it on; the battery may be flat – the phone should be charged up*. Elicit and add to the board how this should appear in a troubleshooting flow chart.

- Set the task for pairwork. Monitor and assist.

- As feedback, ask some students to draw their flow chart on the board.

Answers

Answers depend on the students.

- Now have the students do the Workbook exercises for Unit 12, Lesson 9.

Workbook answers

Reading and writing: flow charts

A What the flow chart shows

Microbial growth inspection and maintenance actions.

B Write bullet-pointed instructions

Answers depend on the students.

Language: sentence structure

A Make changes to the sentences

1 The display should be set to minimum.

2 If the indicator does not light up, you should replace the bulb and recheck it.

3 Warning: ensure that the power supply is switched off before beginning inspection.

4 If equipment is still not working, then call a supervisor.

5 By using the flow chart, the technician can trace the fault without repeating any operations.

6 Inspect the unit for damage and signs of mishandling.

7 Once the unit is running, you can apply the load smoothly.

8 When the temperature reaches 70°C, check the pressure of the outlet valve.

Writing: spelling

A Correct the spelling

1 equipment

2 reinstall

3 correct

4 accessible

5 correct

6 correct

7 clearance

8 signs of

9 correct

10 failure

11 leak

12 correct

Lesson 10 — MRO: Review II

Objectives

To practise interpreting technical drawings.

To practise scan-reading and reading procedural instructions.

Language

Verbs for describing jacking procedure, e.g., *raise*, *adjust*.

Speaking

A Interpret the drawings

- As an introduction, review Unit 11, Lessons 7 and 9, which deal with types of access gear.

- Elicit the answer to question 1 from the whole group. Ask students what type of drawing this is.

- Go over key vocabulary such as *jack*, *jacking point* and *stand*.

Answers

The drawings are orthographic drawings of a light aircraft positioned on three jacks to lift the undercarriage clear of the ground.

B Discuss rationale for equipment layout

- Set for pairwork. Feed back orally.

Answers

The jacks are positioned for stability: the two jacks under the wings are near the centre of gravity of the aircraft; the tail stand completes a safe triangular base.

C Speculate about best procedure

- Elicit examples of how the expressions in the box might be completed, and drill the pronunciation of '*d*. Point out that students need to at least be aware of it so that they recognise it.

- Take the opportunity to elicit verbs such as *lower*, *raise*, *lift*, *pick up*, *move*, which students will probably need to use.

- Set the task for pairwork. Elicit one or two ideas to start the task. Stress that more than one person may be needed for the job.

- Feed back orally. Correct the use of the target language, but accept all reasonable suggestions. Students will read to check their ideas.

Answers

Answers depend on the students.

Vocabulary

A Use the verbs to complete the sentences

- Ensure that students understand all of the words in the box.
- Give students time to read through the sentences 1–10 without writing.
- Set the task for individual work and pairwork checking. Do the first item as a group.
- Feed back orally. Go over additional vocabulary items, such as: *locking pin*, *leg strut*, *centre tube central pillar*, *adapter*.

Answers

1	Raise	**6**	Open
2	Extend	**7**	Continue
3	Lower	**8**	Install
4	Adjust	**9**	Position
5	add	**10**	Insert

Reading

A Scan for vocabulary items

- Set a time limit which prevents students from reading the text word by word. Emphasise that the wording of the phrases and sentences may not be exactly the same as in the Vocabulary section.
- Pairs compare answers. Feed back orally.

Answers

raise	A, B, E, H, I
lower	A, G
add	F
extend	A, B
adjust	C, E, H, I
position	E, H, J

install	A, I, J
open	G
continue	I
insert	H

B Read for details of information

- Set the task for individual work at first. Point out that the first and last two steps of the procedure are already numbered. Clarify that students do not need to understand every word to do the task.
- Monitor and assist, particularly with vocabulary problems.
- When they are ready, students compare answers with a partner. Encourage students to explain and discuss the order they have decided in the event that there is disagreement.
- Feed back onto the board the order(s) that students propose. They will listen to check in the next task.

Answers

No answers at this point.

Listening

A Listen to check ideas (CD3 Track 22)

- Set the task. Play the recording once.
- Pairs compare ideas. Feed back orally.

Answers

1	D	**6**	A
2	H	**7**	I
3	J	**8**	G
4	E	**9**	C
5	B	**10**	F

Extension: writing and speaking

- Refer students back to the drawings in the Speaking section on page 240, and to the words in the Vocabulary box below them.

- From memory, students should make notes, either in the Course Book on the drawings themselves, or in their notebooks, about key aspects of the jacking procedure.

- Insist on the use of reduced note forms. Monitor and assist.

- They then practise asking and answering questions about the jacking procedure with a partner.

- Now have the students do the Workbook exercises for Unit 12, Lesson 10.

Workbook answers

Language: MRO terms

A Complete the table

access equipment	maintenance procedures	documentation	faults
jack	apply oil	airworthy	signs of rust
platform	disassemble	discrepancy	pitting
access docking	go through checks	sign off/OK	scratches
facilities	reinstall	IAW	failure

B Put more words in the table

Answers depend on the students.

Writing: discrepancy reports

A Mark the words *in* or *out*

1	in	4	in	7	in	10	out
2	out	5	out	8	out	11	out
3	in	6	out	9	in	12	in

B Write the third form of the verbs

1	adjusted	5	checked
2	recalibrated	6	reassembled
3	verified	7	found
4	completed	8	tightened

C Complete the discrepancy report form

Answers depend on the students.

Language: word combinations

A Match the words to make common word combinations

1	f	level position
2	k	extension lead
3	l	threaded ball-end
4	a	authorised person
5	i	required level
6	h	jacking point
7	j	approximate height
8	c	access equipment
9	b	locking collar / pin
10	g	central pillar
11	e	tail stand
12	d	hydraulic jack

Unit test

A test for this unit is available at:

http://www.garneteducation.com/reps/ documents/1256/SDT-u12-test.pdf

Contact your local Garnet Education representative for information about how to access these resources.

Glossary

AC (alternating current) (n) electricity which flows through a circuit, first in one direction and then in the opposite direction. The number of changes of direction per second (the frequency) is measured in units called Hertz.

accumulator (n) a device used to store hydraulic fluid under pressure. When a valve is opened, hydraulic fluid is forced out of the accumulator by compressed gas and can be used to operate equipment.

actuator (n) a device which changes hydraulic or electrical energy into mechanical energy in order to do work. Valves can be actuated by electric motors and air brakes by hydraulic pistons.

aerospace industry (n) all the companies that design, manufacture and maintain aircraft, rockets and spacecraft.

aileron (n) a hinged part of the aircraft wing which can be raised or lowered to control the turning and rolling movement of the plane.

air brake (n) a hinged surface, lowered from the aircraft when it is landing, which helps to slow the plane by increasing air resistance.

air duct (n) a channel or tube which carries air. Clean air is circulated through an aircraft by means of air ducts.

air traffic controller (n) (ATCO) someone who makes sure that aircraft in the air are safely separated from each other, both vertically and horizontally.

airframe (n) the main body of the aircraft consisting of the fuselage, wings, tail and skin, but without the engines, instruments or any other internal equipment.

airworthy (adj) in a good enough condition to fly safely and efficiently.

alloy (n) a metal which is made by mixing two or more liquid (molten) metals together. For example, brass is an alloy of copper and zinc. Solder is an alloy of lead and tin.

alternator (n) a type of generator using rotary mechanical motion of an engine to produce an alternating current

altimeter (n) an instrument which measures the vertical distance above sea level (the altitude) of an aircraft by means of air pressure changes or radio signals.

ammeter (n) an instrument which measures the quantity of electric current (the number of *amperes* or *amps*) flowing in a circuit.

ampere (amp) (n) the basic unit of the quantity of electric current. Very small currents are measured in **milliamps** (thousandths of an amp).

amplifier (n) an electronic device which increases the strength of an electrical signal.

analogue multimeter (n) an instrument which can be set to measure different electrical quantities, such as current, voltage and resistance. The position of the pointer on the dial of the analogue multimeter varies directly with the quantity it is measuring.

antenna (aerial) (n) a device connected to a radio transmitter or receiver which is used to send or receive signals, usually mounted on the outside of the plane.

armature (n) part of the rotating shaft of an electric motor or generator which has a coil or coils of wire wound on it.

artificial horizon (n) an instrument which shows a pilot the position of his aircraft relative to the real horizon, without needing to see outside the plane. It consists of a horizontal line which rotates clockwise or anticlockwise as the plane rolls, climbs or descends.

assembly (n) a group of parts connected together (assembled) to make another part, such as the gearbox of a car or the tail section of an aircraft.

ATC (n) an abbreviation of Air Traffic Control.

avionics (n) the electronic communication, navigation and control equipment of an aircraft.

backup system (n) equipment which is only used when the main equipment fails. Aircraft may have one, two or even three backup systems for particular pieces of equipment. For example, if a generator fails, power can be supplied by a backup battery instead.

battery (n) a device used for storing direct current electricity.

bearings (n) these support a moving part in the correct position but allow it to move freely. For example, electric motors have bearings at each end of the shaft, which keep it in position but allow it to rotate easily.

blind rivet (n) A type of *rivet* which can be fixed from one side of the workpiece only. Blind rivets are used when it is impossible to get to the other side of the workpiece, or when there is only one person to do the work.

boost (v) to increase the strength or the supply of something above the normal amount.

booster pump (n) a pump which can supply more fluid or higher pressure than normal.

bracket (n) a piece of metal, usually in the shape of an L or a triangle, which is used to support and hold individual components or complete items of equipment.

brake (v & n) (v) regular verb which means to slow down or stop a moving vehicle; (n) a device used to slow down or stop a moving vehicle.

bridge crane (n) a type of overhead mobile crane used to lift heavy loads in hangars, workshops and factories. The lifting equipment is suspended from a high steel beam (the bridge). It can move along the beam, which may also be able to move within the workplace.

brittleness (n) a property of materials which means they break into pieces easily when hit. Biscuits, glass and cast iron are examples of brittle materials.

bulkhead (n) a wall which divides one part of the aircraft from another in order to provide extra structural strength or safety. Small single-engine planes often have a fire-proof bulkhead between the engine and the cockpit.

burner (n) a device for controlled burning of fuel. Some jet fighter aircraft have afterburners, which inject fuel into the hot exhaust gas, providing sudden extra power.

burrs (n) small, sharp pieces of material forming the rough uneven edge of holes, sheets or workpieces which have not been properly finished and made smooth. Burrs can cause injury and damage.

busbar (n) large conductor with a high current capacity, used to connect specific power supply to a number of pieces of equipment.

camshaft (n) an engine shaft which has irregular-shaped sections (cams) mounted on it which open and close valves as the shaft rotates.

capacitance (n) the ability of a component or system to store a DC static charge. Capacitance can be measured with a capacitance meter in units called Farads.

capacitor (n) an electronic component which stores a DC static electric charge and will conduct AC current.

centre of gravity (n) the point on an aircraft where it is balanced in all directions. On modern aircraft, it is between the nose wheel and the wings. On older aircraft (taildraggers), it was between the wings and the tail.

centrifugal (adj) moving away from the centre of something that is rotating.

chassis (n) structure in the form of a metal box or frame on which components are mounted. In the past, cars were built on a chassis. Electronic equipment is often built on a chassis.

check valve (n) a valve which allows a liquid or gas to flow in one direction only. The simplest types contain a ball and spring or a hinged flap.

circuit board (n) a sheet of insulated material containing a number of connected electronic components. Sometimes, the board will have a circuit already printed on it, ready to receive the components.

circuit diagram (n) a drawing consisting of lines and component symbols which show how an electrical or electronic device works and how its components are connected to each other.

circuit-breaker (n) an automatic or manual emergency switch which disconnects the electrical power supply before equipment becomes overloaded and damaged or dangerous.

CNC (computerised numerical control) machine tools (n) lathes, drills, milling machines, etc., which are controlled by signals from a computer instead of being controlled by the hands of an operator. They can accurately produce standard components very quickly.

cockpit (n) the section of an aircraft which contains the pilot, instruments and flight controls. Training aircraft often have two cockpits.

coefficient of linear expansion (n) a measure of the increase in length of a material when it is heated. Each material has a different C of LE which is used to calculate its expansion at different temperatures. Aluminium expands more than cast iron when heated and has a higher C of LE (aluminium = 0.000022/C0, cast iron = 0.000011/C0).

combustion chamber (n) an enclosed space inside a *turboprop, turbofan* or jet engine where the fuel is burnt. Burning (combustion) of the fuel creates hot, expanding gases which produces *thrust*.

commutator (n) rotating part at the end of the shaft of a DC electric motor or generator which supplies or collects current.

component (n) a separately made part of a device or piece of equipment.

compression stroke (n) the stage in the operation of the internal combustion engine when the piston moves into the cylinder and compresses the fuel-air mixture prior to combustion.

compressor (n) a pump or machine that increases the pressure of a gas or vapour and decreases its volume.

conductivity (n) (adj. conductive) the property of allowing heat or electricity to pass through a material.

conduit (n) a length of pipe, duct or channel used to isolate and protect smaller pipes or cables.

control column (n) the main manual control used by the pilot to control the pitch and roll of an aircraft.

control surfaces (n) moveable sections of the wings and tail of an aircraft which are under the control of the pilot and enable him to change the position and direction of the aircraft.

corrosion (n) a chemical process which can damage or destroy a metal. It is the result of a chemical reaction between the metal and another material, such as acid or water.

cylinder barrel (n) the hollow inside of a cylinder in which a piston moves.

cylinder block (n) part of an internal combustion engine that contains one or more cylinder barrels.

cylinder head gasket (n) a type of sealing washer which ensures a tight seal between the cylinder head and the cylinder block.

decoy (n) a protection device used by fighter aircraft to attract enemy missiles to the wrong place so that they do not hit the fighter.

defogging nozzle (n) the end of a small tube which passes warm, dry air onto the cockpit glass to prevent condensation or fogging in order to maintain clear visibility out of the aircraft.

density (n) (adj. dense) the mass (or weight) of a material contained in a specific volume of space. A cubic centimetre of steel weighs more than a cc of aluminium, and is therefore denser.

digital multimeter (n) an instrument which can be set to measure different electrical quantities, such as current, voltage and resistance. Unlike analogue meters, it has no moving parts, and the results of a measurement are displayed as separate digits.

direct current (DC) (n) an electric current that flows continuously in the same direction from one pole to another.

display panel (n) a separate panel in the *cockpit* which holds flight information instruments and warning lights.

drogue parachute (n) a small parachute which comes out first and helps to pull out the main parachute.

ductility (n) (adj. ductile) the ability of a material to be stretched (drawn) into a thinner and thinner wire. Copper and aluminium are used for electric cable because of their high ductility.

durability (n) (adj. durable) the ability of a material to last for a long time without changing. Glass, gold and stainless steel are all durable materials.

ejection seat (n) a seat in a military aircraft which can be shot quickly out of the aircraft in an emergency, with the crewman still sitting in it.

elasticity (n) (adj. elastic) the ability of a material to regain its original shape after stretching or compression.

electrical arc (n) a spark caused by an electrical discharge between two adjacent terminals or contacts. Arcing in electrical switches is sometimes the cause of fires.

electrical resistivity (resistance) (n) the property of preventing the flow of electricity through a material measured in ohms.

electrolytic capacitor (n) a type of capacitor which consists of electrodes separated by a chemical substance called an electrolyte.

electromagnet (n) a piece of ferrous material which is made magnetic when an electric current is passed through a coil which is wound round it. Relay switches are often operated by electromagnets.

elevators (n) moveable control surfaces, usually mounted on the tail, which are used to vary the pitch by forcing the nose of the aircraft up or down.

engine block (n) the main part of an internal combustion engine, containing the cylinder block.

evaporator (n) a device which enables a liquid to change to a vapour (evaporate) easily.

exhaust (n) the escape of waste gases from the rear of an aircraft or other vehicle.

exhaust duct (n) a type of pipe used to convey *exhaust* gases from the engine to the outside of a vehicle.

external power unit (EPU) (n) equipment which supplies electrical power to an aircraft when it is on the ground and bypasses the aircraft's own generator and battery systems.

fan belt (n) a belt of flexible material which connects the main shaft of an engine with a cooling fan.

filament bulb (n) a traditional type of glass lamp which contains a thin wire which gives off light when a current is passed through it.

filter (n) a device for removing unwanted material from a liquid or gas.

fin (n) the vertical surface on the tail of an aircraft (the vertical stabiliser).

flammable (adj) easy to burn.

flange (n) a protruding edge of a component used to position or connect it.

flashpoint (n) the temperature at which the vapour of a flammable liquid such as kerosene will suddenly start to burn without any external source of ignition.

flexibility (n) the ability to be bent out of position and then back again. The wings of aircraft are designed to move up and down (flex) during flight, reducing stress on the airframe.

flight control systems (n) the electronic, electrical and mechanical equipment and instruments which are used to control the movement of an aircraft in the air.

flight data recorder (n) (black box) a very tough and well-protected container of instruments which continuously records details of a flight so that this information is available in the event of an accident.

flight path (n) the line or course flown by an aircraft through the air.

flywheel (n) a heavy wheel which continues to turn after power is disconnected. Flywheels are used to maintain smooth continuous rotation in engines and other equipment.

foreplane (n) small stabilising wings attached to the nose of jet fighters and missiles.

fuse (n) a protective device consisting of a piece of conducting material which melts and disconnects a circuit when there is too much electric current flowing through it.

fuselage (n) the central body of an aircraft without the wings and the tail.

gas compressor (n) a machine designed to increase the pressure of a gas by using cylinders and pistons or a series of powerful fans.

gasket (n) a piece of strong sealing material held tightly between two connected parts to prevent any leaks of liquid or gas.

GPS (global positioning system) (n) A navigation system which uses the position of satellites in orbit around the earth (the globe) to detect the exact position of an object on the surface of the earth.

grommet (n) a piece of flexible material fitted into a hole to prevent damage to pipes or cables by the sharp edges of the hole.

groundloop (n) a sudden accidental turning movement of an aircraft that is moving along the ground.

gyro-compass (n) a direction-finding instrument which uses a gyroscope instead of a magnetic needle to indicate the direction of travel.

gyroscope (n) a rapidly rotating wheel that maintains the position of its axis of rotation even when its container changes direction.

gyro-system (n) any control or instrument system which is based on the use of a gyroscope.

hand-propping (n) starting an aircraft engine by quickly pulling a propeller blade downwards by hand.

hangar (n) a large building which is used to store, repair or build aircraft.

induction stroke (n) the stage in the operation of an internal combustion engine when the piston moves away from the top of the cylinder and sucks air/fuel mixture into it through a valve.

internal combustion engine (n) any engine that produces power by rapidly burning fuel inside a cylinder containing a moving piston.

isolator (n) a device which is used to cut all mechanical or electrical connections between two parts of a system.

jack (n) a device which uses gears, compressed air or hydraulic fluid to produce a mechanical advantage to lift heavy loads.

jet pump (n) a type of fuel pump without moving parts which uses a liquid stream of fuel to suck more liquid into the stream.

joystick (n) the main control column of an aircraft.

landing gear (n) the extendable wheel and leg assembly usually mounted beneath the wings which, together with the nosewheel, enables it to land safely.

landing gear leg (n) an extending/retracting strut to which a landing wheel is attached.

lead (pronounced 'leed') screw (n) a rotating shaft with a pitched thread which is attached to a lathe. It enables bolts and screws to be made accurately by controlling the speed of the cutter along the workpiece.

lightness (n) the property of not being heavy. The lightness of aluminium is one of the main reasons for using it in aircraft.

locking pin (n) a small metal rod which is inserted into a folding or extending mechanism (such as a landing gear leg or a jack) to make sure that it doesn't change position suddenly.

longeron (n) an important structural beam which extends along the length of the fuselage.

longitudinal axis (n) the centre line which goes through the length of something. The longitudinal axis of a plane runs through the centre of the aircraft, from nose to tail.

lubricity (n) the property of reducing friction so that parts can move easily against each other. Some fuels have a high lubricity and so reduce engine wear.

lug (n) a small piece of a component or device which projects out of the main body and is used to attach it to something else.

malleability (n) the property of being beaten into a thin sheet or a different shape without breaking. Gold and lead are both very malleable metals.

melting point (n) the temperature at which a solid changes into a liquid. Metals used in aircraft engines need to have very high melting points.

missile (n) a pilotless flying weapon which explodes when it hits its target.

monocoque construction (adj & n) a body, such as an egg, which has all the strength in the skin, without any internal reinforcement. Most aircraft have a semi-monocoque design so that the stresses are carried by the internal framework as well as the outer skin.

MRO (maintenance repair and overhaul) (n) The systematic and regular inspection, repair and parts replacement procedures to ensure that aircraft are airworthy.

multi-stage compressor (n) a compressor which consists of two or three compressors which act on the gas one after the other and increase the pressure more each time.

NDI (non-destructive inspection) (adj & n) careful checking of a material or component without taking it to pieces or removing a sample from it. Examples are eddy current examination of metal skins and the use of a special instrument called a boroscope to look inside engines.

nickel-cadmium battery (n) a type of battery made from cadmium and nickel hydroxide. It can be recharged and gives a steady supply of power without any reduction until just before it runs out.

nickel-metal-hydride battery (n) an improvement on the *nickel-cadmium battery*. One of these can store about 40% more power than an NC battery of the same size and weight.

nose-wheel (n) a landing and take-off wheel (often retractable) which is attached to the front part of the aircraft (the nose).

nozzle (n) the projecting end of a tube with a small hole in it which is used to spray liquids or gases.

O ring (n) a flexible circular washer, made of rubber, plastic or silicone, which has a circular cross-section. Often used for sealing high-pressure liquid or gas connections.

OBOGS (onboard oxygen generation system) (n) equipment which enables an aircraft to make its own oxygen as it flies, instead of loading and carrying large, heavy, pressurised containers.

oscillator (n) an electronic device which produces a high-frequency AC signal, either as current in a circuit or as radio waves.

oscilloscope (n) an electronic instrument which displays electronic signals visually on a screen. It is used for inspecting, testing and troubleshooting electronic circuits.

payload (n) the extra weight that an aircraft is carrying in the form of passengers, luggage, cargo or bombs.

piston (n) a solid cylinder that fits inside another cylinder and moves under pressure to push or displace gas or liquid.

pitch (n) 1. the angle at which the nose of an aircraft is pointing up or down; 2. the angle of the thread on a bolt or screw.

plated (adj) having a covering of one metal on the surface of another. Some electronic terminals are gold-plated to improve their electrical conductivity and resist corrosion.

pneumatic piston assembly (n) the cylinder and piston of an air pump.

potential freight load (n) the maximum weight of cargo that it is possible for an aircraft to carry, rather than what it actually carries normally.

power plant (n) the main engine or engines of an aircraft.

power to weight ratio (n) the amount of power an engine can produce compared with its weight. Jet engines have a higher PWR than internal combustion engines.

PPE (personal protective equipment) (n) clothes or equipment worn to prevent injuries or illness due to workplace hazards. Safety helmets, gloves and ear defenders are examples of PPE.

pressure transducer (n) a device which produces small electrical currents when it is put under mechanical pressure.

processor (n) a small electronic component which does calculations and sends and receives information inside a computer.

production cell (n) a small group of workers who are responsible for several stages in the production of a manufactured item such as a car or a plane. Some companies find that production cells work better than each worker having just one small repeated job.

propeller (n) a rotating device with two or more pitched blades which forces an aircraft through the air.

propeller shaft (n) The rotating shaft that the propeller is mounted on.

pushrod (n) a metal rod which connects a cam on the camshaft to a spring-loaded valve. The rotating cam pushes the rod against the valve to open it.

R

radar (radio detection and rangefinding) (n) equipment which sends very high-frequency radio signals to display the location and direction of a flying aircraft on a radar screen.

radiator (n) a form of heat exchanger designed for cooling liquids. A hot liquid circulates through the inside, and the large external surface area transmits unwanted heat into the air.

ram (v & n) (v) to suddenly increase pressure on a liquid solid or gas; (n) a device designed to do this. Air which is collected from the front of a plane is called ram air.

reciprocal lathe (n) an early type of lathe which used backwards and forwards linear motion in the form of a bow, pole or strap to produce rotation of the workpiece.

reciprocating piston (n) a piston which moves constantly up and down in the cylinder of an engine, pump or compressor.

reduction gearbox (n) a set of mechanical gears designed to change the high speed of an engine into a lower speed that is more suitable for attached equipment. The propeller of an aircraft is often connected to a reduction gearbox.

reservoir (n) a container used to store a liquid or gas so that is always available when needed.

resistor (n) an electronic component with a value measured in ohms, which resists the flow of an electric current.

retract chamber (n) the lower part of a hydraulic cylinder. When it is filled with hydraulic fluid, the piston is forced to the top of the cylinder in order to retract an attached device.

rheostat (n) a variable resistor which is used to reduce the flow of electricity to equipment or circuits. Lights can be made dimmer or brighter by using a rheostat.

rib (n) a lateral beam or strut forming part of the framework of an aircraft.

rivet (n) a pin with two heads which is used to fasten metal parts very tightly together.

rocker arm (also called a rocker) (n) a lever in an internal combustion engine that connects a push rod to a valve.

rocket firing unit (n) part of the ejection seat mechanism that forces the seat and pilot out of the aircraft at extremely high speed.

rotary engine (n) an unusual type of internal combustion engine which has major parts that rotate in addition to or as well as the shaft. In the Wankel engine, the pistons rotate, and in the gnome aircraft engine, the engine block rotates around the shaft.

rotor (**n**) any device which turns round a centre point or axis.

runway (**n**) a long piece of ground with a prepared smooth surface which is used by aircraft for taking off and landing.

selector (**n**) a control switch, handle or knob which enables a pilot to choose a particular operation.

selector switch (**n**) an electrical selector.

semi-monocoque (**adj**) an aircraft design in which both the skin and the internal frame of the aircraft carry the stress loads. Also see *monocoque*.

sensor (**n**) an instrument which detects or reacts to a physical condition such as a rise in temperature or the presence of a dangerous gas. Aircraft have many different types of sensor to constantly monitor the engines, the fuel system and the cabin atmosphere.

simple transformer (**n**) a device consisting of a primary and secondary coil which changes the voltage of an AC current.

single-skin construction (**adj & n**) made with only one layer of covering material. Most modern aircraft are made with two skins with an insulating layer between them.

slip rings (**n**) rotating metal rings at the end of the shaft of an alternator which feed current to the supply circuit. They are similar to the commutator in a DC generator or motor.

solenoid (**n**) an *actuator* consisting of a moveable, soft, iron rod surrounded by a coil. When a current flows through the coil, the rod moves in or out to operate another device.

spark plug (**n**) a small device screwed into the top of the cylinder of an internal combustion engine. It produces a high-voltage spark which ignites the fuel mixture in the cylinder.

specific heat capacity (**n**) the amount of heat energy required to increase the temperature of a material one degree. The SHC of aluminium is about 900 jules per kg, whereas iron is about 470 jpk, which means that it takes twice as much energy to increase the temperature of aluminium as it does to increase the temperature of iron.

static charge (**n**) DC electricity which does not flow as a current but is released instantly. A capacitor can build up and store a static charge. Lightning is caused by a static charge of millions of volts.

stiffening (**n**) additional pieces of material fastened to a structural part such as a beam or strut to make it more rigid and less likely to bend under a load.

strainer (**n**) a type of filter designed to separate two or more materials from each other.

stressed skin panel (**n**) a piece of aircraft skin which is not just a cover, but is designed to withstand stress loads and contribute to the general strength of the structure.

stringer (**n**) a thin metal strip which goes along the length of the fuselage.

strut (**n**) a rod or beam which connects one part of a plane with another to increase strength and support. Small light aircraft often have struts between the wings and the fuselage.

sub-assembly (**n**) a group of connected components which are put together to make a part before it is fitted into a larger group (*assembly*) of connected parts.

sump (**n**) a *reservoir* of lubricating oil located at the bottom of an engine.

supercharger (**n**) a special high-speed fan which forces air into an internal combustion engine and increases its power.

taildragger (**n**) an informal name for an aircraft which has a landing wheel on the tail instead of the nose.

take-off selector (**n**) a *selector* used to set the control surfaces to the position required for the plane to leave the ground.

tensile strength (**n**) the ability of a material to withstand being pulled or stretched without breaking.

terminal block (n) a piece of insulating material containing metal connectors that the ends of wires are attached to.

thermal conductivity (n) the property of allowing heat to pass through a material. Copper has a high thermal conductivity, whereas wood and polystyrene do not.

thrust (n) the pushing power of an aircraft engine which makes the plane move forward.

torsion (n) a twisting force which tries to turn something clockwise and anticlockwise at the same time.

tricycle (n) type of landing gear system found on most modern aircraft which consists of a nose wheel and two main wheels or sets of wheels in the middle of the aircraft, behind the centre of gravity.

trim (n) the balance of an aircraft in flight. To keep the aircraft in exactly the right position, small flight control surfaces called trim tabs have to be constantly adjusted.

turbo-fan engine (n) a type of jet engine, used on most large planes today, which is fitted with a large fan that provides most of the thrust.

turboprop engine (n) a type of jet engine which drives an external propeller.

turn coordinator (n) a flight instrument which tells the pilot how effectively and safely he is turning the aircraft in order to change direction.

vacuum (n) space or container where there is an absence of pressure and matter.

valve (n) a device that regulates the flow of gases, liquids or materials through a pipe or other passage.

volatility (n) the ability to change from a liquid to a vapour.

voltage probe (n) a thin conducting rod with a pointed end which can be put in contact with parts of a circuit that are difficult to reach during inspection and fault tracing.

voltage regulator (n) a device in a power supply system which ensures that current is supplied at a constant electromotive force (voltage), without sudden increases (surges) or decreases (drops).

voltmeter (n) an instrument which measures the strength of an electromotive force in volts.

workpiece (n) a piece of material that is being shaped, drilled, cut, filed or worked on in some way.

Electrical symbols

AC voltage source

Aerial/antenna

Autotransformer

Battery

Capacitor   Polarised capacitor   Variable capacitor

Chassis earth

Circuit breaker

Coaxial cable

Diode

Double pole single throw switch

Double shielded conductor

Earth

Fuse

Generator

Inductor   Variable inductor

Iron core transformer

Lamp

Loudspeaker

Motor

Piezoelectric crystal (transducer)

Pressure switch (closes on pressure release)

Relay

Relay switch

Resistor

Rheostat

Rotary switch

Shielded and twisted conductor

Shielded conductor

Single pole double throw normally closed, momentarily open switch

Single pole double throw switch

Single pole single throw switch

Solenoid

Spare conductor with insulated end

Switch

Terminal block

Terminal strip

Transformer

Wires cross (no contact)   Wires cross (tie point)

Word list
Unit by unit

Unit 1

Lesson 1
air cushion vehicle (n)
bagless vacuum cleaner (n)
capacity (n)
gigabyte (n)
height (n)
horsepower (n)
innovative (adj)
kilometre (n)
LCD television (n)
memory (n)
MP3 player (n)
power (n)
research and development (n)
rotary engine (n)
speed (n)
technology (n)
vertical take-off (adj, n)
watt (n)
weight (n)

Lesson 2
aircraft (n)
component (n)
conditions (n)
crankshaft (n)
design (n)
drawing (n)
economics (n)
flywheel (n)
mass-production (n)
materials (n)
military (adj)
model (n)
prototype (n)
regulation (n)
sketch (n)
specification (n)

Lesson 3
alloy (n)
availability (n)
brittleness (n)
chemical stability (n)
coefficient of linear
    expansion (n)
composite (n)
corrode (v)
density (n)
ductility (n)
elasticity (n)
electrical resistivity (n)
hardness (n)
malleability (n)
mechanical (n)
melting point (n)
moulded plastic (n)
non-metallic (adj)
physical (adj)
plated metal alloy (n)
property (n)
range (n)
specific heat capacity (n)
specific operating
    conditions (n)
strength (n)
suitability (n)
synthetic (adj)
thermal conductivity (n)
toughness (n)
transparent (adj)

Lesson 4
adhesive (adj, n)
airliner (n)
application (n)
characteristic (n)
commercial (adj)
fine (adj)

Pascal (n)
radial (adj)
tensile strength (n)

Lesson 5
enable (v)
facilities (n)
geometry (n)
horizontal (adj)
impression (n)
manufacturing costs (n)
optics (n)
orthographic projection (n)
passenger (n)
perspective (n)
pictorial (adj)
point of view (n)
potential freight load (n)
safety (n)
sponsor (n, v)
technique (n)
three-dimensional (adj)
two-/three-point
perspective (n)
two-dimensional (adj)
vanishing point (n)
visualise (v)

Lesson 6
acceleration (n)
cabin (n)
ceiling (n)
circumference (n)
crew (n)
cubic metre (n)
distance (n)
divided (adj)
duration (n)
flight (n)
kilogram (n)

kilometres per hour (n)
length (n)
maximum (adj, n)
minus
multiplied
payload (n)
plus power plant (n)
relative to sphere (n)
surface area (n)
take-off (n)
volume (n)
width (n)

Lesson 7
air brake (n)
decoy (n)
display (adj, n)
ejection seat (n)
foreplane (n)
fuel (n)
fuel tank (n)
fuselage (n)
long-range (adj)
medium-range (adj)
missile (n)
port (adj)
radar (n)
refueling probe (n)
retractable (adj)
short-range (adj)
starboard (adj)
tail (n)
tank (n)
turbofan engine (n)
undercarriage (n)
wing (n)

Lesson 8
brake (n)
economics (n)

funding (n)
hangar (n)
invest (v)
manufacturing (n)
runway (n)
scrapped (adj)
trial (n, v)

Lesson 9

approximately (adv)
area (n)
centimetre (n)
cruise speed (n)
customer (n)
data (n)
dimension (n)
distance (n)
equivalent (adj)
estimate (v)
feet (n)
imperial (adj)
inch (n)
kilowatt (n)
metre (n)
metric (adj)
miles per hour (n)
performance (n)
pound (n)
rough (adj)
square (adj, n)
war plane (n)
wingspan (n)

Unit 2

Lesson 1

bend (v)
blade (n)
calibrate (v)
edge (n)
face (n)
file (n, v)
grip (v)
hammer (n, v)

hand tool (n)
handle (n)
jaw (n)
measure (v)
pliers (n)
pound (v)
power tool (n)
precision (n)
punch (n, v)
saw (n, v)
screw (n, v)
screwdriver (n)
sharp (adj)
sharpen (v)
size (n)
teeth (n)
thread (n)
tighten (v)
vice (n)

Lesson 2

batch (n)
beat (v)
blade (n)
flexible (adj)
fold (n, v)
forge (v)
gap (n)
machine tool (n)
molten (adj)
polish (v)
process (n)
razor-sharp (adj)
sheet (n)
smith (n)
solid (adj)
tie (v)
weld (v)
welding (n)
wrench (n)
wrought iron (n)

Lesson 3

accuracy (n)
axis/axes (n)

CNC machine tool (n)
configuration (n)
cutter (n)
hand-operated (adj)
interchangeable (adj)
key (n)
margin of error (n)
milling machine (n)
operation (n)
repeatability (n)
tailor (v)
tolerance (n)
versatile (adj)

Lesson 4

adjust (v)
air-powered (adj)
align (v)
base (n)
drill bit (n)
feed (n)
gearbox (n)
handwheel (n)
pillar (n)
pneumatic (adj)
transmit (v)
turning tool (n)
work table (n)
workpiece (n)

Lesson 5

bow (n)
crank (n)
gyroscope (n)
internal combustion engine (n)
irrigation (n)
lathe (n)
momentum (n)
pole (n)
reciprocal lathe (n)
rotation (n)
strap (n)
treadle (n)
wood turning (n)

Lesson 6

apron (n)
connect (v)
dial (n)
ensure (v)
feed (n, v)
headstock (n)
isolator (n)
knob (n)
lead screw (n)
lever (n)
longitudinal axis (n)
mains (n)
protection (n)
reverse (adj, n, v)
selector (n)
spindle (n)
thread pitch (n)
transverse axis (n)

Lesson 7

aluminium (n)
asbestos (n)
chromium (n)
coating (n)
conductivity (n)
conductor (n)
construction (n)
copper (n)
corrode (v)
corrosion (n)
durability (n)
electrochemically (n)
gold (n)
lead (n)
lightness (n)
lightweight (adj)
manganese (n)
oxide (n)
resistance (n)
stainless steel (n)
versatile (adj)
zinc (n)

Lesson 8

burr (n)

hole (n)

instructor (n)

malleable (adj)

mark out (v)

radius (n)

scribe (v)

shears (n)

smooth (adj)

Lesson 9

airworthy (adj)

bar (n)

blind rivet (n)

bolt (n)

clip (n)

fastening (n)

insert (v)

man-hour (n)

non-structural part (n)

pin (n)

pinhead (n)

pre-drilled hole (n)

rivet (n)

riveting tool (n)

shear (adj)

shearing strength (n)

standard rivet (n)

structural part (n)

titanium (n)

vibration (n)

washer (n)

Lesson 10

factory (n)

kit (n)

tin snip (n)

Unit 3

Lesson 1

blade (n)

control surface (n)

fin (n)

flap (n)

hinged (adj)

horizontal (adj)

landing gear (n)

longitudinal (adj)

main body (n)

pitched (adj)

power plant (n)

propeller (n)

rod (n)

rudder (n)

skin (n, v)

structure (n)

strut (n)

support (n, v)

surface (n)

vertical (adj)

Lesson 2

backwards (adj, adv)

bank (v)

bending (n)

climb (v)

clockwise (adj, adv)

compression (n)

dive (v)

fall (v)

force (n)

forwards (adj, adv)

gravity (n)

hover (v)

lift (n, v)

roll (v)

rotate (v)

sideways (adj, adv)

stress (n)

tension (n)

torsion (n)

Lesson 3

descent (n)

gain height (v)

grab (v)

manoeuvre (n, v)

pull up (v)

shallow angle (n)

turn around (v)

twist (v)

wingspan (n)

Lesson 4

airframe (n)

function (n)

hollow (adj)

internal [structure] (n)

rib (n)

rigid (adj)

skeleton (n)

withstand (v)

Lesson 5

bulkhead (n)

extend (v)

external (adj)

frame (n)

framework (n)

lateral (adj)

lengthwise (n)

longeron (n)

monocoque (adj)

plate (n)

semi-monocoque (adj)

spar (n)

stiffening (n)

stressed skin panel (n)

stringer (n)

trussed (adj)

Lesson 6

assembly (n)

attach (v)

carry out (v)

electronics (n)

engine (n)

fit (v)

flight control system (n)

flight test (n)

hydraulics (n)

install (v)

main assembly (n)

perform (v)

pre-flight test (n)

production line (n)

splice (v)

weapons system (n)

Lesson 7

bridge crane (n)

hangar (n)

in situ (adv)

part (n)

production site (n)

shift (n)

site (n)

sub-assembly (n)

tarmac (n)

Lesson 8

aerospace industry (n)

automation (n)

contract (n)

inaccessible (adj)

maintenance (n)

manual procedure (n)

manufacturing industries (n)

robot technology (n)

robotic assembly (n)

security (n)

single-skin construction (n)

wing box (n)

Lesson 9

background (n)

draughtsman (n)

engineer (n)

experience (n)

machinist (n)

manual (n)

modification (n)

navigation (n)

qualification (n)

steering (n)

systems engineer (n)

technician (n)

Lesson 10

be accountable for
    [something] (v)

brief [someone] (v)

cellular manufacturing (n)

chain of communication (n)

clarification (n)

collective responsibility (n)

communication (n)

intercom (n)

log (n)

malfunction (n, v)

production cell (n)

relieve [someone] (v)

terminology (n)

Unit 4

Lesson 1

channel (n)

constant (adj)

flap valve (n)

flow rate (n)

gradient (n)

inlet (adj, n)

kinetic energy (n)

level (n)

outlet (adj, n)

pipe (n)

piston (n)

pool (n)

potential energy (n)

pressure (n)

pump (n, v)

reciprocating piston (n)

reservoir (n)

slope (n)

sophisticated (adj)

source (n)

suction (n)

supply (n, v)

Lesson 2

clack valve (n)

cycle (n)

delivery valve (n)

device (n)

fluid (n)

hydraulic (adj)

increase (n, v)

primary flow (n)

ram (n, v)

spring (n)

spring-loaded (adj)

upwards (adv)

valve (n)

Lesson 3

atmospheric (adj)

boiling point (n)

freezing point (n)

hydraulic power (n)

hydraulic system (n)

lifting (n)

linkage (n)

loading (n)

mechanical advantage (n)

tarmac (n)

technological (adj)

towing (n)

Lesson 4

actuate (v)

actuator (n)

control (v)

de-energise (v)

energise (v)

extend (v)

indicate (v)

microswitch (n)

operate (v)

psi [pounds per square
    inch] (n)

relay (n, v)

retract (v)

select (v)

selector switch (n)

solenoid (n)

Lesson 5

absorb [shock] (v)

artificial horizon (n)

descent (n)

detect (v)

determine (v)

distribute [the load] (v)

emergency landing (n)

flexibility (n)

impact (n)

line up (v)

load (n)

minimise (v)

mobility (n)

orientation (n)

precise (adj)

pressurised fluid (n)

shock (n)

simultaneous (adj)

spread (v)

Lesson 6

air resistance (n)

center of gravity (n)

groundloop (n, v)

isolator (n)

leg (n)

nose wheel (n)

rough ground (n)

spin round (v)

stationary (adj)

taildragger (n)

tricycle (n)

visibility (n)

Lesson 7

chamber (n)

control angle (n)

drag (n)

inoperative (adj)

intermediate chamber (n)

land selector (n)

partially/fully extended (adj)

piston head chamber (n)

primary (adj)

redirect (v)

retract chamber (n)

secondary (adj)

spring-loaded (adj)

take-off selector (n)

trainer (n)

Lesson 8

aileron (n)

cable (n)

elevator (n)

joystick (n)

lateral axis (n)

longitudinal axis (n)

pedal (n)

pitch (n, v)

pre-flight check (n)

roll (n, v)

slow down (v)

vertical axis (n)

yaw (n, v)

Lesson 9

buckle (v)

control column (n)

excessive load (n)

groove (n)

hollow (adj)

play (n)

pulley [wheel] (n)

rigid (adj)

rotor (n)

slack (adj)

stiff (adj)

tension (n)

terminal (n)

torque (n)

tube (n)

Unit 5

Lesson 1

alternator (n)
camshaft (n)
cylinder block (n)
engine block (n)
fan (n)
fan belt (n)
filter (n)
generator (n)
pushrod (n)
rocker (n)
rocker arm (n)
spark plug (n)
starter motor (n)
strainer (n)
sump (n)
timing chain/belt (n)

Lesson 2

cruise (v)
gas turbine (n)
grounded (adj)
in-line (adj)
price (n)
reliability (n)
rubber band (n)
simplicity (n)
spare part (n)
streamlined (adj)

Lesson 3

angular (n)
decrease (v)
increase (v)
linear (adj)
mass (n)
overall (adj, adv)
perpendicular (adj)
skater (n)
stable (adj)
straight line (n)
strain (n, v)

top (n)
velocity (n)

Lesson 4

admit (v)
air-cooled (adj)
by hand (adv)
downwards (adv)
exhaust (n)
expand (v)
ignited gas (n)
mixture (n)
power stroke (n)
power-to-weight ratio (n)
radiator (n)
ratio (n)
resist (v)
suck [in] (v)
vacuum (n)

Lesson 5

automatic (adj)
aviation fuel (n)
bearings (n)
bore (n)
breakdown (n)
data (n)
diameter (n)
displacement (n)
gear (n)
high gear (n)
liner (n)
low gear (n)
lubrication (n)
power rating (n)
starter (n)
stroke (n)
supercharger (n)

Lesson 6

assemble (v)
blocked (adj)
brass (n)
clean (v)

cylinder head gasket (n)
damaged (adj)
dirty (adj)
dismantle (v)
flood (n, v)
frame (n)
gasket (n)
inspect (v)
leak (n, v)
loose (adj)
machine (v)
mount (v)
out of alignment (adj)
overhaul (v)
plug (n)
reattach (v)
rebuild (v)
refit (v)
remove (v)
replace (v)
sandblast (v)
strip down (v)
transfer (v)

Lesson 7

air intake (n)
act on (v)
add [to] (v)
combustion chamber (n)
discharge (v)
draw in (v)
drive (v)
exert pressure on (v)
exhaust duct (n)
expel (v)
mix (v)
multistage compressor (n)
propeller shaft (n)
provide [power] (v)
reduction gearbox (n)
spark (n)
turboprop (n)

Lesson 8

altitude (n)
bypass (v)
core (n)
cowling (n)
cruising altitude (n)
exit velocity (n)
extract (v)
fahrenheit (n)
jet fighter (n)
revolutions per minute (n)
statistics (n)
thrust (n)

Lesson 9

chemical reaction (n)
corrosiveness (n)
efficiency (n)
flash point (n)
freezing point (n)
kerosene (n)
liquid (n)
lubricity (n)
net [heat] (adj)
standard specification (n)
sulphur (n)
volatility (n)

Lesson 10

battery (n)
booster pump (n)
centrifuge (n)
collector tank (n)
delivery (n)
flammable (adj)
flange (n)
freeze (v)
gear pump (n)
jet pump (n)
seal (n, v)
transfer (n)
vane (n)
venturi principle (n)

Unit 7

Lesson 1

assess (v)

caution (n, v)

consult (v)

evaluate (v)

first aid (n)

forbid (v)

hazard (n)

maintain (v)

mandatory (adj)

manual (adj)

notice (n)

PPE (n)

prevent (v)

prohibit (v)

risk (n, v)

safety equipment (n)

scald (n, v)

sign (n)

warning (adj, n)

Lesson 2

change (= replace) (v)

clean up (v)

disconnect (v)

dry out (v)

fumes (n)

guard (n)

ignition (n)

switch on/off (v)

well-ventilated (adj)

Lesson 3

blanket (n)

colour-coded (adj)

conduct (v)

emergency (n)

extinguish (v)

film (n)

fire extinguisher (n)

foam (n)

fuel (v)

multi-purpose (adj)

powder (n)

residue (n)

suitable (adj)

versatile (adj)

Lesson 4

breathing equipment (n)

bumpy (adj)

check (n, v)

Civil Aviation Authority (n)

cockpit (n)

confidential (adj)

evacuation (n)

fray (v)

harness (n)

intercom (n)

life jacket (n)

megaphone (n)

oxygen mask (n)

passenger cabin (n)

priority (n)

safety inspector (n)

seat belt (n)

service tag (n)

sick bag (n)

turbulence (n)

twisted (adj)

Lesson 5

bracket (n)

chafing (n)

contamination (n)

cracking (n)

discard (v)

distortion (n)

elongation (n)

evidence (n)

fading (n)

finish (n)

fixing (n)

loosen (v)

mechanism (n)

repair (n, v)

report (v)

restraint (n)

saddle washer (n)

scrutinise (v)

stitch/stitching (n)

strap (n)

tear (n, v)

wear (v)

Lesson 6

agent (n)

atmosphere (n)

burning [process] (adj)

damage (n)

disperse (v)

displace (v)

environment (n)

environmentally friendly (adj)

fraction (n)

halon (n)

harm (n, v)

interrupt (v)

ozone layer (n)

put out [a fire] (v)

react [with chemicals] (v)

spray (n, v)

suffocate (v)

Lesson 7

ballistic threat (n)

built-in (adj)

eliminate (v)

enriched (adj)

ground support (n)

HEI (high explosive
    incendiary) (adj)

incendiary round (n)

inert (adj)

integral inlet (adj)

life cycle (adj, n)

lightning (n)

nitrogen (n)

slosh (v)

static discharge (n)

survivability (n)

threat (n)

ullage (n)

vapour (n)

vulnerability (n)

Lesson 8

back pressure (n)

block (v)

braided (adj)

clamp (n, v)

concentration (n)

concentrator (n)

crimped (adj)

defect (n)

deteriorate (v)

downstream (adv)

hose (n)

hypoxia (n)

lethal (adj)

line [= tube] (n)

LOX converter (n)

message traffic (n)

OBOGS (onboard oxygen
    generating system) (n)

quality assurance (n)

sleeve (n)

Lesson 9

activate (v)

breech (adj, v)

collide (v)

deploy (v)

drogue parachute (n)

ejection gun (n)

initiate (v)

limb-restraint cord (n)

multi-tubed (adj)

navigator (n)

parachute (n)

propel (v)

reinforce (v)

remote (adj)

rocket firing unit (n)

rocket motor (n)

stabilise (v)

static line (n)

telescopic (adj)

Lesson 10

accumulator (n)
backup system (n)
charge (v)
check valve (n)
emergency package (n)
oil head (n)
pressure transducer (n)
release valve (n)
restrictor (n)
survival (n)
take over (n, v)

Unit 8

Lesson 1

bead (n)
bulge (n, v)
bump (v)
burst (v)
crosswind (n)
deflate (v)
flip (v)
foreign body (n)
gauge (n)
groove (n)
inflated (adj)
layer (n)
misalignment (n)
sidewall (n)
skid (v)
tread (n)
tyre (n)

Lesson 2

absorb (v)
annular space (n)
axle (n)
compressed (adj)
cylinder barrel (n)
dampen (n)
displaced (adj)
landing gear leg (n)
non-level (adj)

phase (n)
pneumatic (adj)
reciprocation (n)
recoil (v)
shock (n)
slip ring (n)
uneven (adj)

Lesson 3

basket (n)
burner (n)
Celsius/centigrade (n)
copilot (n)
dense (adj)
envelope (n)
equivalent (adj, adv)
Kelvin (n)
occupy (v)
propane (n)
proportional to (adv)

Lesson 4

arm (v)
bypass (adj, n)
cock (v)
compressor (n)
housing (n)
maintenance manual (n)
oil cooler (n)
pneumatic piston
    assembly (n)
prime (v)
purge (v)
set (v)
shutdown (adj)
tattle tale (n)
vent (adj, n)

Lesson 5

circulate (v)
compression [stroke] (adj, n)
condense (v)
condenser (n)
evaporate (v)
evaporator (n)

freeze (v)
give off (v)
induction [stroke] (adj, n)
liquefy (v)
melt (v)
orifice (n)
piping (n)
refrigerant (n)
solidify (v)
state (v)
stroke (n)
vaporise (v)
waste (v)

Lesson 6

absolute humidity (n)
absorption (n)
atmospheric humidity (n)
bulb (n)
condensation (n)
controlling humidity (n)
humidify (v)
humidity (n)
hygrometer (n)
intermix (v)
measuring humidity (n)
moist (adj)
moisture (n)
molecule (n)
relative humidity (n)
saturated (adj)
saturation point (n)
temperature reading (n)

Lesson 7

AC (air conditioning) (n)
activate (v)
coil (n)
cooler (n)
draw in (v)
filter (n, v)
freshen (v)
hygrostat (n)
recirculate (v)
reheater (n)

thermostat (n)
unit (n)

Lesson 8

consume (v)
contaminant (n)
dilute (v)
equipped (adj)
high-efficiency (adj)
interval (n)
jetliner (n)
replenish (v)
ventilate (v)

Lesson 9

distribute (v)
draw off (v)
duct (n)
exhausted (adj)
grille (n)
heat exchanger (n)
intake (n)
lobe (n)
microscopic particle (n)
outflow valve (n)
overhead [outlet] (adj)
pattern (n)
trap (v)

Lesson 10

bleed (adj, v)
defogging nozzle (n)
ECS [environmental
    control system] (n)
extract (v)
firewall (n)
nozzle (n)
outlet (n)
sensor (n)
separator (n)
turbine (n)

Unit 9

Lesson 1

ambient (adj)

armour (n)

bedding (n)

busbar (n)

cross-sectional area (n)

current (n)

dissipate (v)

filler (n)

heavy-duty (adj)

insulate (v)

insulation sleeve (n)

multi-cored (adj)

sector-shaped (adj)

sheath (n)

tape (n)

voltage (n)

wire (n)

Lesson 2

ampere (n)

bundled [cable] (adj)

circuit (n)

conduit (n)

curve (n)

drop (n, v)

formula (n)

ohm (n)

square root (n)

value (n)

volt (n)

voltage (n)

watt (n)

Lesson 3

cable clamp (n)

drainage hole (n)

grommet (n)

installation (n)

loop (n)

lug (n)

maintenance check (n)

maintenance form (n)

protrude (v)

redo (v)

reposition (v)

ring (n)

slack (adj, n)

terminal block (n)

Lesson 4

lead-acid [battery] (adj, n)

lithium-ion [battery] (adj, n)

nickel-cadmium
[battery] (adj, n)

nickel-metal-hydride (adj, n)

percentage capacity (n)

recharge (v)

self-discharge rate (n)

storage (n)

Lesson 5

charge up [battery] (v)

cut out [engine] (v)

EPU [external power unit] (n)

generator (n)

handpropping (n)

injury (n)

risky (adj)

run down [battery] (v)

taxi off (v)

turn over [engine] (v)

Lesson 6

[radio] static (n)

air gap (n)

armature (n)

brush (n)

brush holder (n)

build up (v)

burn out (v)

capacitance meter (n)

capacitor (n)

commutator (n)

contact (v)

electrical arcing
(sparking) (n)

multi-graph (n)

operating manual (n)

pitted (adj)

scale (n)

shorted (adj)

snag (v)

tangled (adj)

Lesson 7

circuit diagram (n)

diode (n)

electrolytic capacitor (n)

filament bulb (n)

inductor (n)

lamp (n)

resistor (n)

semiconductor (n)

simple transformer (n)

single pole switch (n)

troubleshooting (n)

voltmeter (n)

Lesson 8

AC (alternating current) (n)

analogue multimeter (n)

cable pulling draw tape (n)

crimping tool (n)

DC (direct current) (n)

digital multimeter (n)

electrical tool catalogue (n)

hacksaw (n)

mains (adj, n)

multi-grip (adj)

side cutters (n)

socket set (n)

solder (n, v)

soldering gun (n)

soldering iron (n)

tape measure (n)

voltage probe (n)

WD40 lubricant (n)

wire strippers (n)

Lesson 9

avionics (n)

bring online (v)

display panel (n)

dual output (n)

dual-role (adj)

hertz (n)

inverter (n)

isolate (v)

power socket (n)

starter-generator (n)

transistorised static
inverter (n)

trickle-charge (v)

voltage regulator (n)

Lesson 10

circuit-breaker (n)

common connection
point (n)

electromagnet (n)

flat strip (n)

fuse (n)

hollow (adj)

melt (v)

multiple switch (n)

on-board (adj)

overloaded (adj)

relay (n)

remote switch (n)

Unit 10

Lesson 1

air traffic controller
[ATC] (n)

attitude (n)

collision (n)

departure point (n)

flight data recorder (n)

trim (n)

Lesson 2

airborne (adj)

altimeter (n)

back-up (v)

barometer (n)

button (n)
compass (n)
gyro-compass (n)
panel (n)
recalibrate (v)
stall (v)
turn-coordinator (n)

Lesson 3

ammeter (n)
analogue (adj)
digital (adj)
division (n)
input (n)
instrumentation (n)
needle (n)
output (n)
pointer (n)
solid state (adj)

Lesson 4

algebra (n)
application (n)
arithmetic (n)
calculus (n)
chips (n)
circuit board (n)
coordinate (v)
electromechanical (adj)
fabricate (v)
fine-tune (v)
functional (adj)
hardware (n)
integral (adj)
interpret [data] (v)
junction box (n)
know-how (n)
layout drawing (n)
magneto (n)
malfunction (n)
optimum (adj)
oscilloscope (n)
personnel (adj)
processor (n)
programming (n)

regulate (v)
schematic (n)
set up (v)
software (n)
systemic [problems] (adj)
troubleshoot (v)

Lesson 5

amplifier (n)
chassis (n)
click (n)
gyro-system (n)
hum (n)
interference (n)
rheostat (n)

Lesson 6

fault (n)
integrated circuit (n)
intermittent (adj)
magnifying glass (n)
mishandle (v)
parallel (adj)
PCB (printed circuit
    board) (n)
reading (n)
right angles (adj)
trace [fault] (v)

Lesson 7

aerosol solvent cleaner (n)
foreign matter (n)
grease (n)
heat sink (n)
lacquer (n)
particle (n)
short circuit (n, v)
solder sucker (n)
trim (v)

Lesson 8

aerial (n)
air to ground
    communication (n)
antennas (n)

clear for take-off (v)
dots and dashes (n)
echo (n)
flight path (n)
frequency (n)
identification code (n)
inbound (adj)
landmark (n)
Morse code (n)
navigation chart (n)
non-precision aid (n)
outbound (adj)
path (n)
quadrant (n)
radar [Radio Detection
    and Ranging] (n)
radio-based system (n)
refine (v)
tower [radio] (n)
transmitter receiver (n)
two-way radio (n)
VOR [Very high-frequency
    Omni-directional Radio
    range] (n)

Lesson 9

ELT [emergency locator
    transmitter] (n)
format (v)
GPS [Global Positioning
    System] (n)
grid (n)
icon (n)
latitude (n)
longitude (n)
monitor (n)
radar tracking (n)
satellite (n)
update (n)

Lesson 10

AC-excited (adj)
crack (n)
eddy current (n)
grain size [of metals] (n)

induce (v)
magnetic field (n)
oscillator (n)
permeability (n)
probe (n)
void (n)

Unit 11

Lesson 1

airworthiness (n)
appropriate (adj)
authority (n)
certify (v)
factor (n)
regulatory (adj)
shaft (n)

Lesson 2

compulsory (adj)
deactivate (v)
flatten (v)
live (adj)
personnel (n)
reset (v)
secure (adj)
tags [warning ~] (n)

Lesson 3

complex (adj)
disassemble (v)
nut (n)
o-ring (n)
put back (in, together,
    etc.) (v)
reassemble (v)
reinstall (v)
take apart (v)

Lesson 4

apply (v)
contaminate (v)
improvement (n)

Lesson 5

coning (n)

disc (n)

field conditions (n)

industrial pollution (n)

lining (n)

NDI (non-destructive
    inspection) (n)

restore (v)

sand (v)

sandpaper (n)

schedule (n, v)

scratch (n, v)

thickness (n)

Lesson 6

access door (n)

indicator (n)

lock (n)

locking pin (n)

miss [the problem] (v)

spot [the problem] (v)

verify (v)

Lesson 7

access equipment (n)

adjacent (adj)

clearance (n)

extension lead (n)

jack (n)

platform (n)

remote control unit (n)

servicing equipment (n)

work cage (n)

Lesson 8

docking (n)

domestic water service (n)

fire detection (v)

process water service (n)

steady (v)

take out [of service] (v)

undertake [maintenance] (v)

works [= maintenance] (n)

wrap around (v)

Lesson 9

abrasive (adj)

castor (n)

container (n)

dispose of (v)

drain (n, v)

failure (n)

loss of load (n)

MRO (Maintenance
    Repair and Overhaul)
    saddle (n)

Lesson 10

authorised (adj)

discrepancy (n)

document (v)

IAW (in accordance with)

L/H (left-hand)

R/H (right-hand)

record (v)

sign off (= authorise) (v)

work order (n)

Word list
Alphabetical

abrasive (adj)

absolute humidity (n)

absorb (v)

absorption (n)

AC (air conditioning) (n)

AC (alternating current)
    (n)

acceleration (n)

access door (n)

access equipment (n)

accumulator (n)

accuracy (n)

AC-excited (adj)

act on (v)

activate (v)

actuate (v)

actuator (n)

add [to] (v)

adhesive (adj, n)

adjacent (adj)

adjust (v)

admit (v)

aerial (n)

aerosol solvent cleaner (n)

aerospace industry (n)

agent (n)

aileron (n)

air brake (n)

air cushion vehicle (n)

air gap (n)

air resistance (n)

air to ground
    communication (n)

air traffic controller
    [ATC] (n)

airborne (adj)

air-cooled (adj)

aircraft (n)

airframe (n)

airliner (n)

air-powered (adj)

airworthiness (n)

airworthy (adj)

algebra (n)

align (v)

alloy (n)

alternator (n)

altimeter (n)

altitude (n)

aluminium (n)

ambient (adj)

ammeter (n)

ampere (n)

amplifier (n)

analogue (adj)

analogue multimeter (n)

angular (n)

annular space (n)

antennas (n)

application (n)

apply (v)

appropriate (adj)

approximately (adv)

apron (n)

area (n)

arithmetic (n)

arm (v)

armature (n)

armour (n)

artificial horizon (n)

asbestos (n)

assemble (v)

assembly (n)

assess (v)

atmosphere (n)

atmospheric (adj)

atmospheric humidity (n)

attach (v)

attitude (n)

authorised (adj)

authority (n)

automatic (adj)

automation (n)

availability (n)

aviation fuel (n)

avionics (n)

axis/axes (n)

axle (n)

back pressure (n)

background (n)

back up (v)

backup system (n)

backwards (adj, adv)

bagless vacuum cleaner
    (n)

ballistic threat (n)

bank (v)

bar (n)

barometer (n)

base (n)

basket (n)

batch (n)

battery (n)

be accountable for
    [something] (v)

bead (n)

bearings (n)

beat (v)

bedding (n)

bend (v)

bending (n)

blade (n)

blanket (n)

bleed (adj, v)

blind rivet (n)

block (v)

blocked (adj)

boiling point (n)

bolt (n)

booster pump (n)

bore (n)

bow (n)

bracket (n)

braided (adj)

brake (n)

brass (n)

breakdown (n)

breathing equipment (n)

breech (adj, v)

bridge crane (n)

brief [someone] (v)

bring online (v)

brittleness (n)

brush (n)

brush holder (n)

buckle (v)

build up (v)

built-in (adj)

bulb (n)

bulge (n, v)

bulkhead (n)

bump (v)

bumpy (adj)

bundled [cable] (adj)

burn out (v)

burner (n)

burning [process] (adj)

burr (n)

burst (v)

busbar (n)

button (n)

by hand (adv)

bypass (adj, n, v)

cabin (n)

cable (n)

cable pulling draw tape (n)

calculus (n)

calibrate (v)

camshaft (n)

capacitance meter (n)

capacitor (n)

capacity (n)

carry out (v)

castor (n)

caution (n, v)

ceiling (n)

cellular manufacturing (n)

Celsius/centigrade (n)

center of gravity (n)

centimetre (n)

centrifuge (n)

certify (v)

chafing (n)

chain of communication (n)

chamber (n)

change (= replace) (v)

channel (n)

characteristic (n)

charge (v)

charge up [battery] (v)

chassis (n)

check (n, v)

check valve (n)

chemical reaction (n)

chemical stability (n)

chips (n)

chromium (n)

circuit (n)

circuit board (n)

circuit diagram (n)

circuit-breaker (n)

circulate (v)

circumference (n)

Civil Aviation Authority (n)

clack valve (n)

clamp (n, v)

clarification (n)

clean (v)

clean up (v)

clear for take-off (v)

clearance (n)

click (n)

climb (v)

clip (n)

clockwise (adj, adv)

CNC machine tool (n)

coating (n)

cock (v)

cockpit (n)

coefficient of linear expansion (n)

coil (n)

collective responsibility (n)

collector tank (n)

collide (v)

collision (n)

colour-coded (adj)

combustion chamber (n)

commercial (adj)

common connection point (n)

communication (n)

commutator (n)

compass (n)

complex (adj)

component (n)

composite (n)

compressed (adj)

compression (n)

compression [stroke] (adj, n)

compressor (n)

compulsory (adj)

concentration (n)

concentrator (n)

condensation (n)

condense (v)

condenser (n)

conditions (n)

conduct (v)

conductivity (n)

conductor (n)

conduit (n)

confidential (adj)

configuration (n)

coning (n)

connect (v)

constant (adj)

construction (n)

consult (v)

consume (v)

contact (v)

container (n)

contaminant (n)

contaminate (v)

contamination (n)

contract (n)

control (v)

control angle (n)

control column (n)

control surface (n)

controlling humidity (n)

cooler (n)

coordinate (v)

copilot (n)

copper (n)

core (n)

corrode (v)

corrosion (n)

corrosiveness (n)

cowling (n)

crack (n)

cracking (n)

crank (n)

crankshaft (n)

crew (n)

crimped (adj)

crimping tool (n)

cross-sectional area (n)

crosswind (n)

cruise (v)

cruise speed (n)

cruising altitude (n)

cubic metre (n)

current (n)

curve (n)

customer (n)

cut out [engine] (v)

cutter (n)

cycle (n)

cylinder barrel (n)

cylinder block (n)

cylinder head gasket (n)

D

damage (n)

damaged (adj)

dampen (n)

data (n)

DC (direct current) (n)

deactivate (v)

decoy (n)

decrease (v)

de-energise (v)

defect (n)

deflate(v)

defogging nozzle (n)

delivery (n)

delivery valve (n)

dense (adj)

density (n)

departure point (n)

deploy (v)

descent (n)

design (n)

detect (v)

deteriorate (v)

determine (v)

device (n)

dial (n)

diameter (n)

digital (adj)

digital multimeter (n)

dilute (v)

dimension (n)

diode (n)

dirty (adj)

disassemble (v)

disc (n)

discard (v)

discharge (v)

disconnect (v)

discrepancy (n)

dismantle (v)

disperse (v)

displace (v)

displaced (adj)

displacement (n)

display (adj, n)

display panel (n)

dispose of (v)

dissipate (v)

distance (n)

distortion (n)

distribute (v)

distribute [the load] (v)

dive (v)

divided (adj)

division (n)

docking (n)

document (v)

domestic water service (n)

dots and dashes (n)

downstream (adv)

downwards (adv)

drag (n)

drain (n, v)

drainage hole (n)

draughtsman (n)

draw in (v)

draw off (v)

drawing (n)

drill bit (n)

drive (v)

drogue parachute (n)

drop (n, v)

dry out (v)

dual output (n)

dual-role (adj)

duct (n)

ductility (n)

durability (n)

duration (n)

echo (n)

economics (n)

ECS [environmental
    control system] (n)

eddy current (n)

edge (n)

efficiency (n)

ejection gun (n)

ejection seat (n)

elasticity (n)

electrical arcing
    (sparking) (n)

electrical resistivity (n)

electrical tool catalogue (n)

electrochemically
    electrolytic capacitor (n)

electromagnet (n)

electromechanical (adj)

electronics (n)

elevator (n)

eliminate (v)

elongation (n)

ELT [emergency locator
    transmitter] (n)

emergency (n)

emergency landing (n)

emergency package (n)

enable (v)

energise (v)

engine (n)

engine block (n)

engineer (n)

enriched (adj)

ensure (v)

envelope (n)

environment (n)

environmentally friendly (adj)

EPU [external power unit] (n)

equipped (adj)

equivalent (adj, adv)

estimate (v)

evacuation (n)

evaluate (v)

evaporate (v)

evaporator (n)

evidence (n)

excessive load (n)

exert pressure on (v)

exhaust (n)

exhaust duct (n)

exhausted (adj)

exit velocity (n)

expand (v)

expel (v)

experience (n)

extend (v)

extension lead (n)

external (adj)

extinguish (v)

extract (v)

fabricate (v)

face (n)

facilities (n)

factor (n)

factory (n)

fading (n)

fahrenheit (n)

failure (n)

fall (v)

fan (n)

fan belt (n)

fastening (n)

fault (n)

feed (n, v)

feet (n)

field conditions (n)

filament bulb (n)

file (n, v)

filler (n)

film (n)

filter (n, v)

fin (n)

fine (adj)

fine-tune (v)

finish (n)

fire detection (v)

fire extinguisher (n)

firewall (n)

first aid (n)

fit (v)

fixing (n)

flammable (adj)

flange (n)

flap (n)

flap valve (n)

flash point (n)

flat strip (n)

flatten (v)

flexibility (n)

flexible (adj)

flight (n)

flight control system (n)

flight data recorder (n)

flight path (n)

flight test (n)

flip (v)

flood (n, v)

flow rate (n)

fluid (n)

flywheel (n)

foam (n)

fold (n, v)

forbid (v)

force (n)

foreign body (n)

foreign matter (n)

foreplane (n)

forge (v)

format (v)

formula (n)

forwards (adj, adv)

fraction (n)

frame (n)

framework (n)

fray (v)

freeze (v)

freezing point (n)

frequency (n)

freshen (v)

fuel (v)

fuel tank (n)

fumes (n)

function (n)

functional (adj)

funding (n)

fuse (n)

fuselage (n)

gain height (v)

gap (n)

gas turbine (n)

gasket (n)

gauge (n)

gear (n)

gear pump (n)

gearbox (n)

generator (n)

geometry (n)

gigabyte (n)

give off (v)

gold (n)

GPS [Global Positioning
    System] (n)

grab (v)

gradient (n)

grain size [of metals] (n)

gravity (n)

grease (n)

grid (n)

grille (n)

grip (v)

grommet (n)

groove (n)

ground support (n)

grounded (adj)

groundloop (n, v)

guard (n)

gyro-compass (n)

gyroscope (n)

gyro-system (n)

hacksaw (n)

halon (n)

hammer (n, v)

hand tool (n)

handle (n)

hand-operated (adj)

hand-propping (n)

handwheel (n)

hangar (n)

hardness (n)

hardware (n)

harm (n, v)

harness (n)

hazard (n)

headstock (n)

heat exchanger (n)

heat sink (n)

heavy-duty (adj)

HEI (high explosive
    incendiary) (adj)

height (n)

hertz (n)

high gear (n)

high-efficiency (adj)

hinged (adj)

hole (n)

hollow (adj)

horizontal (adj)

horsepower (n)

hose (n)

housing (n)

hover (v)

hum (n)

humidify (v)

humidity (n)

hydraulic (adj)

hydraulic power (n)

hydraulic system (n)

hydraulics (n)

hygrometer (n)

hygrostat (n)

hypoxia (n)

IAW (in accordance with)

icon (n)

identification code (n)

ignited gas (n)

ignition (n)

impact (n)

imperial (adj)

impression (n)

improvement (n)

in situ (adv)

inaccessible (adj)

inbound (adj)

incendiary round (n)

inch (n)

increase (n, v)

indicate (v)

indicator (n)

induce (v)

induction [stroke] (adj, n)

inductor (n)

industrial pollution (n)

inert (adj)

inflated (adj)

initiate (v)

injury (n)

inlet (adj, n)

in-line (adj)

innovative (adj)

inoperative (adj)

input (n)

insert (v)

inspect (v)

install (v)

installation (n)

instructor (n)

instrumentation (n)

insulate (v)

insulation sleeve (n)

intake (n)

integral (adj)

integral inlet (adj)

integrated circuit (n)

interchangeable (adj)

intercom (n)

interference (n)

intermediate chamber (n)

intermittent (adj)

intermix (v)

internal [structure] (n)

internal combustion
    engine (n)

interpret [data] (v)

interrupt (v)

interval (n)

inverter (n)

invest (v)

irrigation (n)

isolate (v)

isolator (n)

jack (n)

jaw (n)

jet fighter (n)

jet pump (n)

jetliner (n)

joystick (n)

junction box (n)

Kelvin (n)

kerosene (n)

key (n)

kilogram (n)

kilometre (n)

kilometres per hour (n)

kilowatt (n)

kinetic energy (n)

kit (n)

knob (n)

know-how (n)

L/H (left-hand)

lacquer (n)

lamp (n)

land selector (n)

landing gear (n)

landing gear leg (n)

landmark (n)

lateral (adj)

lateral axis (n)

lathe (n)

latitude (n)

layer (n)

layout drawing (n)

LCD television (n)

lead (n)

lead screw (n)

lead-acid [battery] (adj, n)

leak (n, v)

leg (n)

length (n)

lengthwise (n)

lethal (adj)

level (n)

lever (n)

life cycle (adj, n)

life jacket (n)
lift (n, v)
lifting (n)
lightness (n)
lightning (n)
lightweight (adj)
limb-restraint cord (n)
line [= tube] (n)
line up (v)
linear (adj)
liner (n)
lining (n)
linkage (n)
liquefy (v)
liquid (n)
lithium-ion [battery] (adj, n)
live (adj)
load (n)
loading (n)
lobe (n)
lock (n)
locking pin (n)
log (n)
longeron (n)
longitude (n)
longitudinal (adj)
longitudinal axis (n)
long-range (adj)
loop (n)
loose (adj)
loosen (n)
loss of load (n)
low gear (n)
LOX converter (n)
lubrication (n)
lubricity (n)
lug (n)

machine (v)
machine tool (n)
machinist (n)
magnetic field (n)
magneto (n)
magnifying glass (n)
main assembly (n)

main body (n)
mains (adj, n)
maintain (v)
maintenance (n)
maintenance check (n)
maintenance form (n)
maintenance manual (n)
malfunction (n, v)
malleability (n)
malleable (adj)
mandatory (adj)
manganese (n)
man-hour (n)
manoeuvre (n, v)
manual (adj, n)
manual procedure (n)
manufacturing (n)
manufacturing costs (n)
manufacturing industries (n)
margin of error (n)
mark out (v)
mass (n)
mass-production (n)
materials (n)
maximum (adj, n)
measure (v)
measuring humidity (n)
mechanical (n)
mechanical advantage (n)
mechanism (n)
medium-range (adj)
megaphone (n)
melt (v)
melting point (n)
memory (n)
message traffic (n)
metre (n)
metric (adj)
microscopic particle (n)
microswitch (n)
miles per hour (n)
military (adj)
milling machine (n)
minimise (v)
minus

misalignment (n)
mishandle (v)
miss [the problem] (v)
missile (n)
mix (v)
mixture (n)
mobility (n)
model (n)
modification (n)
moist (adj)
moisture (n)
molecule (n)
molten (adj)
momentum (n)
monitor (n)
monocoque (adj)
Morse code (n)
moulded plastic (n)
mount (v)
MP3 player (n)
MRO (Maintenance
    Repair and Overhaul) (n)
multi-cored (adj)
multi-graph (n)
multi-grip (adj)
multiple switch (n)
multiplied (v)
multi-purpose (adj)
multistage compressor (n)
multi-tubed (adj)

navigation (n)
navigation chart (n)
navigator (n)
NDI (non-destructive
    inspection) (n)
needle (n)
net [heat] (adj)
nickel-cadmium [battery]
    (adj, n)
nickel-metal-hydride (adj, n)
nitrogen (n)
non-level (adj)
non-metallic (adj)
non-precision aid (n)

non-structural part (n)
nose wheel (n)
notice (n)
nozzle (n)
nut (n)

OBOGS (onboard oxygen
    generating system) (n)
occupy (v)
ohm (n)
oil cooler (n)
oil head (n)
on-board (adj)
operate (v)
operating manual (n)
operation (n)
optics (n)
optimum (adj)
orientation (n)
orifice (n)
o-ring (n)
orthographic projection (n)
oscillator (n)
oscilloscope (n)
out of alignment (adj)
outbound (adj)
outflow valve (n)
outlet (adj, n)
output (n)
overall (adj, adv)
overhaul (v)
overhead [outlet] (adj)
overloaded (adj)
oxide (n)
oxygen mask (n)
ozone layer (n)

panel (n)
parachute (n)
parallel (adj)
part (n)
partially/fully extended (adj)
particle (n)
Pascal (n)
passenger (n)

passenger cabin (n)

path (n)

pattern (n)

payload (n)

PCB (printed circuit
  board) (n)

pedal (n)

percentage capacity (n)

perform (v)

performance (n)

permeability (n)

perpendicular (adj)

personnel (adj, n)

perspective (n)

phase (n)

physical (adj)

pictorial (adj)

pillar (n)

pin (n)

pinhead (n)

pipe (n)

piping (n)

piston (n)

piston head chamber (n)

pitch (n, v)

pitched (adj)

pitted (adj)

plate (n)

plated metal alloy (n)

platform (n)

play (n)

pliers (n)

plug (n)

plus (v)

pneumatic (adj)

pneumatic piston
  assembly (n)

point of view (n)

pointer (n)

pole (n)

polish (v)

pool (n)

port (adj)

potential energy (n)

potential freight load (n)

pound (n, v)

powder (n)

power (n)

power plant (n)

power rating (n)

power socket (n)

power stroke (n)

power tool (n)

power-to-weight ratio (n)

PPE (n)

precise (adj)

precision (n)

pre-drilled hole (n)

pre-flight check (n)

pre-flight test (n)

pressure (n)

pressure transducer (n)

pressurised fluid (n)

prevent (v)

price (n)

primary (adj)

primary flow (n)

prime (v)

priority (n)

probe (n)

process (n)

process water service (n)

processor (n)

production cell (n)

production line (n)

production site (n)

programming (n)

prohibit (v)

propane (n)

propel (v)

propeller (n)

propeller shaft (n)

property (n)

proportional to (adv)

protection (n)

prototype (n)

protrude (v)

provide [power] (v)

psi (pounds per square
  inch) (n)

pull up (v)

pulley [wheel] (n)

pump (n, v)

punch (n, v)

purge (v)

pushrod (n)

put back (in, together,
  etc.) (v)

put out [a fire] (v)

Quadrant (n)

qualification (n)

quality assurance (n)

R/H (right-hand)

radar (n)

RADAR [Radio Detection
  and Ranging] (n)

radar tracking (n)

radial (adj)

radiator (n)

radio static (n)

radio-based system (n)

radius (n)

ram (n, v)

range (n)

ratio (n)

razor-sharp (adj)

react [with chemicals] (v)

reading (n)

reassemble (v)

reattach (v)

rebuild (v)

recalibrate (v)

recharge (v)

reciprocal lathe (n)

reciprocating piston (n)

reciprocation (n)

recirculate (v)

recoil (v)

record (v)

redirect (v)

redo (v)

reduction gearbox (n)

refine (v)

refit (v)

refrigerant (n)

refueling probe (n)

regulate (v)

regulation (n)

regulatory (adj)

reheater (n)

reinforce (v)

reinstall (v)

relative humidity (n)

relative to (v)

relay (n, v)

release valve (n)

reliability (n)

relieve [someone] (v)

remote (adj)

remote control unit (n)

remote switch (n)

remove (v)

repair (n, v)

repeatability (n)

replace (v)

replenish (v)

report (v)

reposition (v)

research and development (n)

reservoir (n)

reset (v)

residue (n)

resist (v)

resistance (n)

resistor (n)

restore (v)

restraint (n)

restrictor (n)

retract (v)

retract chamber (n)

retractable (adj)

reverse (adj, n, v)

revolutions per minute (n)

rheostat (n)

rib (n)

right angles (adj)

rigid (adj)

ring (n)

risk (n, v)

risky (adj)

rivet (n)

riveting tool (n)

robot technology (n)

robotic assembly (n)

rocker (n)

rocker arm (n)

rocket firing unit (n)

rocket motor (n)

rod (n)

roll (n, v)

rotary engine (n)

rotate (v)

rotation (n)

rotor (n)

rough (adj)

rough ground (n)

rubber band (n)

rudder (n)

run down [battery] (v)

runway (n)

S

saddle (n)

saddle washer (n)

safety (n)

safety equipment (n)

safety inspector (n)

sand (v)

sandblast (v)

sandpaper (n)

satellite (n)

saturated (adj)

saturation point (n)

saw (n, v)

scald (n, v)

scale (n)

schedule (n, v)

schematic (n)

scrapped (adj)

scratch (n, v)

screw (n, v)

screwdriver (n)

scribe (v)

scrutinise (n)

seal (n, v)

seat belt (n)

secondary (adj)

sector-shaped (adj)

secure (adj)

security (n)

select (v)

selector (n)

selector switch (n)

self-discharge rate (n)

semiconductor (n)

semi-monocoque (adj)

sensor (n)

separator (n)

service tag (n)

servicing equipment (n)

set (v)

set up (v)

shaft (n)

shallow angle (n)

sharp (adj)

sharpen (v)

shear (adj)

shearing strength (n)

shears (n)

sheath (n)

sheet (n)

shift (n)

shock (n)

short circuit (n, v)

shorted (adj)

short-range (adj)

shutdown (adj)

sick bag (n)

side cutters (n)

sidewall (n)

sideways (adj, adv)

sign (n)

sign off (= authorise) (v)

simple transformer (n)

simplicity (n)

simultaneous (adj)

single pole switch (n)

single-skin construction (n)

site (n)

size (n)

skater (n)

skeleton (n)

sketch (n)

skid (v)

skin (n, v)

slack (adj, n)

sleeve (n)

slip ring (n)

slope (n)

slosh (v)

slow down (v)

smith (n)

smooth (adj)

snag (v)

socket set (n)

software (n)

solder (n, v)

solder sucker (n)

soldering gun (n)

soldering iron (n)

solenoid (n)

solid (adj)

solid state (adj)

solidify (v)

sophisticated (adj)

source (n)

spar (n)

spare part (n)

spark (n)

spark plug (n)

specific heat capacity (n)

specific operating
    conditions (n)

specification (n)

speed (n)

sphere (n)

spin round (v)

spindle (n)

splice (v)

sponsor (n, v)

spot [the problem] (v)

spray (n, v)

spread (v)

spring (n)

spring-loaded (adj)

square (adj, n)

square root (n)

stabilise (v)

stable (adj)

stainless steel (n)

stall (v)

standard rivet (n)

standard specification (n)

starboard (adj)

starter (n)

starter motor (n)

starter-generator (n)

state (v)

static discharge (n)

static line (n)

stationary (adj)

statistics (n)

steady (v)

steering (n)

stiff (adj)

stiffening (n)

stitch/stitching (n)

storage (n)

straight line (n)

strain (n, v)

strainer (n)

strap (n)

streamlined (adj)

strength (n)

stress (n)

stressed skin panel (n)

stringer (n)

strip down (v)

stroke (n)

structural part (n)

structure (n)

strut (n)

sub-assembly (n)

suck [in] (v)

suction (n)

suffocate (v)

suitability (n)

suitable (adj)

sulphur (n)

sump (n)

supercharger (n)

supply (n, v)

support (n, v)

surface (n)

surface area (n)

survivability (n)

survival (n)

switch on/off (v)

synthetic (adj)

systemic [problems] (adj)

systems engineer (n)

tags [warning ~] (n)

tail (n)

taildragger (n)

tailor (v)

take apart (v)

take out [of service] (v)

take over (n, v)

take-off (n)

take-off selector (n)

tangled (adj)

tank (n)

tape (n)

tape measure (n)

tarmac (n)

tattle tale (n)

taxi off (v)

tear (n, v)

technician (n)

technique (n)

technological (adj)

technology (n)

teeth (n)

telescopic (adj)

temperature reading (n)

tensile strength (n)

tension (n)

terminal (n)

terminal block (n)

terminology (n)

thermal conductivity (n)

thermostat (n)

thickness (n)

thread (n)

thread pitch (n)

threat (n)

three-dimensional (adj)

thrust (n)

tie (v)

tighten (v)

timing chain/belt (n)

tin snip (n)

titanium (n)

tolerance (n)

top (n)

torque (n)

torsion (n)

toughness (n)

tower [radio] (n)

towing (n)

trace [fault] (v)

trainer (n)

transfer (n, v)

transistorised static
    inverter (n)

transmit (v)

transmitter receiver (n)

transparent (adj)

transverse axis (n)

trap (v)

tread (n)

treadle (n)

trial (n, v)

trickle-charge (v)

tricycle (n)

trim (n, v)

troubleshoot (v)

troubleshooting (n)

trussed (adj)

tube (n)

turbine (n)

turbofan engine (n)

turboprop (n)

turbulence (n)

turn around (v)

turn over [engine] (v)

turn-coordinator (n)

turning tool (n)

twist (v)

twisted (adj)

two-/three-point
    perspective (n)

two-dimensional (adj)

two-way radio (n)

tyre (n)

ullage (n)

undercarriage (n)

undertake [maintenance] (v)

uneven (adj)

unit (n)

update (n)

upwards (adv)

vacuum (n)

value (n)

valve (n)

vane (n)

vanishing point (n)

vaporise (v)

vapour (n)

velocity (n)

vent (adj, n)

ventilate (v)

venturi principle (n)

verify (v)

versatile (adj)

vertical (adj)

vertical axis (n)

vertical take-off (adj, n)

vibration (n)

vice (n)

visibility (n)

visualise (v)

void (n)

volatility (n)

volt (n)

voltage (n)

voltage probe (n)

voltage regulator (n)

voltmeter (n)

volume (n)

VOR [Very high-frequency

Omni-directional Radio
    range] (n)

vulnerability (n)

war plane (n)

warning (adj, n)

washer (n)

waste (v)

watt (n)

WD40 lubricant (n)

weapons system (n)

wear (v)

weight (n)

weld (v)

welding (n)

well-ventilated (adj)

width (n)

wing (n)

wing box (n)

wingspan (n)

wire (n)

wire strippers (n)

withstand (v)

wood turning (n)

work cage (n)

work order (n)

work table (n)

workpiece (n)

works [= maintenance] (n)

wrap around (v)

wrench (n)

wrought iron (n)

yaw (n, v)

zinc (n)

Tapescript

Unit 1

CD1 Track 1: Unit 1, Lesson 1

1.

A: So this is your new telly then?

B: Yep – what do you think?

A: Brilliant picture ... and it's so thin.

B: And it's really light ... only about 18 kilos ...
    I can lift it easily!

2.

A: Have you ever been on a hovercraft?

B: Yes. We went across to France on one last year.

A: What was it like?

B: Amazing! I didn't know if I was on a boat or a
    plane. We were doing nearly a hundred and
    twenty klicks and we were only about a metre
    above the water. Fantastic!

3.

A: Look, this is the one I'd recommend Bob. It's got
    a massive memory.

B: Really?

A: Yeah, 60 gigabytes. You'll be able to play your
    horrible music
    for hours!

B: Ha ha.

4.

A: Whoa, what was that?

B: Oh, it's one of the Harriers from the air base
    along the coast. It's probably doing a thousand
    kilometres an hour.

5.

A: Can I help you, sir?

B: Yes. I'm interested in this machine here. Can you
    tell me something about it?

A: Sure ... Well, of course, there's no bag to worry
    about and it's got a big strong motor, 1,400 watts

in fact ... And a two-litre bin ... So you don't
have to empty it too often.

6.

A: That's the new Mazda RX-8, isn't it?

B: Yeah. It's a lovely motor. 230 brake horsepower.
    That's far more than you usually get from
    smallish cars like that. I bet it goes like a rocket
    on the motorway!

CD1 Track 2: Unit 1, Lesson 4

Host: Good evening and welcome to *Amazing Animals*.
       This evening, we are going to hear about one
       of nature's strongest materials, made by some of
       nature's smallest creatures, the spiders. To tell us
       more about these remarkable little insects is Dr
       Donald Parsons, the director of Hopewell Zoo ...
       Good evening, Doctor Parsons.

Dr P: Good evening, Robin, and ... er ... before we go
       any further, can I, er, just correct one small
       mistake ...?

Host: Oh ... yes, of course.

Dr P: Yes ... well, you see the thing is, spiders are not
       actually insects.

Host: Oh, really?

Dr P: No, they actually belong to a group of creatures
       called arachnids, which have eight legs, not six,
       like insects, and they are more related to crabs
       and scorpions than they are to flies and beetles.

Host: Oh, well, thank you for putting me straight about
       that ... Now, I believe that you have got some
       interesting information for us about the silk that
       spiders use to make their webs with.

Dr P: Yes, that's right. And not just to make webs
       either. They use the silk to wrap up the small
       creatures they catch in their webs, to make
       shelters where they can hide from their enemies
       and even as lifelines to help them escape when
       they are being chased.

Host: So this spider silk must be a pretty amazing material. I know I get spider webs on my car wing mirrors in the morning and they don't blow off even when I'm driving fast.

Dr P: That's right. The silk often contains a strong adhesive; it's very, very sticky. Like chewing gum … but also extremely, extremely strong. Did you know that someone once calculated that a length of spider silk only the thickness of a pencil could stop a Boeing 747 airliner, without breaking.

Host: Incredible!

Dr P: Yes, weight for weight, it's much stronger than steel … up to five times as strong under the right conditions. If I can give you a few technical details …

Host: Well, I don't …

Dr P: Yes, there's something called Young's modulus of elasticity, which is a measurement of the tensile strength, how much you can stretch something before it breaks really . . and this is measured in units called Pascals. The radial thread of a spider's web …

Host: Radial thread?

Dr P: Yes … the line that goes from the centre of the web to the edge.

Host: Uh huh …

Dr P: Yes, well, that radial thread has a tensile strength of well over a thousand Pascals, whereas mild steel, by contrast, comes out at about 400 Pascals.

Host: Hmm. That's a big difference. Does it have any other interesting properties?

Dr P: It certainly does. It is very, very light stuff, 25% lighter than synthetic plastics made from oil. And it keeps its strength at very low temperatures – down to minus 40, in fact. At that temperature, a lot of materials become very brittle and really quite useless. It is also fairly resistant to moisture – more so than the silk that is used to make clothes …

Host: But if it's so marvellous, why don't we see it being used more often? It sounds like an almost perfect material.

Dr P: Well, the problem is, it's almost impossible to collect it in any useful quantities. The silk is very, very fine – finer than a human hair – and you need an awful lot of it to make a useable amount. And you can't farm spiders like silkworms. They don't do as they are told. But seriously, spider silk has occasionally been used in the past, to make the cross hairs in instruments and gun sights, for example. However, there may be a chance of producing more of it in the future.

Host: Really?

Dr P: Yes, apparently, some biologists and chemists in Canada are trying to produce a type of spider silk from goat's milk.

Host: Goat's milk?

Dr P: Yes, it's really most interesting. Their idea is to mix the genes of the …

Unit 2

CD1 Track 3: Unit 2, Lesson 2

PM: Welcome to another edition of *Collector's Corner*. I'm Phillip Martin. In today's programme, we'll meet someone who claims to have the world's largest collection of bicycles, someone with a wonderful collection of Arab coffee pots and lastly a chap who collects … wait for this … you're not going to believe it … the labels from tea bags – yes, tea bag labels are now collectable! You heard it here first.

But to start today's programme we're going to talk to Richard Bolton about that traditional weapon of the Japanese warrior, the katana sword.

Hello, Richard and welcome to the programme. Now you've been collecting Japanese swords for quite a number of years, I believe?

RB: Yes, that's right. It all started when I went to Japan on a business trip about 15 years ago. I was lucky enough to be taken to see a Japanese swordsmith at work while I was there. I found it so fascinating that I became completely hooked … and … well, I've just been collecting Japanese armour and weapons ever since.

PM: How many swords have you got altogether?

RB: Seven.

PM: Is that all? Forgive me … uh … it's just that it doesn't seem like very many.

RB: I know … but you have to remember that it's quality, not quantity, that counts! These weapons are very, very expensive and each one has been produced by a different master craftsman … and a couple of them are well over a hundred years old … But each one is special and different in its own way.

PM: Can you tell us a little about how they are made?

RB: Well, in two words really … heating and beating.

PM: Is that it … surely not?

RB: Well, I am making it a bit simple … but you see … the uh … technique developed because originally the Japanese were only able to produce rather impure metal.

PM: Why was that?

RB: Because Japan had very little high-quality coal so it was difficult to heat metal to very high temperatures in the furnaces.

PM: Right … so the work that might have been done by the heat had to be done by the craftsmen.

RB: Yes, that's it. Basically, the swordmaker repeatedly heats, folds and hammers the metal. This repeated folding and beating eliminates any air bubbles or gaps in the metal and makes sure that the carbon content of the steel is uniform – it homogenised it … and the continual heating, cooling and reheating helped get rid of … to eliminate many of the impurities in the rather low-quality steel.

PM: And what is the end result of all this work? What kind of sword do you get at the end of it all?

RB: Quite an unusual one, actually. For a start, it's rigid, not very flexible … designed to be used edge-on like a guillotine. And there is only one sharp-cutting edge, not two …

PM: Ugh … don't go on … it sounds nasty.

RB: Yes … sure … but did you know that the techniques used in making swords were also used to make gardening and woodworking tools, planes, chisels, cutters, knives and so on. A lot of …

PM: Really … how interesting … and presumably there was some special armour made to protect soldiers from these deadly blades?

RB: Armour … oh yes, that's another subject in itself … you know most people … think of heavy metal when they hear the word 'armour' … but in fact, the Japanese armour consisted mainly of bamboo wood, leather and quilted cotton which was ….

CD1 Track 4: Unit 2, Lesson 4

a)   The drill bit can be moved in more than one axis. (1)

b)   The handle is usually made of plastic. (1)

c)   Power is transmitted to the drill bit by the gears. (BOTH)

d)   Its wide heavy base helps to prevent unwanted movement. (2)

e)   The pillar supports the work table. (2)

f)   The drill is powered by compressed air. (1)

g)   It's unsuitable for drilling teeth. (2)

h)   The drill bit can be accurately positioned. (BOTH)

i)   The position of the bit can be controlled by the handwheel. (2)

CD1 Track 5: Unit 2, Lesson 6

… so you might wonder why it is important to have standardised screws and bolts in the modern world. The answer is simple. Industry is international. This is not as recent development as you might think, though. Brought on by the 2nd World War, screws and bolts first became standardised during the 1940s. In order to produce these standardised screws and bolts, the modern lathe has become an essential piece of machinery.

CD1 Track 6: Unit 2, Lesson 6

The lead screw is a long, threaded rod that carries a tool along the axis of a rotating workpiece. It ensures that the workpiece moves at a constant, even speed so that threads can be cut into it. The relationship between the longitudinal speed of the tool and the rotational speed of the workpiece can be varied by means of a gearbox.

CD1 Track 7: Unit 2, Lesson 8

T: John, Martin, can you come over here? OK, good ... now have you done ... what I asked?

J: Yes, we've cut two square sheets of this metal you gave us, hang on, here you are, is that all right?

T: Hmm, it looks OK but, uh ... Martin, run your thumb along the edge there, yes, ... that one.

M: Ouch, it's a bit sharp.

T: Yes, I thought so ... it's really important to smooth off the edge of any metal you cut. You know why?

J: Yes, cos somebody could get hurt.

T: That's right, safety is always important, but there's another reason to do with this particular metal.

M: What, you mean aluminium sheet?

T: Well, this isn't ordinary aluminium, though, it's an alloy, which has been through a process called alcladding. It's had a skin of pure aluminium applied to it on both sides.

M: What's the point of that?

T: It's to stop corrosion ... The aluminium forms a protective oxide skin. But this means that you have to handle it and work with it very carefully because ...

J: ... you might break the skin and then the metal underneath ... you could get corrosion in underneath the skin.

T: Exactly ... and in an aeroplane, thousands of feet up in the sky, that's dangerous ... very dangerous.

M: Does this stuff have a special name?

T: This particular sheet is called 2024-T3 ... It's used a lot in aircraft.

J: So how exactly do you have to be careful when you work with it?

T: Well, think for a moment ... What are the four main operations you do on sheet metal?

J: ... er ... cutting, drilling, bending ... er ...

M: ... and marking out ...

T: ... exactly, ... marking out. Before you do anything else ... and that's when you have to start being careful ... from the beginning.

M: ... when you're scribing the lines.

T: Ah ... but that's the first thing to remember. You mustn't use a scriber to make the lines and points. Scribers can cut too much into the oxide surface, which allows corrosion in ... Also, where you cut into the metal, it makes it a little weaker along those lines ... and if ... there is vibration and stress in the metal ... it can fracture there. It's a bit like breaking the pieces off a bar of chocolate. The chocolate breaks along the lines marked in it. So you should always use a special marker pen like this ... It's called a sharpie ... you can get them in different colours, but I prefer blue, it seems to show up better.

J: What about cutting the sheet? Are there any special things to remember?

T: Yes, three things: finishing, finishing and finishing. You must make sure that any cut edges are rounded off and completely smooth. Rough and jagged edges are like small cracks in the metal ... And these can get bigger and bigger. The best test is to check that you can't see the marks of the saw teeth ... or the guillotine blade if you use the treadle shears. The same thing when you drill a hole ... you must remove all the burrs and make sure the inside of the hole is completely smooth.

M: And what about making bends?

T: Now that's quite a complicated subject ... But the basic rule is that you mustn't put too much stress on the metal. You mustn't bend it over too far or with a bend radius that is too small for the metal you are working with.

M: How do you know if you've done that?

T: You make sure you don't by looking at the special tables ... no ... no, I think that's enough talking for today. We'll leave it until tomorrow. What I want you two to do now is mark out these two sheets, and drill the holes here and here, just like you can see on this drawing ... you see ... here and here ... and then bring them back to me. And remember you must remove all the burrs and smooth all the sharp edges. You should be able to finish it in about half an hour. But don't worry if you can't. You don't have to finish it today. The important thing is to do it correctly …

J &M: Right … OK …

T: You can use that bench over there.

CD1 Track 8: Unit 2, Lesson 10

I: So, Robert, this is the plane you built yourself?

R: Yes, this is my baby, what do you think of her?

I: Well, to be honest, I'm amazed! It looks just like a new plane from a factory. It's so well finished. You must have done a lot of studying and technical training before you started building it!

R: No, not really. Of course, I've always been interested in planes and I've had a pilot's license for over ten years now. But my normal job is in a bank … and … apart from a few household repairs, I don't have much hands-on technical experience. At least, I didn't before I started building her.

I: And how long did it take you from start to finish?

R: Hmm … let's see now … this is August and the kit was delivered in April the year before last … yes, so just over two years.

I: And how much of that time did you spend on the project?

R: Practically every weekend really, as well as a lot of evenings after work. I'm afraid my family didn't see very much of me. I suppose it was about 800 hours altogether.

I: You said it was a kit?

R: Oh yes, I couldn't have built it from scratch. A lot of the difficult stuff had already been done at the factory.

I: And where did you actually make it?

R: In my garage.

I: Your garage?

R: Yes, of course. I had to move the car out and put a workbench in!

I: I bet you did! You must have a pretty big garage.

R: Hmm, biggish but not enormous. What is it … Yes … It's about 24 feet long, nine feet wide and nine feet high. Of course, I had to fit the wings on outside. But I did most of the construction inside.

I: You must have needed a lot of special equipment.

R: No, not really, although I did buy myself a kit of good hand tools.

I: Such as?

R: Oh, you know, the usual thing, screwdrivers, pliers, saws, spanners and so on. Oh yes, and a pair of tin snips for cutting the sheet metal.

I: So just those hand tools, then?

R: Not exactly. I did get myself a new electric drill … the old one was on its last legs. And a hand rivet gun for the blind rivets.

I: 'Blind' rivets?

R: Yes, sorry, it just means a kind of rivet that one person can fit easily on their own into any part of the plane. You only need to work from one side.

I: Right … and tell me … how much did all this cost, apart from your time, of course?

R: Just under 14,000 pounds for the kit, plus about 200 for the tools. Of course, I did have to take my family away for a special holiday to make up for all the time they didn't see me. That was a couple of thousand for all of us!

I: And one last question, if you don't mind?

R: Sure.

I: Why did you do it?

R: Ah. Well, partly the cost, it was a lot cheaper than buying one ready-made. But really, I suppose it was also because I enjoy a challenge. And I wanted to do something that was completely different to my work. You know, even if I never build anything else, I'll always be proud of this … it's given me a lot of satisfaction.

Unit 3

CD1 Track 9: Unit 3, Lesson 3

L: No, it's no good … I can't see it. My eyes just aren't as good as yours. There's not enough light for me yet.

G: Don't worry. Be patient. Close your eyes and then look across the lake again. He's on a branch about halfway up that very tall tree on the left of the rocks. Start at the bottom of the tree, and move your eye slowly up until …

L: Ah yes, there he is! My gosh, look at him … if he just stays still long enough while I get my binoculars … There, yes, I can see him clearly … every feather. He's a big chap.

G: Yes, fully grown … and he's going to start hunting in a minute, I think. Don't use your binoculars; you won't be able to follow him. He's much too fast. Do you see those circles in the water over to your left? That's a big fish just coming up to feed on the early morning insects. I'm sure the eagle must have seen him and he'll … yes, there he goes.

CD1 Track 10: Unit 3, Lesson 3

L: There he goes, he's taken off from the branch. Will you look at those wings – the wingspan must be at least a metre. He's not going for the fish though, he's climbing away. What's he doing? He's a couple of hundred metres beyond where the fish are feeding. Ah, now he's turning back, banking towards us.

G: Yes … he's got the height he needed to start his descent. He'll dive at quite a shallow angle so the fish won't see him coming …

L: … and what a speed, too! He must be doing at least 60 km an hour.

L: Got it! Wow, he just picked the fish up as if he was in a supermarket … wait a second, he seems to be having some trouble – he's just dragging it along the surface of the water.

G: Yes, he can't … the fish is too big and heavy. He can't produce enough lift from his wings … no, I think he's OK. He's beginning to climb. But he's certainly a lot slower … Oh, I think you are a lucky man – quickly, look above him and to the right … you see?

L: Woooh! It's another eagle – it's diving straight down towards him. What's it doing?

G: Wait, you'll see. It's a female … they are usually bigger and heavier.

L: I don't believe it! Is it trying to steal the fish?

G: Yes. If the first bird drops it, the other one will dive down and grab it in midair. They are famous thieves, these eagles. But it's unusual to see one fish eagle steal from another; they usually attack other birds.

L: What a sight! They've both got their talons into the fish and they're flapping away … he really doesn't want to let go of his breakfast. Is she trying to pull him up higher? She seems to be. Why's that?

G: Yes. She is. If she can gain enough height, she will try a special manoeuvre … yes … watch.

L: Ah, look at that, she's rolling and falling at the same time, and the male is having real trouble keeping up.

G: Yes. Because the female is bigger and stronger, she will try to twist the fish away from him or make him drop it, which he may have to do in the end to avoid falling into the lake. There … he's dropped it.

L: Woooah … and she's shot down past him and grabbed it … ha! … and she's off with his fish. She's climbing really fast … she must have some power in those wings. Fantastic. She's away. Well, what a lot of excitement – and the day has only just started! I wonder what we'll see next.

G: Well, it's your lucky day for sure. Perhaps a lion will come down to drink … there's still time.

CD1 Track 11: Unit 3, Lesson 6

The final production assembly line for the German version of the Eurofighter is located in the South German town of Manching. It is here, that as well as equipping the main fuselage, the engineers and technicians put together more than 300 pieces of equipment, sub-assemblies and assemblies to produce the finished aircraft.

At the first stage, the three main assemblies are fitted or spliced together. These are the centre fuselage, the rear fuselage and the cockpit. Next, the flight control surfaces such as the wings, flaps and fins are attached to the fuselage. At this point, the fighter starts to look like a real plane. At the third stage, all the electrical cables are thoroughly tested and then the aircraft is moved on to Station 4, where the mechanical, electrical and hydraulic systems are subjected to rigorous testing with detached computerised equipment. Following satisfactory completion of these system tests, the aircraft is ready to have its engines and weapons systems fitted.

Following this, the flight control and navigation systems are installed and the plane is now ready for its pre-flight tests. Once these have been carried out, the aircraft is tested in flight, before finally being moved to the paint shop to be painted in the colours of the German Air Force.

CD1 Track 12: Unit 3, Lesson 9

A: I left university 18 months ago with a degree in maths and physics and an MSc in metallurgy. Since then, I've been working for a testing laboratory in London. It's quite a small firm so we all have to deal directly with customers face to face, especially if we have an urgent job on … which is quite often! We've recently had quite a few jobs for one of the big aerospace companies, mainly on stress fractures as well as routine structural loading tests. That's how I became interested specifically in the aircraft industry. I know I'll have to start at the bottom and work my way up. That's no problem. And I'm not married yet, so I don't really mind where I work.

B: I've lived in the UK all my life. All my family's here. I'd never want to move away. I originally wanted to go to art school. I was always drawing and painting as a kid. I still do in my spare time. Thing is, there's no money in it … so anyway, I took my dad's advice and got an engineering degree. Luckily, I was good at maths and science, too! My job combines both, art and engineering. Mind you, I'm having to use computer-aided design more and more these days. I don't stay long with any one company. I prefer short-term contracts … I think the longest I've done is eight months. I work away, sometimes. You know … find a comfy bed and breakfast for Monday to Friday and then head back home at the weekends.

C: I was in the air force for 20 years after I finished my apprenticeship … had a wonderful time and went all over the place … Germany … Cyprus … Malta. I worked on helicopters and Hercules transports mainly, skin fabrication mainly. That's my speciality, I suppose you could say. Since I came out of the air force, I've been working locally as a machinist, but to tell the truth, I'm getting a bit bored. It's the same thing day in, day out, and really I'm a sheet metal basher at heart. I'd like to go somewhere sunny for a few years, preferably the Middle East, and try something new. I'd be able to save a bit more towards my pension, too!

D: Mathematics is my chosen field. I've always loved it. I did a degree in maths and physics and then an MSc in avionics and control systems. I've been working on a doctorate and teaching at the university at the same time. I've also done quite a few short contract jobs for the air force. I'm looking for something long-term, a proper career in the systems and navigation field. I'd really like to work on a project from the early stages right through to completion. My family is from Italy originally. They came to South Wales to work in the coal mines during the last century. But I love it here – such a beautiful place – I would never move away.

Unit 4

CD1 Track 13: Unit 4, Lesson 2

T: Right … OK … if I could have your attention, please. Good. Now, before the break, we were talking about various different ways that farmers could pump water onto their land. We looked at very hot, dry countries where there isn't much water anyway. Now I want us to study a device which can be used when you don't have a shortage of water, but when the problem is that the water is in the wrong place. Can anybody think of a situation like that? Yes, John.

J: Well ... in countries where they get a lot of heavy rain one part of the year and then it goes very dry. You might want to fill up a reservoir or some storage tanks ... so you have water all the year round.

T: Exactly. Good. Yes, Peter?

P: Or maybe you need to supply animals with drinking water in fields which are high up ... or water crops, maybe?

T: Yes, that first point is a good one. Any other ideas? No. OK, right. Now I'll just put this diagram up on the board. There. Now, the basic idea is this. You see here where the water comes down through the pipe at point A. This is the primary flow, water flowing through a pipe on a slope – so, what force is making the water flow, anybody?

Student: Gravity.

T: Excellent! So gravity is our source of energy for this machine. And of course, gravity is free! Now, the water flows down through the pipe towards this valve at the bottom. And what do you think will happen as it hits the valve?

J: It will force the valve closed.

T: Yes, good – and this is why it's called a ram; because it pushes, or rams, the valve shut. But now – and this is the really clever bit – because the moving water has nowhere to go, there is a sudden increase in pressure. This is called 'water hammer'. You get a reaction in the opposite direction to the flow. To help you understand this, think about what happens if you throw a stone at a wall. The stone is moving and then it is suddenly stopped. Does it just drop down vertically?

Various students: No/it bounces/it comes back/it rebounds.

T: Exactly. So when this happens to the water, there is this backwards pressure in the pump, and this is enough to open the delivery valve here in the centre of the diagram, and push water upwards through the outlet pipe. Of course, then the pressure inside the pump goes down again, and as it returns to normal, our spring-loaded valve at the bottom – it's sometimes called a 'clack'

valve because it makes a clacking sound like someone hitting two pieces of metal together ... actually, sometimes it doesn't have a spring, just a weight – anyway, this valve opens again, and of course the – what happens to the delivery valve up here?

A: It, er, closes again, because the pressure inside the delivery pipe has dropped.

T: Well done. Yes, and so the whole cycle starts again: first, gravity moves the water – the moving water has momentum – which is what, anybody?

Various students: Mass multiplied by velocity.

T: That's right. Mass times velocity. And when the water hits the closed valve, this energy of momentum is changed into pressure – is changed into work. It lifts the water up the outlet pipe. So there you have it, a three-part cycle: momentum, pressure, work; momentum, pressure, work, and so on, about 30 to 60 times a minute. As long as there's water coming down the pipe, the pump just keeps going and you can move the water up, to any other place you need it. It only has two moving parts and it doesn't need an electric motor or an engine to power it. Now, has anybody got any questions?

CD1 Track 14: Unit 4, Lesson 5

So when this happens to the water, there is this backwards pressure in the pump, and this is enough to open the delivery valve here in the centre of the diagram, and push water upwards through the outlet pipe. Of course, then the pressure inside the pump goes down again, and as it returns to normal, our spring-loaded valve at the bottom – it's sometimes called a 'clack' valve because it makes a clacking sound like someone hitting two pieces of metal together ... actually, sometimes it doesn't have a spring, just a weight – anyway, this valve opens again, and of course the – what happens to the delivery valve up here?

CD1 Track 15: Unit 4, Lesson 5

... the use of hydraulics in matters of life or death. If I can take an example that everybody's familiar with – the domestic cat. Cats are good climbers and will often jump from a tree at a bird, or sometimes just slip and fall. But, a fall that would kill or badly injure a human or another animal may have little effect on a cat, because they have two characteristics which are essential for a safe landing, and which the aviation industry, among others, makes use of.

Firstly, cats have a small but extremely sensitive fluid-filled organ in their inner ear. When they move, so does the fluid, and so they know precisely their position relative to the ground at all times, and in a landing situation, of course, that's vital information. The equivalent in an aircraft is an instrument called the artificial horizon, which tells the pilot his orientation. As soon as a fall – or a descent, in the case of an aircraft – starts, this detector enables the animal, or pilot, to determine their orientation quickly so that they can prepare for the landing impact.

That's one important feature. Then, secondly, to help prepare the body itself for landing, the physical design needs to enable the falling body to easily get into the best position. Cats benefit from an extremely flexible skeleton. They don't have collar bones, and the bones in their backs have an especially high mobility. This enables them to twist and turn their bodies quickly and easily in an emergency landing, so that by the time they hit the ground they are oriented in the best position to absorb the shock. Aircraft don't have anything like that structural mobility, although there is some flexibility built into the airframe.

But what a pilot can do is to use the aircraft's control surfaces – the surfaces that move – like the rudder, flaps, etc. He can use them to slow the aircraft's descent – something that a cat can't do. This will minimise the shock of landing and give him time to manoeuvre. The control surfaces also allow him to … to manoeuvre into the best position very accurately.

The cat looks at the ground to help here; so, of course, does the pilot. This is where hydraulics comes in. To distribute the load, cats always land on all four feet simultaneously; aircraft, however, land just on their two main wheels or wheel groups. Cats curve their backs,

which aircraft can't do, of course, but where animals use muscles to minimise the shock of the impact, aircraft have to make use of hydraulics: as we saw with the air brake, a hydraulic chamber is filled with pressurised fluid.

When the aircraft hits the ground, the landing gear assembly basically acts as the piston inside the chamber, pushing against the fluid, which absorbs the force of the landing.

All being well, both the animal and the aircraft will land safely – although, hopefully, the aircraft will have many more than just nine lives …

CD1 Track 16: Unit 4, Lesson 6

J: Hello, George, haven't seen you for ages … must be what … seven or eight months? We thought you'd left the club but forgotten to say goodbye.

G: No … no. I've just been so busy at work … there just hasn't been enough time to fit the flying in. Anyway, how are you, Jack? How's the family?

J: Fine … fine, thanks. We're all well. My eldest son's just gone off to university. And you and yours?

G: Yep, all fine. As I say, I haven't had much free time, but things are a bit quieter now so I'm looking forward to getting some more flying in.

J: I see you're still flying with that taildragger landing gear. You should really move with the times.

G: No, thanks. I'm quite happy with this layout, thanks. I know it's more difficult to take off and land, but I enjoy the excitement and it keeps my flying skills up to scratch. Anyway, there are some real advantages to this taildragger gear, you know?

J: Oh yes, such as …?

G: Well, for a start, it's much easier to land on rough and uneven ground. There's no danger of breaking the nose wheel in a hole or digging up the field with the propeller when the nose tips forward. I'd much rather make an emergency landing in a field in my plane than in yours.

J: Hmmm … that's true, I suppose. But what about on the runway? I mean, the centre of gravity is behind the wheels, so the moment you get out of line, the plane's going to start trying to spin round … and … the cockpit is pointing up at the sky. You can't see properly.

G: Oh, you get used to the visibility problem, you just have to pay more attention. But I agree there is a danger of spinning round during take-off and landing. In fact, I have done a couple of ground loops myself. But that was when I was a beginner. If you know your plane and pay attention to the wind speed and direction, it's not a problem.

J: But what about when you're parked on the runway? There's much more danger of the plane being damaged by sudden gusts of wind, with the wings angled upwards like that. You remember what happened when we had that bad storm a couple of years ago. David's plane was a write-off.

G: Yes, that's true. I admit that can be a problem. But you just have to make sure you tie the machine down well or put it in the hangar.

J: And another thing, I can get in and out of my plane far more easily than you can, and it's much easier and quicker to load stuff in because everything is level with the ground.

G: Ah yes, but I can fly faster when I'm in the air because my landing gear is smaller. It weighs less and it doesn't have as much air resistance. I save quite a bit on fuel as well. If I had the tricycle gear fitted, I'd be spending about ten per cent more on fuel, I reckon.

J: Hmmm, well, I can see we're never going to agree. Brrr, it's getting cold out here. Let's get to the clubhouse and grab some hot coffee.

G: Now that I do agree with!

CD1 Track 17: Unit 4, Lesson 8

M: Hi, I'm back!

F: How did it go?

M: Fantastic! I actually got to fly the plane by myself!

W: You mean they let you go up alone?

M: No, of course not. That's against the law. Dennis – the instructor – was there all the time. But I did actually steer the plane and made it go up and down. It's got dual controls.

B: What's that?

M: It means you have your own set of controls but the instructor can take over from you instantly if he needs to. Anyway, I'm going to have another lesson as soon as I've saved enough money. It's just such a wonderful feeling up there!

F: Slow down, slow down. Why don't you tell us all about it from the beginning?

M: Sorry, yes, well I got to the airfield and Mr Saunders – Dennis – he began by showing me the main controls in the cockpit and explained what they do.

D: Now, before you get in, I'm just going to show you the main controls and how they relate to the flight control surfaces. Those are what we call the bits of the plane that move about. So if you stand just there while I get in …

D: OK. First, here's the control column. That's the correct name for it, but it's sometimes called the joystick, or even just … just the 'stick'. Now I can move it in the longitudinal axis, backwards and forwards … and I can move it laterally, side to side. Now if I push it forward like this – look at the back of the plane. You can see the elevators moving up and down – so it also controls movement in the vertical axis.

M: Oh yes, I see.

D: Now, when I pull the stick back, the elevators move up. And it's this control we use for what's known as the pitch of the plane – raising and lowering the front end, the nose.

M: Right. So you can use it to increase or decrease your flying height.

D: Yes, that's it. Now, as I said, we can also move the stick from side to side. Look, watch what happens.

M: Ah yes, those flaps on the wings are moving up and down.

D: Well, actually they aren't flaps. The flaps are actually further in, near the fuselage, you see?

M: OK. What do they do?

D: Well, we mainly use those for take-off and landing. You won't be doing either of those things today, so don't worry about it for now. No, these control surfaces are called ailerons and they control the rolling movement of the plane, sometimes called banking. You see, if I want the plane to roll or bank to the left, I move the stick to the left. If I want to bank to the right, I have to move the stick laterally to the right.

M: OK, yes, I've got it.

D: And lastly, there are these two pedals down here. Now if I press them down with …

M: Ah yes, it's like a ship's rudder.

D: Yep, that's it. It is the rudder. We use it to swing or turn the nose of the plane to the right or left. It's called yaw. Y-A-W: yaw.

M: So that's the control we have to use if we want to turn the plane round.

D: Well, in fact, not exactly – you mainly use the ailerons for that – but yes, you still need the rudder to help you control the turn.

M: How are the cockpit controls connected to the surfaces? Is it electrical or hydraulic?

D: In this little plane, neither! All the linkages are mechanical. It's all rods and cables, so you can really feel directly how the plane is behaving.

M: Right.

D: So. Now you know about the primary control system, I think that's enough theory for the moment. Now we're going to do all the pre-flight checks and then we're off.

M: Great! So what checks do you have to do?

CD1 Track 18: Unit 4, Lesson 10

Commentator: Hello and welcome to our radio listeners to the Spanlow Air Show. As you can probably hear, there's a large, expectant crowd here today and we're all waiting for the opening event, which is of course, a 20-minute display by the famous Red Arrows aerobatic team … And I've just been told that we're about to start … and yes, here they come. All nine planes are flying horizontally towards us and they turn together, greeting us in the famous Diamond Bend formation, which usually starts the Red Arrows show. Fabulous.

Now three of the planes have split off and … ah yes, two of them are climbing. Now, they cross at the top, bank and dive towards each other to make a Heart shape with their smoke. Oh, and here comes the third aircraft right through the middle to complete the picture … fantastic!

The three are joined by a third aircraft now as they fly away from us. I wonder what – ahaa, I think I can see what – yes, now this really is amazing, ladies and gentlemen. Watch closely. Two of them are rolling in long horizontal loops in a kind of spiral round each other, just like a corkscrew, while the other two fly straight and flat inside. They're so close to each other, each pilot is only three or four metres away from the pilot next to him during this manoeuvre, and there's no computerised flying here, let me tell you – every movement of the control surfaces is under the direct manual control of the pilot. That manoeuvre is called the Corkscrew, in fact. Very demanding.

The Red Arrows, of course, spend the whole of the winter each year practising for the display season – six or seven months in all. Every pilot always flies in the same position in the formation. That group has headed off and been replaced by another group of four and they're climbing … climbing … climbing. Now, they're looping over, and down they come, diving down vertically … that's extraordinary. I really don't know how to describe the shape they've drawn … it's like straight vertical lines with loops at the bottom. The team have given it the equally strange name 'Twizzle'. I bet that one takes a lot of practice!

What's next, I wonder? Here we've a bigger group, flying horizontally straight at us again, two, four, six, seven planes flying close together – wooah – and they suddenly break apart. What a beautiful sight, all the coloured smoke trails rushing away from each other. The pilots experience 7 G of gravitational force, ladies and gentlemen, in that turn – the Vixen Break. The Hawk aircraft itself has a structural limit of 8 G!

And now a smaller group is climbing for the Opposition Loop. Up, up. And they turn together in very tight formation … accelerating downwards. Down they go. Beautiful. And they spread apart just a little but still stay close to each other. These pilots really are exceptional.

Unit 5

CD1 Track 19: Unit 5, Lesson 2

R: Now you've flown all kinds of planes, haven't you, Tom?

T: Yes, it's nearly 30 years since I started, and I hope I'll be doing it for another 20!

R: What changes have you seen?

T: Well, I suppose the biggest change really is in engines. Aircraft body design hasn't altered so much, but engine technology's another matter. And it's the power source that drives the machine along, so it's central to the whole thing. Without that, it's just a glider.

R: That's true! What would you say are the important factors in engine design?

T: I'd say probably the most important thing, above all others, is reliability. You have to remember that a plane can't just stop in the air if something goes wrong, like a ship or a car can do. You need to be absolutely sure that the engine will keep working from the time the plane takes off to the time it lands. The thing is that aircraft engines have to run at very high power most of the time. When it takes off, of course, it's using maximum power, but even when it's just cruising along happily it may well need between 65 and 75% of its power, whereas a car only uses that percentage of power for about 20% of the journey time.

The next most important factor in my book is weight: the lighter you can make the engine, the better. If the plane is carrying less weight, less power is needed and less fuel is required. So reliability and weight, yes, those are my top two. In the early days, engines tended to be pretty heavy, but these days, with lightweight metals like aluminium, and other technical improvements, they are much, much lighter. They've got some incredible materials these days.

R: What else would be on your list apart from those two?

T: Well, the next priority I suppose would have to be power. The more power in an engine, the better, because then the weight and the air resistance can be overcome easily. Plus, it's always good to have some power in reserve in case of an emergency situation. And lastly, I guess, well, size makes a difference. The smaller it is, the less drag you get again, the less air resistance it has, and if you can streamline the shape, that helps too.

R: Sure. And wh …

T: Actually, there is one more thing, sorry, if it can possibly be done. It's nice if the engine is kept as simple as possible, with spare parts that are easy to get hold of and reasonably cheap. There's nothing worse than having your plane grounded because some high-priced part has to be sent from the other side of the world.

CD1 Track 20: Unit 5, Lesson 2

R: Which engine is your favourite?

T: My favourite … well, there are three that I've had good experience of. The twin-cylinder in-line, you know, one behind the other, is a nice engine. It's a fairly simple design and, of course, a very well streamlined arrangement, and as I said before, that's always a good thing, although it doesn't have much power compared to some other types. Then there's the radial with the cylinders set in a circle. That's a bit more complicated. More spare parts, of course, but it is very powerful and it runs very smoothly. And I guess the other would have to be the gas turbine. A turbine is always expensive but it's powerful, and again, it's a very good shape for fitting in a plane. No cylinders, of course – the whole thing is very streamlined, so drag is kept to a minimum.

R: Which was your first engine?

T: Oh, none of those! No, I started like most people of my age with a rubber band! I didn't have the money to buy a proper engine. No, I just turned the propeller with my finger a couple of hundred times and then let the plane go! Ah ... happy days.

CD1 Track 21: Unit 5, Lesson 6

I've always been interested in aircraft and in engines particularly, so when I retired, I decided to see if I could find an old aircraft engine to work on. Well, I advertised in local newspapers as well as a couple of aircraft magazines, and nothing happened for a month or so. And then, out of the blue, when I thought I wasn't going to get anywhere, I got a phone call from a film company. Apparently, they had an old Rolls Royce Merlin engine in working condition. They were using it to produce wind in the studio – winds of up to 400 miles per hour, they said. But it was unreliable as well as being noisy and smelly, so they were going to replace it with more modern equipment. They didn't want to just throw the engine away, so I could have it free if I paid for the transport. I didn't hesitate. And a week later, I was the proud owner of a working – well, sort of working – Merlin MK 20 aero engine.

But they weren't joking when they said it was unreliable. In fact, it was really on its last legs and should have had a complete overhaul. Anyway, the first time I started it up, steam came hissing out at the front, the two front cylinder heads, because of a water leak. Luckily it was quite easy to fix. I just fitted new cylinder head gaskets. The next problem was another leak. This time oil, not water. There are these brass tubes that carry oil back to the crankcase and they were all leaking ... not much, but enough to be a problem. They all had to be replaced, and that was quite expensive, because I had to have them specially made. Spare parts for 50-year-old engines aren't that easy to get hold of.

Things went all right for a while ... until I started to get problems with the carburettor. It was flooding, you know, filling up with too much fuel, so of course the engine wasn't running properly, because the fuel was staying in the carburettor, you see, instead of going to the combustion chambers. Anyway, that's what I thought. Well, I stripped down the carburettor, inspected it, it looked OK, but I cleaned it anyway, and then put it all back together again, started the engine, and it was just the same! It still wasn't firing on all cylinders. So I removed the rocker cover and what did I find? One valve completely missing and half the valve springs broken. So I replaced all of those, including the unbroken ones, of course. After that, I didn't have any trouble for about a year, and then the two front cylinder heads started leaking again. This time, I completely dismantled the engine. And it was then that I discovered that it had been in some sort of crash and the front cylinder blocks were slightly out of alignment. Anyway, they had to be machined so that they fitted properly. While I had the engine in pieces, I completely overhauled and replaced anything I could. Since then, she's been running perfectly. Listen ...

CD1 Track 22: Unit 5, Lesson 8

It is easy to forget what a technological marvel modern flight is. When a Boeing 747-400 is cruising at 35,000 feet, each of the four engines is generating 12,000 pounds of thrust, and continues to do so for many hours at a stretch.

To do this, the engine draws in 700 pounds of air every second, although in fact as much as about 80 per cent of the air which goes in actually bypasses the core. The 120 pounds of air that does enter the core is pressurised to more than 150 pounds per square inch and, at the same time, heated to more than 850°F in the compression section.

When the air is compressed inside the engine core, one and three-quarter pounds of fuel is injected into it and burned to heat the air/fuel mixture of combustion gas to more than 2,000°F. The turbines extract enough energy from these gases to turn the fan at about 3,300 rpm and the compressor much faster still, at around 9,500 rpm. When the gas mixture exits the turbine section, it is moving at a velocity of 1,400 feet per second and is still very hot – over 1,000°F.

All this goes on every second that the engine is cruising up there, hour after hour.

CD1 Track 23: Unit 5, Lesson 10

I: OK, now we talked about fuel and fuel tanks in small aircraft yesterday. Can anyone remind us what the two main problems are with aircraft fuel? Yes, Ali.

A: It's highly flammable, so there is always a danger of, maybe, fire.

I: Yes, that's right. But of course, the trouble is if it wasn't highly flammable, it wouldn't be much use ... But you're right, you have to design everything to get round that problem. What was the other main problem – there's something else about fuel which can cause problems if it isn't allowed for. Yes, Abdelhakim?

A: It's a liquid, so it's a bit difficult to store. It can run around the plane.

I: Yep, that's it. Not exactly around the whole plane, of course, but round –

A: In the tanks.

I: Yes, you've got it. That was it. The fuel can move around inside the fuel tanks when the plane turns, you just try running round the canteen with a bowl of soup in your hands. It'll go all over the place, especially if you suddenly stop, start or turn. OK, fine, so how do you think this causes a problem for a pilot flying a plane?

A: It makes the plane difficult to keep balanced because the weight of the fuel keeps moving in position. It's like all the passengers suddenly change seats.

I: Exactly. And the plane becomes very difficult for the pilot to control. So what do you think the answer is?

A: You have to keep the fuel balanced all the time you are flying.

I: That's right. And how can you do that? Ali?

A: A lot of smaller fuel tanks?

I: Well OK, yes. Quite hard to build though. Other ideas? Mohammed.

M: Move the fuel from one tank to the other while flying.

I: That's right. Well done. How?

M: A pump?

I: By pumping it from one tank to another ... in any part of the plane, in any of the tanks, and so you maintain the balance of the plane – the 'trim', as it's called. For example, you don't want a right wing tank empty and the left one full. And of course, the bigger the aircraft, the more important this is ...

CD1 Track 24: Unit 5, Lesson 10

I: ... and what kind of pump, can anyone suggest ... I mean, there are different kinds, so which is best for moving fuel in and out of tanks? Yes, Yusef?

Y: It has to be very safe.

I: That's true. Well ... have a look at these pictures. Now, these five are pumps from a trainer plane, and they all do different jobs. Let's look at the first one – pump number 1. Can anyone tell me what type of pump this is?

Y: A gear pump.

I: Well done, Yusef. In fact, it's the high-pressure pump, which supplies the combustion chamber of the engine. You can see it's got a filter on the outlet to make sure that the fuel is clean when it goes into the engine. Yes, Abdelhakim?

A: Where does the power come from?

I: From the engine. It's connected to the engine. As the engine turns, the pump rotates and supplies more fuel.

Y: But how can it start if the engine isn't turning, before the plane takes off?

I: Good question. And the answer is this second kind of pump. This doesn't need any engine power. It's a 28-volt booster pump which runs from the battery until the engine takes over and powers the high-pressure pump. And it's a centrifugal pump. In the PC9, there is one in each collector tank. Those are the two fuel tanks in the centre of the plane. You can see from the flange at the bottom of this pump that it's fixed to the bottom of the tank. Yes, Ali?

A: Isn't it dangerous to have an electric pump in a jet fuel tank?

I: Well, no, not really. You see, it is the fuel vapour that is most dangerous, and these booster pumps are immersed in the liquid fuel. Also, they are carefully sealed and they are not used for long periods. Now, the pump that is used for long periods is this third one here, the main fuel pump. You could call it the 'heart' of the fuel system.

Y: It looks like a vane pump.

I: Yes, that's right, that's exactly what it is. It's got rotating vanes in the centre here, which push the fuel through. It's powered by the engine, like the gear pump.

A: What's the valve for?

I: The one on the right, you mean, Abdelhakim?

A: Yes.

I: Ahh, that's to relieve the pressure if it is too high on the outlet side.

A: I see …

I: This vane pump does two jobs. It not only supplies fuel to the first pump, the high-pressure gear pump. It also drives these other two pumps, numbers 4 and 5. As you can see, these two are quite similar. Let's look at the simplest one first, number 4. What do you notice about it?

A: There's nothing in it, just a tube.

I: Right, and pump number 5 is the same, except for this flap valve at one end to make sure that the fuel can only go in one direction. They are both called jet pumps and they work on the 'venturi' principle. Now then, can anyone remind us what that is? Yes, Yusef?

Y: When a high-speed liquid goes through a smaller space it makes a vacuum behind it.

I: That's it. You see the main pump, number 3 here, drives the fuel, in what's called the motive flow, through the main tube of the jet pumps, and then more fuel is sucked up by vacuum. This type, number 4, is used to transfer fuel from the wing tanks to the collector tanks. It's called a transfer jet pump. And number 5, this one with the valve, called the delivery jet pump, supplies fuel continuously from the collector tank.

A: And they work all the time as the plane is flying?

I: They do. No moving parts, not much to go wrong and they can keep working indefinitely, non-stop. So the fuel is always moving from the wings to the centre and keeping the plane properly balanced. Yes, Abdelhakim?

A: Can you show us one of each type of pump?

I: Yes. In fact, we're going to strip a couple down in the workshop tomorrow.

Unit 6

CD2 Track 1: Unit 6, Lesson 4

Part 1

Mr G: Hello, John. How are you getting on with drilling out those two sheets of aluminium?

J: Fine, thanks, Mr Green. We've drilled out most of the holes. I'm doing the outside holes along the edge on the machine here, but we can't do the centre holes on the machine. Frank's doing those with a hand drill over there on the bench behind the guillotine.

J: Frank, Frank … are you OK? You're as white as a sheet. What have you done to your hand?

F: No, I'm not OK … I think I've broken my little finger …. The chu-… aaghhh, the chuck was still in the drill … oooohhh … after I changed the bit, … I didn't notice …

Mr G: You've cut it quite badly as well. Right, let's get the first aid kit and get the finger covered and then off to hospital ASAP. You're going to be off work for a couple of days, even if it isn't broken. I'd like you to go with John. Report back to me when you get back, OK?

J: Right. Shall I …?

F: So stupid … aaaa … I just didn't check … I looked away for a second.

Mr G: OK, OK, we can talk about it later, but now let's get that finger treated. That's the first thing.

Part 2

Mr G: Come in.
Ah, hello, John. How did it go at the hospital?

J: He was quite right, the finger is broken, but it's a nice clean break. The doctor said he'll need six weeks off work though, because he works with his hands. It would be different if he was working in an office.

Mr G: Right … well, we'd better do this form. Just check it with me. So, reference number. Workshop 2, wasn't it?

J: Yes.

Mr G: So, W2.9 – dear me, the ninth accident this year – 06. OK, … full name. Could you just read off his personnel data up on the screen there?

J: Yeah, Frank … no, sorry, his full name's Francis – F-R-A-N-C-I-S. Yes, Francis Robert – I never knew that. Robert … – the surname's Day … address: 245 Bartlett Street.

Mr G: Is that double t at the end?

J: Yes. Bart – B-A-R-T and lett – L-E-double T.

Mr G: Occupation … fitter. Age?

J: 28.

Mr G: And he works directly for the company. Type of injury … fracture. First aid … 'yes'.

J: Yeah, we put a loose bandage round the whole hand.

Mr G: OK … agent of injury … that's clear … and injury site … left finger. Date and time … 26th November, and it was, what … about 11 o'clock?

J: 11:15.

Mr G: Yes, that's right, and you and I were the witnesses, so I'll put my name and address. Now, cause of incident. Hmmmm … I don't think we can put that in until I've spoken to Frank a bit more. Just to make sure we get the facts straight. Now, brief description of accident. I'm going to put 'operative injured by chuck key left in hand drill'.

J: Yeah, that seems to be what happened.

Mr G: Hmm … recommendations … No, again, I'll wait until Frank is back before I do that. Now, date of incapacitation … same as today's. Date of return … can't say yet. I think that's all we can do for the moment. Oh yes, property damage.

J: Well, the alclad sheet he was working on can't be used again.

Mr G: No, you're right … I'll talk to the Accounts Department about the cost of that. Right, that's it for now. If you can just put your name and address in the top right-hand box and I'll send this across to the safety superintendent …

CD2 Track 2: Unit 6, Lesson 8

Mr A: Hello, 201 70973.

Ted: Is that Mr Armstrong?

Mr A: Yes, speaking.

Ted: Oh, hello sir. It's Ted from Sankey's Garage here. Just ringing to let you know your car is ready for collection. We've done the full 20,000-mile service as you asked.

Mr A: Everything OK?

Ted: Yes, all pretty good, really. I'm just looking at the job sheet now … Ah, yes … except for – you said the brakes didn't feel right … Yes, we had to do quite a lot on those as it turned out.

Mr A: Oh, so that's going to be quite expensive.

Ted: Well, I'm afraid it won't be cheap. But the rest of the service was fine.

Mr A: So, what was the problem?

Ted: There were a couple of things, actually. First of all, there was a leak in the master cylinder, brake fluid was getting out and air was getting in.

Mr A: So, did you repair it?

Ted: Oh no, now it would be expensive to repair the old one. No, it's cheaper to replace them. So we did that … we also found a slight leak in the fluid line to the left-hand front brake. So, obviously we had to empty all the fluid out of the system to replace the damaged tube.

Mr A: And you refilled it with new fluid.

Ted: Yes, and I'm afraid that we also found the surface of the disc on the right front wheel was damaged – it was slightly out of alignment.

Mr A: Why was that?

Ted: Hard to tell. It's usually caused by excessive heat, you know, if the brakes are having to do too much work. When we inspected it, it looked as if – have you been doing any hard or fast driving recently?

Mr A: Yes, we spent the week before last driving in the mountains. That's when I noticed there was the vibration and the car was pulling to one side when I pressed the brake pedal. You have to do a lot of braking on those roads.

Ted: Sure. Yes, I should think that front wheel brake unit was doing most of the work because the other brake wasn't getting sufficient pressure. Anyway it's all fine now. We machined the damaged disc surface and straightened it again. We've checked and cleaned the other three discs and they're all OK, so it's ready for you to collect and working perfectly.

Mr A: Mm ... that's the good news. What's the bad news?

Ted: Ha ha ... well, yes, the main service was the standard £120 plus parts, but then the extra work on the brakes is going to ...

Unit 7

CD2 Track 3: Unit 7, Lesson 2

1.

A: Do you know where Jack is? I've been waiting for him for 20 minutes.

B: Last time I saw him, he was going off to wash his hands. He's been doing some painting ... Oh, here he is now. All right, Jack? Why the bandage?

J: I've scalded myself with that water. I should have put some cold in first. It serves me right for ignoring the notice! Lucky it's only one hand.

A: You won't make that mistake again in a hurry.

2.

Foreman: How's that pump running now, Barry?

Barry: Fine, I've replaced the front main bearing, so I'm just going to give it a bit more lubrication, clean it up and then it's ready to refit.

Foreman: OK ... but make sure it's switched off when you do it! We don't want another accident like last week.

3.

A: Phyeeugh ... look at it, this overall's ruined and I've got some down my neck, I ... can feel it. Ugh, its all sticky.

B: And some on your face and hair, too. Look at the state of you. What happened?

A: The sump plug fell out while I was underneath. It's not funny!

C: It is from where I'm standing!

B: Well, you're completely covered in the stuff. Off you go to the washroom and clean yourself up thoroughly, and make sure you change that overall!

4.

A: Ah ... yes, I think this is the right stuff for cleaning all that paint away.

B: Better check first, though. What does it say on the label?

A: Hmmm ... harmful, do not swallow. Only use in a well-ventilated area. Wash hands immediately after use. No smoking when using this product. Do not breathe fumes. Keep in a cool place well away from sources of ignition ... Here we are ... To use as a paint solvent, apply to a cotton cloth.

5.

A: Have you finished the block yet?

B: Yes, I'm just going to finish it off on the grinder ... get it nice and shiny and smooth.

A: OK, but make sure you don't try to use the grinder over there. There's no protection on it at all. It got broken yesterday afternoon and hasn't been replaced yet. In fact, I'll go over and remove the wheel so it can't be used.

B: Might be an idea to disconnect it from the power supply as well.

A: Yep ... I'll do that.

6.

A: So, this is where the new milling machines are going to go.

B: Yes, they're being installed next week ... once this place has dried out.

A: Hmmm ... yeah, it's a pretty strong smell. They could do with a few more windows open. So what's going to go in these cupboards, do you know? Look out, mind that door.

B: Oh no! This is a new jacket. There should be a sign here.

CD2 Track 4: Unit 7, Lesson 4

A: Hello and welcome to *I Could Do That*, the programme in which we give young people a chance to ask questions about interesting and unusual jobs, and today in the studio my guests are Martin Robbins. You're how old, Martin?

M: 12.

A: … and Terry Gardner from the Civil Aviation Authority. I won't ask you how old you are, Terry.

T: Thank you.

A: Now you are, I think I've got the title right, a civil aviation operations cabin inspector.

T: That's correct.

A: And can you give us an idea of what you do, in a few words?

T: Well, as the job title suggests, I inspect aircraft cabins and make sure that the correct procedures are followed by the crew before, during and after a flight.

A: Right. Now I think it's over to you, Martin. Let's have your questions.

M: Can you tell me – what exactly do you actually inspect?

T: Most of the things that you would expect to find inside an aircraft really.

M: Like the engines and the cockpit and stuff?

T: No, no, I'm not an aircraft engineer. Those sorts of thing are checked by the technical maintenance people. No, my main priority, the most important thing, I mean, is the safety of the aircraft passengers – making sure that everything in the passenger cabin is OK and that the cabin attendants do all the right things throughout the flight. For example, one of the first things I check when I go on board is that there are legible signs in all the correct places. You know the sort of thing – no smoking, exit signs, emergency equipment. To make sure the signs are there and that everyone can read them easily, especially in case of an emergency.

M: What sort of emergency equipment do you look for?

T: Well, every model of plane is different, of course, there's a different specification for each one. But I need to make sure that there are the right number of fire extinguishers, that they are in the right place, and that they have an up-to-date service tag.

M: What's that?

T: The service tag? That's a little label which tells you when the equipment was last checked. And of course, there's a lot of other safety equipment. Aircraft – or airlines – are not allowed to fly if they don't have all the right safety equipment. Things such as oxygen bottles and oxygen masks – protective breathing equipment in case there is smoke in the cabin. It's important to check that they drop correctly down from the ceiling. Then there's first aid kits … and the passenger information cards, of course … the flashlights – the torches – I need to make sure there are enough of them – in case the lights go out and, of course, I check that the batteries are charged.

M: Are the information cards the ones in the back of the seat with the pictures on, with the magazine and the sick bags?

T: Yes, that's it, and those are important, those sick bags, especially if there's a lot of turbulence – you know, if it gets bumpy – for everyone's comfort, not only the person who feels sick, and to keep the cabin clean.

M: Yuck. Is that the lot?

T: No! There are dozens of other things … Can you think of anything I might have missed?

M: What about life jackets?

T: Well done. I have to check that there's one under each seat. And again I check the condition … oh, yes, of course there has to be at least one megaphone.

M: In case the intercom on the plane doesn't work.

T: Yes, for that, but mainly for use if the plane has to be evacuated and the crew need to give instructions to the passengers.

M: So after you've checked everything is there, the plane takes off?

T: Not quite. I inspect the cabin before the passengers come on board. And before they do, there's another very important thing I have to check. I can't really do it with the passengers on board – any ideas?

M: Uhh … mmmmmm … oh, I know – the seats.

T: Exactly. I check that they're working, that they can be adjusted up and down, and that the seat belts work properly and that they're not twisted or frayed. I also check the cabin crew seats and their safety harnesses as well. There really is a lot to look at. Then, when I'm satisfied that everything's shipshape, the passengers are allowed to board.

M: Do you sometimes go on a flight without the crew knowing that you are there?

T: Well … actually, I'm afraid that's confidential. It's the one thing I'm not allowed to tell you.

A: Ha ha. I think we can take that as a yes. OK, now perhaps you could tell us a bit more about the next stages in the inspection.

T: Well, as the passengers come on board, I make sure that the cabin crew show them to …

CD2 Track 5: Unit 7, Lesson 8

During my last two years in the US Navy, I served on board an aircraft carrier as a Quality Assurance Representative. For those of you who don't know, the job of the QAR is to keep an eye on the standard of the maintenance, servicing and repair of aircraft.

When I first joined the ship, I was pleased to discover that I wasn't going to be dealing with LOX converters. I don't know if you know, but these are a specialised kind of cylinder assembly designed to store and convert liquid oxygen into gaseous oxygen for aircrews during flight. Hence the name LOX: liquid oxygen. Well, there were a lot of maintenance problems associated with that system, and I was glad that the Hornet fighters based on the ship were all in fact fitted with the new OBOGS oxygen system. OBOGS stands for Onboard Oxygen Generating System, and it was known to be much more reliable. It works by concentrating oxygen from the air during flight and supplying it to the pilot at high altitude: there is no need for the aircraft to carry the oxygen supply.

Unfortunately, a few weeks after I joined the ship, some of the pilots began to report experiencing symptoms of altitude sickness – such as nausea – on a couple of the Hornets, and the OBOGS system was obviously the first thing that was looked at. I consulted all the relevant maintenance manuals carefully, read about similar incidents, and of course questioned the pilots who'd reported the problem. One of the ship's technicians inspected the OBOGS systems and found that on deck it was working perfectly. So, knowing the system worked fine on the ship's deck, but not at altitude, we decided to test the concentrator to see if it deteriorated at high altitude. But no, it was fine. Absolutely no defects.

However, some time after this, one of the aircraft mechanics came to see me and showed me an OBOGS line he had removed. A closer look revealed the hose's exterior metal braided sleeve had separated from the plastic tube inside, which meant that we could see the true condition of the internal hose. We could see that it was almost completely closed off, blocking the free flow of oxygen. The damage was hidden from casual examination because of the protective metal sleeve on the hose. We quickly recognised this was likely to be the source of the problem, so I then went to maintenance control and asked to inspect all the other aircraft. We wanted to determine if this was an isolated case or a general problem. After inspecting 12 aircraft, I found nine OBOGS lines bent or crimped in some way, possibly causing low oxygen.

We now tried to find out the reason for the crimping. I followed our mechanics out to an aircraft and watched them remove an OBOGS concentrator. I quickly discovered that their technique didn't appear correct. The service manual said to move the product line to the side, which allowed space to remove the concentrator. However, this technique resulted in the mechanics treating the OBOGS lines like the old LOX lines: they just moved them out of the way quite carelessly. The difference was that the LOX lines, which we could move aside quite easily, were very flexible, whereas the new OBOGS inner lines were made of a much more rigid plastic. So when the mechanics moved the OBOGS lines quite forcefully, they were getting crimped – and they were crushing the inner plastic hose. The mechanics thought that the line was as tough as the LOX line and couldn't be damaged.

The result of the crimping was that there was some back pressure within the OBOGS concentrator. So it formed a blockage – which reduced the amount of oxygen going to the pilot. The reduced amount of oxygen didn't affect a pilot's performance until the aircraft reached very high altitudes, which is why the system seemed fine on the ground. A final piece of this puzzle could have resulted in death! In those early days, the OBOGS oxygen monitor only detected the oxygen concentration within the concentrator, which meant that a warning light would not go on because the level inside the concentrator was normal. An oxygen monitor didn't exist downstream, so the pilot never got a warning of the oxygen decrease.

Well, what we learned from this was to stop treating OBOGS lines like LOX lines and to be much more careful with them when they were moved. And I'm sure we saved the lives of a few pilots as a result.

Unit 8

CD2 Track 6: Unit 8, Lesson 1

A: Right, let's have a look at this then.

B: So why did they request a tyre inspection? We don't normally do it every time the plane lands, do we? Usually one of the crew does a quick visual.

A: No, I know, but apparently she had a bad landing because of these sudden crosswinds. Came down with quite a bump. And skidded a bit more than usual. That can sometimes do a lot of damage when the plane comes down quite hard.

B: OK, so what's the main thing to look for?

A: We'll do it by the book. Have you got the tread gauge there?

B: Here you are.

A: Thanks. Hmmm .. Right, let's see. Just put it in the grooves – there … and there … and there. Yep, that's good. There's plenty of wear left on these treads. So let's just stand back and look at it from the front. Ah yes, small problem there, can you see?

B: Oh yes, the treads look a bit more worn on the right-hand side than on the left. Does that mean replacing it?

A: Not necessarily. If there's no other problem, we can just flip it.

B: Flip it?

A: Demount it and turn it round so that the wear is evened out. But we'll need to report it, so that the nose wheel gear is checked.

B: So that might be the cause of the problem. Gear misalignment.

A: Yes, and it's probably causing a bit of nose wheel vibration as well. Right, now we need to look for anything stuck in the tyre. No … looks OK. Can you see anything? Any foreign matter?

B: No. Looks all clear. What happens if you find something stuck in the tyre? Do we try to remove it?

A: Not while the tyre is inflated, you don't. You could burst it and really hurt yourself. No, if there's anything noticeable, it has to come off and be repaired in the workshop. Now, let's have a look at the beads and the sidewalls.

B: What are we looking for?

A: The obvious things really – heat damage on the beads, they're the hottest parts of the tyre. Cracking or bulges.

B: Like this, you mean?

A: Ah yes. Well spotted. It's only about the size of your thumbnail, but it's a definite bulge all right. Well, that's it then, off it comes. The shop'll have to inspect it, no question.

B: But it doesn't look too bad.

A: Not yet, it doesn't, but it's a bad sign. It means that there's some damage or separation between the layers. Every time it's used, it'll get worse. Right, I'll let the flight crew know what's happening and then we'll get started.

CD2 Track 7: Unit 8, Lesson 3

F: Good morning!

B: Mr Sturgis?

F: Yes, that's right. We've got a flight booked for today – it's my son's birthday.

B: Hello, James. Happy birthday. I'm Bob.

J: Hello.

F: I've been promising him a balloon flight for ages – a couple went over our house a few months ago and he hasn't stopped talking about it since.

B: Yes, they're a pretty impressive sight. It's the size of course – and the colourful designs.

F: How big are they, actually?

B: Well, the one we're going in today is 10 meters across and about 20 meters high, complete with the basket and gas burner.

J: It doesn't look very big.

B: Not yet, but you wait till it starts to fill. Ah, there's my co-pilot Simon. He's just going to start filling it. He'll start with a petrol-engined fan to start off, and then when the mouth of the balloon is open a bit wider, he'll light the propane burner.

J: Look Dad, that blue one's starting to go up. Wow, it's enormous!

B: Now, while Simon's inflating the envelope, the main balloon, are there any questions you'd like to ask?

F: How does it actually work?

B: Well, I don't know if you remember something from school science called Charles' Law.

F: Hmmm … vaguely. Science wasn't really my strong subject at school.

B: Right, there goes the burner. Shouldn't be too long now. Yes, Charles was an 18th-century French scientist who made the discovery that the volume of a gas is directly proportional to its temperature, so his law states quite simply that if you increase the temperature of a gas – which is what young Simon is doing over there – then the volume will increase, as you can see starting to happen.

J: Oh yes, look Dad, it's starting to fill and move up. It's nearly upright.

B: That's because a certain amount of hot air occupies a larger volume than the same quantity of cold air. And since it occupies a larger volume, it must be less dense than the same amount of cooler air. And less dense means lighter. Yeah? Does that make sense?

F: Um.

B: So, when the gas in a balloon gets hot enough, the weight of the balloon with this hot air is less than the weight of an equivalent volume of cold air, and the balloon starts to rise. Yes, another ten minutes or so and we should be able to get on board. The four of us'll fit in the basket.

B: Right. Everybody ready?

F: Yes.

J: Yeaaah.

B: Good. We just release the ropes that hold the basket down, and – off we go.

F: Woah. It's so strange. The ground is getting smaller, but it doesn't feel like we're really moving.

J: All the people are getting smaller.

J: This is fantastic. Dad, have we got to come down yet?

F: I'm afraid so. We only booked a half-hour trip.

J: How do you make it go down again?

B: Well, of course, I stop using the burners. That way, the air in the balloon cools down, so it contracts and becomes more dense – heavier, if you like, and so we start to sink. Exactly the reverse of the process that made us rise.

F: How long does that take?

B: Well, it depends on the difference between the temperature inside and outside the balloon. The bigger the difference, the longer it takes. But I can speed things up a bit if necessary if it's taking too long. If I pull this cord, you can see that flap at the top of the balloon opens – and it lets some of the warm air out, so the volume and the temperature are both reduced suddenly, so we'll start descending more quickly.

J: I can hear the birds again.

B: And here we are.

J: Fantastic. Can we come again another time? Please Dad.

F: Well, I'll have to think about that. It's not cheap, you know.

CD2 Track 8: Unit 8, Lesson 5

Ali: Excuse me.

Mr B: Yes, Ali?

Ali: I'm looking for information about heat pumps. Could you tell me something about how they work?

Mr. B: Of course. How much do you already know?

Ali: Well, I know that a heat pump transfers heat from one place to another ... but the process isn't very clear to me.

Mr. B: Well, OK, basically a heat pump uses the fact that when a liquid evaporates, it needs to absorb heat from its surroundings, and when a vapour condenses, it gives off heat. The best example is an ordinary domestic refrigerator. The inside is cold, but the grill at the back is warm. The heat is removed from inside the refrigerator and then given off into the room. Which is why you can't cool down a room by leaving the refrigerator door open.

Ali: Because when you leave the door open, the refrigerator works hard to try to cool the room, and at the same time, it's giving off the same heat back into the room, heating the place up.

Mr. B: That's right. The heat can't just disappear. It's transferred from one place to the other – and the room doesn't get any cooler. You just waste a lot of electrical energy.

Ali: But I'm not sure what the main parts are, or how they are joined together.

Mr. B: OK, well, in a refrigerator, or any heat pump, you need a compressor, an evaporator, a condenser, a control valve and a special fluid called a refrigerant. Look, there's a drawing in the book here.

Ali: The evaporator must be the part where the refrigerant changes from a liquid to a vapour.

Mr. B: Yes ... here it is – and remember that it needs to absorb heat to change from the one state to the other, from liquid to gas, and it absorbs the necessary heat from inside the refrigerator, and the inside of the refrigerator gets colder.

Ali: So the condenser is where the refrigerant changes back from vapour to a liquid, and gives off the heat again.

Mr. B: Yes, that's right. It sort of gives the heat back – in a different place – at the back of the fridge. The heat from inside the fridge ends up in the room.

Ali: OK. But what about these valves on the pump? Are these valves V1 and V2?

Mr. B: Yes. These valves operate alternately. When the piston in the pump moves inwards, on the compression stroke, valve two opens and valve one closes.

Ali: Is it correct that when the piston moves inwards, the vapour is compressed into a liquid and forced through valve two into the condenser?

Mr. B: Yes, it is, and on the induction stroke, when the piston moves outwards, you get the opposite – valve one opens, and valve two closes. So the pump keeps the refrigerant circulating by opening and closing these valves. Of course, the refrigerant has to be a special fluid that can change from liquid to vapour and back again quickly and easily.

Ali: And what does the control valve do?

Mr. B: OK, think about the pressures in the evaporator and the condenser. Which will be higher do you think?

Ali: Ah, the pressure will be higher in the condenser than in the evaporator, because it's gas in there.

Mr B: Exactly. And it's the control valve, sometimes called an expansion valve, which maintains the difference. Basically, it's a small orifice, a small hole, which forces the liquid to change into a vapour in order to flow through.

Ali: So it's too small for a liquid to go through, but big enough to allow a vapour through?

Mr B: That's it.

Ali: Hmm.

Mr.B: Is it all a bit clearer now?

Ali: Yes ... Yes. Thank you very much.

Mr B: You're welcome.

CD2 Track 9: Unit 8, Lesson 5

Well OK/ basically a heat pump uses the fact that when a liquid evaporates,/ it needs to absorb heat from its surroundings/ and when a vapour condenses,/ it gives off heat./ The best example is an ordinary domestic refrigerator./ The inside is cold,/ but the grill at the back is warm./ The heat is removed from inside the refrigerator/ and then given off into the room,/ which is why you can't cool down a room by leaving the refrigerator door open./ Because when you leave the door open,/ the refrigerator works hard to try to cool the room/ and at the same time it's giving off the same heat back into the room,/ heating the place up.

CD2 Track 10: Unit 8, Lesson 9

OK, I think we're ready to start. OK, what I'll do is to give you an overview of the system and then go back into it in more detail, with some numbers and statistics.

OK, so, pressurised air for the cabin comes from the compressor in the aircraft's jet engines. Moving through the compressor, the outside air gets very hot as it becomes pressurised. The part drawn off for the passenger cabin is first cooled by heat exchangers in the engine struts and then flows through ducting in the wing. After that, it's further cooled by the main air conditioning units under the cabin floor.

The cooled air then flows to a chamber where it is mixed with an approximately equal amount of highly filtered air from the passenger cabin. The combined outside and filtered air is taken to the cabin and distributed through overhead outlets.

Inside the cabin, the air flows in a circular pattern and exits through floor grilles on either side of the cabin or, on some airplanes, through overhead intakes. The exiting air goes below the cabin floor into the lower lobe of the fuselage. The airflow is continuous and quickly dilutes odours while also maintaining a comfortable cabin temperature.

About half of the air exiting the cabin is immediately exhausted from the back of the airplane through an outflow valve in the lower lobe, which also controls the cabin pressure. The other half is drawn by fans through special filters under the cabin floor, and then is mixed with the outside air coming in from the engine compressors. These high-efficiency filters are similar to the filters used to keep the air clean in hospitals. They are very effective at trapping microscopic particles as small as bacteria and viruses. It is estimated that between 94 and 99.9 per cent of the airborne microbes reaching these filters are captured.

OK, now as I said, some more detailed numbers …

Unit 9

CD2 Track 11: Unit 9, Lesson 3

T: Mr Martin?

M: Hello. Yes.

T: How do you do? I'm Gerry Townsend. I'm the inspecting technician for the electrical system in your plane. I've just been having a look.

M: Oh right, how was it? Everything in order, I should think.

T: Well, yes and no. 95% is OK, but I noticed that you've recently had some new equipment installed by another company.

M: Yes, that's right. Last month, I had a new radio unit fitted. Works very well, so they must have done a good job, surely?

T: Oh, everything is connected up the right way. But to be honest with you, I'm not really happy about the standard of the wiring. There are a few things that could be a safety problem.

M: Really?

T: Yes, I'm afraid so. I can go through it with you here, but it'd be better if we went out to the plane so I can point things out to you.

M: OK, fair enough, let's go. I'll just get my coat. It's a bit nippy out there.

T: Right here we are then. Let's go point by point. Now, they have used the right cable, no problem there. You can see from the identification number marked along the outside sheath … this number tells me the size … the cross-sectional area of the conductor, and this number at the end tells me it's stranded copper with vinyl insulation, and that's correct for equipment with this sort of power load.

M: Goooood, so far.

T: And I'm quite happy about the route of the cables. They're tucked out of the way, so nobody's likely to walk into them or hang on to them by mistake. They've also remembered to fit them above the hydraulic fluid line, just in case there's a leak. So that's all fine.

M: OK, so what is the problem then?

T: Well, if you look at the way the connection has been made to the terminal block …

M: Looks very neat and tidy.

T: … yes, and that's actually the problem. They haven't left a loop. There should be an extra six-inch diameter loop of wire to allow for any sudden tension in the cables. It also makes everything much easier to take out and put back if there's extra wire available.

M: Oh dear.

T: These are also the wrong type of lugs to connect to the terminal block. They should be the closed-ring type, not these open-ended ones – these can slip out too easily if there's a lot of vibration.

M: Yeah, that makes sense.

T: Now the wires have been passed through this hole at the side and that's fine – the hole is plenty big enough. But there's no grommet to stop chafing.

M: And what about where it comes through the other side? That clamp looks a bit too tight to me.

T: Yes, you're right. In fact, it's actually pinching one of the cables. See, it could eventually cut through the sheath and into the insulation sleeve, and look here where the cables are hanging down between the clamps. There's far too much slack in there. I can move it up and down at least a couple of inches – should only be about half an inch at most.

M: So what needs to be done?

T: I'm afraid it'll have to be redone with new cable.

M: Can't we use the existing stuff?

T: No, there isn't enough slack to make a loop at the end.

M: OK, if you say so. Is there anything else?

T: Well, just a couple of points about this new plastic conduit over here.

M: Oh yes, I had the old one replaced when the radio was put in.

T: Right, well, it's OK except for two things. First, these drainage holes at the bottom are a bit roughly finished – there are quite a few burrs on them. And the other thing is the position. I think it could be moved further back. At the moment, it's sticking out too much. I can see your passenger grabbing hold of it by mistake.

M: So is that difficult to fix?

T: No. I'd say it can be removed, the edges of the hole smoothed, and then repositioned in a safer spot, no problem.

M: Oh well, that's something, I suppose. How much do you think it's likely to cost?

CD2 Track 12: Unit 9, Lesson 5

B: Hello, this is Bill Williams and welcome to the 12 o'clock news spot on Radio 310, where local news comes first. Well, it was a very unlucky day for several people out at Wageroo Airfield this morning when Mike Grigson lost control of his plane on the ground and it went charging off all over the tarmac, finally ending up in the hangar, where the air taxi just happened to be in the way. And we've put a picture of it on the website if you can bear to see the results. How did he lose control, you want to know. Well, it seems Mike wasn't actually in the plane at the time – you couldn't make it up if you tried. Mike's on the line right now in fact. G'day, Mike?

M: Yeah, hello Bill.

B: How are you feeling about all this?

M: Er, most of all, I'm just so relieved that no one was hurt. And of course, I feel like a complete idiot. It just goes to show you're never too old to make silly mistakes.

B: Tell us what exactly happened.

M: Well, I was taking my young brother up for a ride this morning – lovely flying weather it was. We've got a two-seater single-engined thing. Light aircraft …

B: It's a beautiful day, yeah.

M: … yeah, and … but, the engine wouldn't start on the electric starter. I'd run a couple of electrical checks and discovered it was the battery. I hadn't been up for about six weeks, so the battery had self-discharged – it was flat, there wasn't enough juice in it to turn the engine over. It's been really hot the last couple of weeks and that makes them run down faster … my own fault really. I should have taken the battery out yesterday and charged it up overnight. There's a good charging set in the hanger. Anyway, I hadn't, so …

B: Right, so you decided to try a bit of hand-propping then?

M: Yeah, I mean it's not ideal, but without an EPU, an electric, an external power source – um, unit – and once you get the engine started, the electric starter motor turns into a generator and starts charging the battery up very quickly. So when you start the engine next time – no problem. Yeah, so I left my brother in the cockpit in charge of the brakes and gave the propeller a pull.

B: He's a qualified pilot, too, is he?

M: Well, no, he's had a few lessons, but he hasn't got a licence.

B: OK, so it was always going to be a risky procedure then.

M: Yeah, well it is, you just grab the prop and swing it round by hand to turn the engine over.

B: You got the engine started though – obviously?

M: Yeah, first time. I think that was the problem actually. Tom wasn't expecting the engine to start so quickly. He got a bit scared and forgot what I'd told him and basically lost control, and I just jumped clear and he taxied off into the hangar. Next thing I heard was this terrible grinding noise.

B: Which was your plane running into and chewing up the other one?

M: Yeah. Then the engine cut out and there was this awful silence.

B: And nobody hurt – that's amazing really.

M: Yeah, only my bank account. It's going to cost me a lot to pay for all the damage – and my pride.

B: It does look a bit like a sliced loaf in the photo. Nice plane, too.

M: Don't remind me. Twin-engined four-seater. Rear fuselage and starboard wing …

B: So have you got any advice for any light aircraft pilots who might be listening?

M: Yes, two things. First, cancel your flight, rather than hand-propping. It's not worth the risk. And second, get an alternator fitted to your plane if you haven't got one.

B: Why's that?

M: They're much better at charging the battery up fully than standard generators like I had.

B: Oh?

M: Well yes, without getting too technical, a generator can charge up the battery, but alternators are better because they charge it up even when the engine's just ticking over. You're less likely to have starting problems with an alternator.

B: OK. Well, I'm no mechanic but that sounds like good advice. Mike, best of luck with all that, thanks for talking to us. Mike Grigson – not a lucky man. On now to a story from the other end of the district …

CD2 Track 13: Unit 9, Lesson 5

a) The engine didn't start on the electric starter.

b) It's been really hot the last couple of weeks and that makes them run down faster.

c) The electric starter motor turns into a generator and starts charging the battery up.

d) Get an alternator fitted to your plane if you haven't got one.

e) I hadn't been up for weeks so the battery had self-discharged – it was flat.

f) We've got a two-seater single-engined thing. Light aircraft …

CD2 Track 14: Unit 9, Lesson 7

1: This figure shows the actual layout of the copper conducting tracks on a printed circuit board, with the holes for component leads to be inserted, viewed from the underside of the board. It is used in the production of the printed circuit board.

2: This figure shows the actual components viewed from the top of the board, as well as indicating the track routing on the underside of the board, as if the PCB was transparent. It would be used by service technicians in identifying the location of components.

3: This figure is a three-dimensional picture of an actual circuit board, showing the components used for this circuit.

4: This figure shows the electrical connections and component values in a way that makes it clear how the electronic system operates. It is used by technicians troubleshooting the board.

5: This figure shows the system as a whole broken down into its main functional sub-systems or blocks. This indicates the way that the sub-systems relate to the whole and gives a simplified idea of the operation of the system.

CD2 Track 15: Unit 9, Lesson 8

A: Jacksons …

B: Oh, hello … I'm just phoning to see if I can order some things from your catalogue, but I need a bit of advice if you can, you wouldn't mind explaining a few things.

A: We can't take orders on the phone, I'm afraid, but I'll certainly try and give you any help you need.

B: Oh, I see. So how do I go about ordering then?

A: Online. Or by post. Do you need a copy of the catalogue?

B: No, I've got the website up on the computer here.

A: Oh good, that's fine. Well, if you click on 'products'.

B: Yep … OK, done that.

A: Now, you've got two options – 'stock list' and 'images'.

B: Right, so 'stock list'?

A: No, 'images' is a better bet if you're not sure exactly what you want.

B: OK, right, I've got a picture of each item with a number next to it.

A: Yeah, that's it. And to order, you just double-click on the picture and it goes in a basket. You fill in your details at the end.

B: OK, well that's easy enough.

A: So, how can I help?

B: Well, I've got to do some electrical work on a small second-hand plane. I've got some general tools, but I'm going to need a few more to do the job properly.

A: OK, what kind of current and voltage are we talking about?

B: It's 28 volts DC for the heavy work, plus an AC system for the avionics.

A: What voltage?

B: 115.

A: OK, well for a start, I'd recommend a voltage tester. 115 volts can give you a pretty nasty shock. Before you touch anything, you can just whip it out of your top pocket and check. They're dirt cheap and it could save your life. We do them for 115 volts, as well as 220-volt mains.

B: OK, so double-click on that – the red one, yes?

A: That's it … and while you're at it, I'd recommend you get a cheap analogue multimeter, too, if you don't already have one. They aren't quite as accurate as the good digital ones, but they're very simple, and you can always buy another one if you lose it. They're so cheap. We do the small blue one there. It comes with a separate set of leads.

B: Good idea, yeah, I can borrow a decent digital one if I need to. OK, and I need a good-quality tape measure – mine hasn't been the same since I stepped on it.

A: I suppose not. Well, we do a wide range, but I'd go for the combi – you can read off the tape directly or use the digital reading at the side.

B: Oh, yes, I see it. Good, yes. One of those.

A: They are very damage resistant, too.

B: OK.

A: And remember the old saying.

B: What's that?

A: Measure twice …

B: … cut once, yes. Absolutely. Right, now let's see. Yes, soldering.

A: Are you going to be doing much?

B: Well, as far as I can tell from the manual, it's just a few joints. Most of the connections are mechanical.

A: Nuts and washers, crimped terminals …

B: Yes, that's it.

A: OK, well, you're probably OK with the simple standard one then, the small one with the red handle there, and you really need the safety stand just to avoid accidents. You'd be amazed how many happen like that. Plus a reel of solder wire.

B: Right, there we go. Erm … actually, I don't know. I think I might be better off with this solder gun next to it. Is that a set of different-sized bits with it?

A: Yes – it does give you that flexibility – probably a bit safer, too, for working inside a plane.

B: Right, so do I just unclick those two? OK, right – and … click on that one.

A: For the terminal nuts, if you've got a lot of them to tighten I'd consider a socket set. It'll save you hours of work and you get a good close contact.

B: I was going to try using the ones I've got for the car. What do you think?

A: No, too big. I'd recommend the small set we do. That'll cover all the sizes you need.

B: Right, OK, got that. Now, I've got an ordinary pair of pliers, but …

A: Are you going to be working alone?

B: Well, yes, most of the time. Why?

A: Then, I'd really urge you to get some locking multi-grip pliers. It's like having a second pair of hands. You can grip something and then free both hands to work on it. They're brilliant – I use them all the time.

B: Are they the thing that looks like a tin opener with black jaws down at the bottom?

A: That's it. And I'd recommend the single-side cutters and the long-nose pliers. They're both vital really, and the single ones are better quality than the ones that come in sets.

B: OK, so that's these three 1 … 2 … and 3. OK, next on my list is a really good crimping tool. I'm going to be doing a lot of that.

A: Do you see the one bottom left with the orange handles? That's your best bet. It's pretty much the one everybody uses. People swear by it – lasts for years. And if you need wire strippers, I'd go for the automatic ones – they're just next to the crimper.

B: So those … and those. Now, what about cable and conduit work? Some of it's quite specialised.

Unit 10

CD3 Track 1: Unit 10, Lesson 2

A: OK. So you've been through the controls and you seem happy with that. We can run through them again if you like.

B: No … no … I think I've got it, thanks.

A: Good, good … OK. So now, before we actually go up in the air, I need to go through the flying instruments with you.

B: Right. There aren't very many, are there? I thought there would be more.

A: Well, you can have more, of course, a big airliner has hundreds – but these are what we call the Basic 6, plus this magnetic compass over here. In a small plane like this you can fly perfectly well with just these.

B: Right.

A: Now the first thing to say is that what you see, the layout of the panel's likely to be the same in any small plane you fly.

B: So it's easier to adapt if you change to flying a different plane?

A: Exactly. And then the second thing to realise is that part of the skill in being a pilot is remembering to check all your instruments continuously. Just the same as in a car you keep looking at the mirrors, the speedometer then the road and other traffic and back to the mirrors, and so on. You can't look at an instrument and then switch off.

B: OK. So that's another good reason for having the same configuration in different planes.

A: Yep. It makes it less likely you'll mistake one instrument for another. Anyway, let's start on the top right here. That's the altimeter. And what this is, basically, is a barometer marked in thousands of feet rather than inches of mercury. It measures the air pressure and that tells you how high above sea level you are flying.

B: How does it work?

A: Well, it just weighs the amount of air above the plane. Obviously, the higher you go, the less air there will be, so the weight of air – the air pressure – will be less.

B: Well, that seems pretty clear.

A: Hmm. The key thing to remember is that the altimeter has to be set before you take off; local sea level air pressure can change from day to day, so all the pilots in that area need to have their instruments set to the same correct sea level pressure.

B: So if I fly into a different airspace, I might need to recalibrate it.

A: Yep, you've got it. Air traffic control will keep you informed on that.
       OK, now this one in the middle, this is the attitude indicator … and that line is the artificial horizon. If you're in clouds or for some other reason you can't see out of the cockpit, well, this can be a life-saver.

B: So the centre line here is the horizon?

A: Yep, that's it. And when that white bar is below it, that means your nose is pointing down, and when it turns clockwise or anticlockwise, it shows the position of the wings relative to the horizon.

B: And this one on the left is the airspeed indicator?

A: Yep. It measures the pressure of the moving air pushing against the front of the plane, and that tells you how fast you're going through the air, which is important because if you go too slow you'll stall, the plane won't stay up, and if you go too fast you'll damage the plane – you'll shake it and break it.

B: Wooah … so it's pretty important then.

A: Certainly is. OK, now … this one here at the bottom on the left …

B: You mean under the air speed indicator?

A: Yes, that's right. It's the turn coordinator. You can use this as a back-up if the attitude indicator fails. But it's really for checking that you're turning the plane properly so that you don't slip down sideways when you bank to the left or right.

B: Sorry, I don't quite get that.

A: Don't worry, I'll give you more explanation when we're airborne if it's still not clear. It'll make more sense then.

B: And this one in the middle looks a bit like a compass.

A: Yeah. It's the direction indicator, it's a gyro compass, in fact – uses an internal gyroscope instead of a magnet. Magnetic compasses aren't very accurate when you're changing direction and speed or you have rough weather. When you're back on a straight level flight, you check it against the magnetic compass and recalibrate it with this knob here.

B: OK, I see.

A: And then finally, this last one on the right is the vertical speed indicator. This measures the aircraft's rate of climb or descent. In other words, it tells how fast you are going up or down. It measures the pressure change.

B: Like the altimeter?

A: Yes, that's the idea. Now, have you got any questions before we take off?

B: No, I'm pretty clear, I think.

CD3 Track 2: Unit 10, Lesson 6

a)   The layout of the panel is likely to be the same in any small plane you fly.

b)   You can't look at an instrument and then switch off.

c)   How does it work?

d)   You'll shake it and break it.

e)   This measures the aircraft's rate of climb or descent.

CD3 Track 3: Unit 10, Lesson 6

P: Right now you've each got a circuit board in front of you and each one has a different fault. I want you to trace the faults. It's the same circuit for each board by the way, but as you can see …

C: Yeah, mine looks a real mess. I'm not surprised there's a fault. Whoever made this up had his eyes shut, I think.

P: You're right, it's all over the place. And that's one of the most important things to remember when you're making up a board, or repairing one

– keep everything neat and tidy, components either parallel or at right angles to each other – much easier to work with. Now, there are basically two kinds of fault that we find on PCBs. Any suggestions? Rasheed?

R: One of the components, or maybe more, maybe don't work properly and, the other problem, sometimes the board not working right.

P: How do you mean?

R: Well, copper path can get broken.

P: Yes, that's right. If the board's been mishandled or is put under a lot of stress or vibration, then it can crack. You can get the same problem if a component hasn't been soldered onto the board properly – the component may be fine but the electrical connection isn't secure.

C: I bet that's the problem with my board.

P: You might well be right, Carlos. We'll see in a minute. But one of the problems with insecure connections like this is that the fault is often intermittent.

C&R: Intermittent?

P: Yes, that means it comes and goes, so sometimes the equipment works fine, sometimes not.

R: Well, how can you find the fault then, if the board is working fine?

P: The first thing is a visual inspection. Look at the board and components under a good strong light. Use a magnifying glass as well. If there aren't any obvious cracks or bad connections, the best way is to connect up the board to a multimeter and then press down with a pencil on the component side of the board in different places. If the board itself seems OK, then put some very gentle pressure on each component. That usually works.

C&R: Right/OK.

P: Now I'll leave you to it for 10 or 15 minutes and see what you come up with. Here's the circuit diagram and a list of the readings you should get at different points on the board. OK. Any questions?

R: No, that's OK.

P: Right. I'll see you a bit later.

P: So, how did you get on? Rasheed?

R: Yeah, this resistor here is the problem. I should get a reading of 25K ohms across it, but there's nothing. It isn't loose, I've checked, but there's no current. So it's definitely faulty.

P: Well done. The problem is with a component. So that's got to be replaced then. And what about you, Carlos?

C: Well, no surprise really. Is cracked copper track. I think cos the board is so badly made. You see, this resistor has been pushed in too far and the copper connection has broken away.

P: Good. Well spotted. So what are you going to do?

R: Throw it in the rubbish – it's a piece of junk. It's difficult to find faults in a badly arranged board.

C: Yeah! Good idea.

P: Yes, normally I'd agree with you, that is the best solution. Sometimes though, we might not have a replacement available. We would need to get it working again then.

C: Well, then, I remove the resistor, repair the track, I test the resistor again, then it's OK again, I think.

P: OK, that would be fine. But this time, we'll just replace it. They're not expensive, and you save a lot of time. Time's money, remember. And make sure you fit the new one neatly.

C: OK. I can do better than this one.

P: Good, I'm glad to hear it. Right, now there are some important rules you both need to know about for doing this kind of work. I'll go through them with you before you start. First of all …

CD3 Track 4: Unit 10, Lesson 6

a) If the board is put under a lot of stress, it might crack.

b) One of the problems with insecure connections is that the fault is often intermittent.

c) Throw it in the rubbish.

d) Press down with a pencil in different places.

e) If the board itself seems OK, put very gentle pressure on each component.

f) Here is the circuit diagram and a list of the readings.

CD3 Track 5: Unit 10, Lesson 8

L: ..., which brings us to radio navigation aids. Now in fact, radio navigation aids were developed around the same time as the mechanical aids we've been talking about – in 1926, successful two-way radio air-to-ground communication began, and the first transmitter/receiver went into mass production in 1928. The earliest radio navigation aid was called the four-course radio range, which was first used in 1929. How the system worked was this – you can see here the layout and how it works. Four towers were set in a square like this, each transmitting the letter A or N in Morse code. A pilot flying toward the square along one of those four paths would hear only A or N in Morse code, in dashes and dots. The dashes and dots got louder or more faint as he flew, depending whether he was flying toward or away from one of the transmitters, which gave him, effectively, a kind of distance reading. Then, if he turned right or left, he would soon start to hear the other letter being transmitted, telling him which quadrant he had now entered. So he also had a reading for his direction.

The first radio-equipped airport control tower was built in Cleveland, Ohio, in 1930, with a range of 15 miles (24 kilometres). By 1935, about 20 more towers had been built. Now, based on continuous radio reports from the pilot, a controller – an air traffic controller – could follow each plane with written notes on a position map. The controller would be able to clear an aircraft for take-off or landing, although the pilot still decided on the best flight path for himself.

Now with regard to the types of radio used, until World War II, which didn't start until 1939 of course, radio navigation relied on low frequencies similar to those of an AM radio. After the war, higher frequency transmitters, called the Very High Frequency Omni-directional Radio range (or VOR), further refined the early 'four-tower' concept of allowing pilots to fly inbound or outbound along a certain quadrant on a line called a radial. These VOR transmitter locations are all printed, with their frequencies and identification Morse codes, on navigation charts.

Before World War II – although as I say, pilots were in constant radio contact with a control tower – the Civil Aeronautics Administration also relied on pilots to radio their position relative to known navigation landmarks to keep aircraft safely at a safe distance from each other in flight. During the war, Radio Detection And Ranging (RADAR) was tested. Radar's primary intent was, and still is, to keep airplanes safely separated – as you know, radar uses the delay between sending a radio ping and receiving the bounced echo to calculate the distance to the other object. And it worked, still works, very well indeed for that. But it was not designed to guide aircraft to a specific point. Because one important thing about these various radio-based systems is not that they are sufficient for navigating between airports, but that they are called non-precision aids because they are not accurate enough, and don't provide enough information, to allow a pilot to land. So the question then is how do pilots get enough information to land safely? Well, of course, the principal considerations for landing an aircraft are ...

CD3 Track 6: Unit 10, Lesson 10

H: Are you still working for the engineering company?

G: No, no, I moved in the end to an aircraft maintenance firm at the airport. It's smaller and friendlier than the last place, and I'm in charge of all the ...

D: Dad! Dad, look what I got.

H: That looks interesting.

D: It's an old coin. I think it might be really old – Roman or Greek or something.

H: David's trying out his new metal detector. David, this is an old friend of mine, George Martin.

G: Hello, David. Pleased to meet you.

D: Hello.

G: Is that a new machine?

D: Yes. I got it for my birthday.

G: What frequency does it operate on?

D: Uuuhmm … I don't really know how it works.

G: You know what an alternating current is?

D: Yes, electricity that goes backwards and forwards in opposite directions. We did that in physics at school.

G: OK, well, that's exactly what your metal detector does. It's got an oscillator circuit that makes a high-frequency AC signal that goes to the coil at the end of the rod, the search coil.

D: Is that what I can hear in the headphones – the oscillator?

G: Yes, that's the noise you hear. And when you pass that coil over a piece of metal, the electromagnetic field in the coil induces a current in the metal. It's called an eddy current. And that makes another magnetic field, which your detector picks up. Your machine just listens for electromagnetic fields.

H: And it changes the sound he hears in the headphones?

G: Yes, that's it, and that tells you there's something metallic nearby. We use something very similar at work. We've got instruments that work in the same way.

H: Metal detectors?

G: Not exactly, although metal detectors like this are what they use for security at airports, of course. No, but we use various instruments to check for faults – cracks or corrosion or even invisible voids inside the metal parts of the planes. Particularly bits that get a lot of wear and tear – failure would be very dangerous.

H: Like the landing gear …

G: … or the engine mountings, yes.

D: Why can't you just look for them?

G: Well, because quite often they're hairline cracks, very tiny and often hidden under paint or dirt.

H: So how do you detect the crack?

G: Well, if there is a crack or some other kind of damage to the structure of the metal, the current won't flow through it as easily as it does through a normal piece of the same metal.

D: So, you get a strange sound through the headphones?

G: Well, our instruments – they're called eddy current meters – have a display rather than headphones, and the instrument we use out on the airfield is quite small, about the size of a very big mobile phone, but yes, you get a different reading. You use a ref – what's called a reference sample, a piece of metal you know is sound, and take a reading from it, and then see if there's a difference between the reading from the part you're testing and the reference reading. Do you see what I mean? If there's a difference, it means there's some kind of fault in the metal.

H: So, someone can go and fix it.

G: Or replace it, usually. It's quicker.

H: Right … so metal detectors aren't just …

D: Dad … I'm just going to do a bit more before it starts raining again. There might be some more coins over there.

H: Righto, see you later. We'll take that coin to the museum this afternoon.

G: Good luck, David!

D: Thanks. See you later, Dad.

Unit 11

CD3 Track 7: Unit 11, Lesson 4

M: Badly designed maintenance documents can be a cause of mistakes. Written procedures that can have more than one meaning, or are long, wordy and repetitive, are likely to cause confusion. Nowadays, many aviation documents are being written in a special kind of English language called Simplified English to make texts as short, simple and clear as possible. This makes the language of maintenance documentation more accessible, particularly for personnel who use English as a second language. Even small improvements in page layout, diagrams and warnings can help to reduce errors. For example, many companies print maintenance documentation in upper case (capital letters), even though it has been known for many years that such text is more difficult to read than text written in the usual mixture of lower and upper case. Replacing blocks of upper case text with normal mixed-case text can increase reading speed by more than 10%.

CD3 Track 8: Unit 11, Lesson 6

Report A:

We took off without any trouble, until we tried to bring up the landing gear. The left-hand gear wouldn't come up – you know why? Maintenance had forgotten to remove the pin before they moved the plane from the hangar to the gate. Anyway, we had to divert to another airport nearby because of heavy traffic. We landed again OK and the pin was removed. Now, maintenance checked it and didn't see it – the pilot did a pre-flight and missed it, and even the push crew who moved them away from the gate failed to spot it. Now … my guess is that there was no warning flag on the pin. I'm sure someone would have noticed if there had been. So that was two mistakes – or three if you count the inspection failures as well.

Report B:

We were flying from our maintenance base to the airport to take the plane back into service. As part of the cockpit check, I called for a fire warning system check and the captain said "complete". Everything seemed fine. It was an uneventful flight and we handed over to a relief flight crew at our destination. But they discovered that the fire extinguisher indicator lights weren't illuminated – and in fact, it turned out that the bottles weren't actually connected! They were still unplugged from when the plane was in maintenance. They'd forgotten to hook up the bottles again! I can't believe that the captain missed it first time round … but I suppose that's what must have happened.

Report C:

As we pushed back from the terminal at Dallas Airport, we got a report that we'd hit another aircraft with our tail. Maintenance was called and they gave us the all clear. Anyway, we flew on down to Mexico City, arriving there at just after midnight local time in the pouring rain. I did the pre-flight with a torch, just before we took off at 2 a.m. – still raining hard – all looked OK. We arrived in Chicago five hours later, and when the aircraft was being inspected in the hangar they found tail damage. Now I'm in trouble with management because they say I didn't do a proper inspection and I shouldn't have taken off from Mexico. It's crazy. Maintenance have all the equipment in the world and they didn't spot it first time round. I'm supposed to see it at 2 a.m. in the pouring rain, in the dark, with a flashlight. It's ridiculous.

Report D:

I accepted the aircraft for a test flight after it had had a couple of weeks in maintenance for fuel tank leaks. The mechanic in charge told me that they'd put in 1,200 lbs of fuel per wing. The pre-flight was fine and take-off normal, and the short flight went OK, although I noticed the fuel transfer light came on just before landing and the fuel gauges indicated zero intermittently, but I suspected a faulty connection problem. I took it up again for a longer run, and about three minutes into the flight, the right engine flamed out, followed 90 seconds later by the left. The engines wouldn't restart and I was forced to make an off-runway landing on a bit of empty highway. Luckily, it isn't used very much. Of course, it turned out that both main tanks were empty. I've learnt my lesson. In the future, I will watch the fuelling being done, look in the tanks myself, or insist that I have an authorised fuel delivery document.

CD3 Track 9: Unit 11, Lesson 8

The efficient maintenance of aircraft depends on the right kind of tools, equipment and facilities. Heavy maintenance, which requires the plane being taken out of service, is usually done inside a hangar. A well-designed maintenance hangar should have the facilities you can see here on the screen, some of them perhaps obvious, I think:

- heating, ventilation and air conditioning systems
- lighting, including emergency lighting
- main, sub-main and small power
- fire detection and alarm systems
- fire protection systems
- domestic and process water services
- process ventilation
- compressed air
- lightning protection and main earthing
- energy management

However, the overall efficiency of a major maintenance hangar – whether it provides aircraft overhaul, heavy maintenance or aircraft paint spraying – is dependent upon the specialist access and lifting equipment provided within the hangar. After the aircraft has been positioned in the hangar, the access docking must quickly wrap around, so as to enable maintenance to commence immediately.

The provision of undercarriage lifting platforms allows removal and testing of the undercarriage equipment without jacking of the aircraft. It is also used to level the aircraft with the access docking before maintenance operations can begin. After the operation is complete, the aircraft is supported by jacks along the fuselage and steadied by the wings. The platforms can then be lowered to undertake maintenance of the undercarriage. Works to all parts of the aircraft can continue to be carried out uninterrupted at a level that is unaltered throughout.

OK, so moving on to some specific pieces of equipment …

CD3 Track 10: Unit 11, Lesson 10

1. 'supervisor
2. 'damaged
3. 'ordered
4. signed 'off
5. co'nnectors
6. e'lectrics
7. 'structural
8. 'generator
9. 'battery
10. re'placed

CD3 Track 11: Unit 11, Lesson 10

1. So I've ordered a new pump.
2. I just replaced it. Put a new one in.
3. One of the O-ring seals was damaged and there was a leak.
4. It wasn't charging the battery properly.
5. No corrosion, no sign of any structural weakness.
6. I refitted it with new connectors.
7. The boss has signed that off, too, so that's OK.
8. Oh yes, problem with the starter generator.

CD3 Track 12: Unit 11, Lesson 10

M: Hello, Fieldings, Maintenance. Mike Armstrong speaking.

D: Hi, David Greenhill here. I'm phoning to see how things are going on my plane.

M: Oh yeah. Which one was it? We've got a few in at the moment.

D: A Skybird 406 – single-engine four-seater.

M: Ah, yes, the red one.

D: That's it.

M: Yep, I did a bit of work on the electrics and instruments … just hang on a minute. I'll get the job file out. Right, OK, let's have a look. Oh yes, problem with the starter generator.

D: That's right … it wasn't charging the battery properly.

M: Yes, actually it turned out it was just a loose connection. Anyway, I took it right out, cleaned it and refitted it with new connectors. All charging up nicely now.

D: What about the magnetic compass? I think it was broken.

M: Yeah, it was. I just replaced it, put a new one in.

D: OK, good.

M: Yes, John Maddox has okayed the avionics and the electrics. Airframe inspection's also been done. No problems there – no corrosion and no sign of any structural weakness … the boss has signed that off, too, so that's OK.

D: That's good news. John Maddox is your supervisor, is he?

M: Yeah. Saved yourself quite a bit of money there.

D: Yes. Did you do the airframe inspection?

M: No, it was Bob, Bob Higgins. It's down here he's also done the nosewheel.

D: Yes, there's been a bit of vibration from that the last few times I've been up. I should have put it in for the service earlier really.

M: Yes, the problem was the damper hydraulic fluid. One of the O-ring seals was damaged and there was a leak. Not much, just a weep really …

D: Enough to make the landing feel pretty strange, I can tell you.

M: I bet, yeah. That hasn't been okayed yet, though. It'll have to do flight tests first.

D: Sure. And what about the engine and fuel system?

M: Right, let's see. Lutfi Tarhoni did that, but … there's something not quite – hang on, he's just come into the office. I'll hand you over, he can give you a better idea. Lutfi, can you have a word with this gentleman?

L: Who is it?

M: Mr Greenhill – the Skybird 406. 100-hour.

L: Oh yes … Hello, Mr Greenhill?

D: Hello.

L: Lutfi Tarhoni with you … Yes, sir, no problem with the engine itself. I did the standard overhaul on that as per the manual … but the fuel pump, well, you can say it's still working, but only 60 or 70%, and to be honest with you, not worth repairing. So I've ordered a new pump which should be here Tuesday next, I'm afraid.

D: OK, well, that's alright, I'm not in a hurry. I'll call back next week then, say Thursday?

L: Yes, sure, I think that will be fine. If we finish it before then, we'll call you.

D: OK, thanks very much. Talk to you next week. Cheers.

L: Goodbye.

Unit 12

CD3 Track 13: Unit 12, Lesson 2

Just to see how effective these agents are, we compared halon with CO_2 and a dry chemical bottle. Um, now this was purely backyard pyromania – we claim no scientific basis for our tests – but the results were impressive. For a test bed, we glued upholstery fabric to foam backing somewhat similar to the material used for aircraft interior panels. We dabbed on a small glass full of gasoline and ignited the panel.

Our halon extinguisher put out this blazing mess in minimum time – with just a couple of squirts. There was no reignition, uhmm, after only about half of the two-and-a-half pound bottle was used. And we were happy that the extinguisher had enough for another go if needed.

Then using a fresh, identical panel, we next tried the CO_2. It took about twice as long as halon to put out the blaze – about three or four squirts. But much worse than that, once the fire appeared to be out, it flamed right back again and needed to be smothered again with the CO_2, so that's maybe eight or 10 seconds overall, which can be a long time in an emergency, of course. By the end of the second try, the bottle was nearly empty, and this fire was not particularly large.

The several types of dry chemicals we tried were nearly as effective as halon in extinguishing the fire – they all did the job after a couple of squirts, using maybe two-thirds of the content. But the mess, the resulting mess was a sight to behold, there was powder swirling in the breeze and coating everything in sight. The air was full of a biting, sour-tasting white dust. We could only imagine trying to fly in a closed cabin with this stuff in the air; parachutes would be preferable.

CD3 Track 14: Unit 12, Lesson 3

O: But we will make our first stop in about 15 minutes at two old houses. Now, is everyone comfortable, not too hot, not too cold?

All: Yeah/Lovely/Fine

O: Good. That's because this coach has modern air conditioning. The temperature outside today is quite high – more than 30 degrees for sure – and because we are not far from the sea, there is high humidity, but we are all comfortable … yes … because the AC takes air from outside, mixes it with recycled air from on board, cools and dries it. And what should we do if the AC breaks down?

S1: Open the windows, of course.

O: For sure, because then we get air flowing through fast because the bus is moving and the air cools us by a process of evaporation from our skin, a bit like an electric fan. But if the bus stops, we are in trouble – we'll soon get hot and humid again.

S2: So how did people stand it in the days before electricity and petrol engines?

O: Well, exactly. Any ideas, anyone …?

CD3 Track 15: Unit 12, Lesson 3

O: OK, are we all here? Right, now you'll see that several of the suggestions you made actually were quite right. Now as you see, this first house has the main building here, the garden and then at the end of the garden, a tall tower. That is called a wind tower, and it's actually an air intake. It's built as tall as possible because the higher you go, the greater the pressure, the force, of the wind, and the lower the air temperature. And the relatively cool, fast-moving wind is ducted down the inside of that tower and under the ground through a tunnel which, compared with the outside air, is quite damp. As the wind passes the tunnel walls, you get evaporation, the air takes heat from the tunnel, so it cools – the tunnel cools in the same way that your skin cools. This means the air coming behind will be cooled and flows into the house and out of the open windows, creating a comfortable environment. And the cycle repeats: the wind flow cools the damp tunnel through evaporation, and the tunnel then takes heat from the air following behind. You don't look convinced. Come inside and you'll see.

O: There, what did I say? Much cooler, isn't it?

S3: That's amazing! How much cooler is it in here? 10 degrees difference?

O: Yes, something like that, somewhere between five and 10 degrees. Low-tech, but very effective. Now we're going to go up to the top of the house and onto the roof so we can see another AC system in operation in the buildings around this one. Be careful of the stairs. They are quite narrow.

CD3 Track 16: Unit 12, Lesson 3

We'll just wait for the last few … OK. If you look over here, you can see that building opposite with the high round roof, and at the top, there is a small dome shape with holes around it. Now what happens in this case is that the wind flows over the top of the dome and, as it does so, it creates low pressure, like the airflow over the top of an aeroplane wing. And because of the lower pressure over the roof, the warm air inside the house is pulled upwards and out of those holes. New air flows into the bottom of the house, and you get a constant airflow through the interior.

S4: But they didn't try to cool the air like in this house?

O: Well, that house also has a big tank, a reservoir, of water under part of the main floor, to help keep the ambient temperature low. Like with the water in the tunnel in the case of this house, evaporation keeps the air temperature down. Unfortunately, they are doing maintenance work at the moment, so we can't go inside. But go and have a look at the wind tower. It's a great lesson in simplicity – the best ideas are often based on simple concepts. Do your sketches and make some notes, and in half an hour we'll go back to the coach and, yes, we should have time to go to the souk before we go back to …

CD3 Track 17: Unit 12, Lesson 3

a) environment / average / conditioning / contaminant

b) system / pressure / falls / intake

c) distribution / reservoir / temperature / ventilate

d)   rise / ducting / moist / cloud

e)   volume / cool / mixture / moisture

f)   exhausted / contracting / compression / constant

g)   humidity / saturate / evaporate / equivalent

CD3 Track 18: Unit 12, Lesson 6

Can you see that OK? Now the three aircraft here, the Sunseeker, the Hughes 300 helicopter and the BAe Hawk, are unusual because they are powered by electric motors rather than by combustion engines. Of course, only this one is big enough to carry a pilot. These other two are working models, but electric motors have several advantages. They are not as noisy as other engines, they produce no dangerous sparks and, of course, they are cleaner. Not only that, they have a longer life and are far more reliable than combustion engines. And – as well as all that – they are extremely efficient and can convert much more electrical energy into mechanical energy than conventional engines.

CD3 Track 19: Unit 12, Lesson 6

Now the fact that it is now possible to fly these aircraft with electric motors is partly due to the development of the brushless DC motor. Look at this. This is a cross-section. In standard brushed motors, as we saw before, the electromagnetic armature on the outside here rotates past static magnets as it is supplied with power through the brushes. But in this BLDC motor, the brushless DC motor is like a brushed motor turned inside out. The electromagnets, in orange here, remain static, and the magnets – in pink – rotate on this central shaft, the black shaft. There is no direct physical connection between the power supply and the rotor around the shaft, the rotor is the grey section here. Instead, the brush and commutator are replaced by an electronic controller.

CD3 Track 20: Unit 12, Lesson 8

JB: ... looking at changes in the manufacturing industry and the aircraft manufacturing industry in particular. With us is Alan Bowden, chief engineer of Apex Aircraft, who produce small passenger jets and turboprop trainers. Is that right, Alan?

AB: Yes, quite correct, we've just started production of our new advanced training aircraft, the D30.

JB: Yes, I'd like you to tell us more about that. But let's go straight to our first caller, Denise Smith, who's on the line from Manchester. Denise, what's your question?

DS: Good afternoon.

AB: Hello, good afternoon.

DS: You were talking about the new plane ...

AB: The D30.

DS: Yeah. Well, I'm really interested in planes and I want to do engineering at college, and I heard something about it on the radio, and they said it had a glass cockpit, as if that was something special, and I was curious. I mean, I thought all cockpits were made of glass, at least the windows are, so I'm assuming perhaps this is something different ...

AB: No, no, you're right. The term 'glass cockpit' doesn't refer to the windows. It refers to the flight instrument display which the pilot uses to get information as he flies the plane.

DS: So, it's a computer system ...

AB: Well, yes, there's certainly a computer – sometimes several – involved. But the most important feature is the way that the six traditional flight instruments are replaced by something called the primary flight display. The PFD for short. Instead of the pilot having to look at six different instruments at once, all the basic flight information is grouped together on the same screen, which means there is less chance of him making a mistake because of misreading his instruments at critical times, like coming in to land.

DS: OK, and the PFD is made of glass.

AB: Well, actually, well, that's where the name comes from, but in fact, it's an LCD computer screen with an image of the information on it. It's called glass because the first screens were made of glass, like an old television screen.

DS: Right, I see, and all information is on there ...

AB: Yes. The centre of the display shows the earth in brown and the sky in blue, with a red cross where the centre of the aircraft is. This does the job of the attitude indicator, which has been present in aircraft since the earliest days of flying: that red cross moves up and down a white scale to show degrees of pitch up or down of the nose. The attitude indicator is in the middle because it's still considered the most important instrument. Then, there is a tick horizontal black line on each side of the centre representing the wings of the aircraft. Those black lines rotate to show roll – whether the aircraft is flying level or not, for turning, in a turning situation, the pilot refers to these black wing symbols to tell him his position in the turn. Now, around that central screen, on either side of the display, are what are known as the 'tapes', which take the form of vertical grey panels. On the left side you have the airspeed tape, and on the right side the altitude tape, which has a further tape on its right showing the rate of climb or descent. That's the vertical speed indicator. The tapes scroll up or down depending on speed and altitude, and in the centre of each tape there is a small box highlighting current airspeed – on the left – and current altitude on the right. And you've got a gyrocompass at the bottom of the screen to give you your heading.

CD3 Track 21: Unit 12, Lesson 8

JB: So, it's a lot like the old Basic Six really?

AB: Yes, only easier to use. And finally, when the aircraft is getting near the ground, when it is landing, the pilot will see a visual indication of the runway on his PFD. It shows him the position of the instrument landing system on the ground. This, and signals coming from the ground, allow him to land using this instrument alone – for example, in bad weather when he can't see much outside the aircraft.

DS: And what else is there besides the PFD?

AB: Well, there is a lot of information about the aircraft systems, of course, which is displayed electronically, but the next most important instrument, I suppose, is the navigation display, or ND as we call it. It combines the functions of radio navigation instruments such as the ADF indicator with a moving map which can be made bigger or smaller.

JB: A satnav sort of thing?

AB: Yes, in a way, plus it can also show the shape of the land you are flying over as well as landing approach maps, weather maps and three-dimensional vertical displays, and 3D navigation images. And the great thing is, the pilot can control a lot of the information, he can change the way it is displayed, or even remove it sometimes. The glass cockpit is actually a result of the same technology that goes into satnavs, digital radios and home computers. It's a combination of the three things really. Global Satellite Technology for navigation information, lightweight LCD display screens and powerful small electronic computers to link everything up. It's a real improvement.

JB: Right. Thank you for that. I hope that answers your question, Denise?

DS: Yes, thanks very much. I think I got the general idea.

JB: Good. Now, Alan, you've been doing a lot of work on the new D30 design. Tell us …

CD3 Track 22: Unit 12, Lesson 10

T: OK, there aren't any obstructions around the aircraft.

E: Right. What's next?

T: Number two: 'Position the tail stand under the tail jacking point. Raise the stand central pillar to the approximate height of the jacking point and insert the locking pin through the centre tube and pillar. Adjust the tail stand threaded ball-end until it contacts the jacking point, ensuring that all three legs of the stand are firmly in contact with the ground.

E: OK, good, that's fine. Now as it says, we need a jacking adapter on top of each jack, which we've just done, and then position them under your jacking points. Make sure the hydraulic valves are closed for now. OK, closed? And the other one. Right, raise the jack … that's it … until the adapter just touches the jacking point. Good. Stop there. Now, line up the adapter and your jacking point properly. Right. And do the same on the other one.

T: It says 'slowly and evenly' – is this OK?

E: That's fine. Keep going. Till you see the wheels are just off the ground.

E: Right, now it's not all that heavy, but we need two pairs of hands for this. Don't try and do this on your own, the manual says. I'll lift the tail and you adjust the stand again to the right height and put a locking pin in. Then I'll put the tail down on it and we'll check if it's at the right height. OK, ready?

T: Yes.

E: Don't take too long over it, mind you – it's not that light either.

T: Right, so the two main jacks have got their locking collars on, so ... 'Raise the aircraft tail', yes, I wasn't sure about this – 'finally adjusting it to the required level by use ... of the threaded ball-end.'

E: Yes, that's important. The ball-end is – look here at the tail stand. See this ...

E: That's it, slowly does it. If you let it down too fast and there's a problem, it can be hard to do anything about it. It's ... it's on the collars now. Close the hydraulic valve – let me see you do it. OK. And the other side. Remember you have to do both jacks. That locks it all up. Do you see how the locking collars lock the jacks mechanically, and the valves lock them hydraulically? That level of security's mandatory. Better safe than sorry.

E: What do you think of the tail height now?

T: It looks level to me.

E: Good, yes, it's OK as it is.

T: I was wondering though, where do we find the forty kilos of ballast. Is there a standard ready-made weight? This is the suspension point ...?

E: We'll get to that in a moment. The first thing is that that is the ballast point, yes.